PROJECTORS AND PROJECTION METHODS

Advances in Mathematics

Volume 6

PROJECTORS AND PROJECTION METHODS

AURÉL GALÁNTAI
Institute of Mathematics
University of Miskolc, Hungary

Kluwer Academic Publishers
Boston/Dordrecht/London

Distributors for North, Central and South America:
Kluwer Academic Publishers
101 Philip Drive
Assinippi Park
Norwell, Massachusetts 02061 USA
Telephone (781) 871-6600
Fax (781) 871-6528
E-Mail <kluwer@wkap.com>

Distributors for all other countries:
Kluwer Academic Publishers Group
Post Office Box 322
3300 AH Dordrecht, THE NETHERLANDS
Telephone 31 78 6576 000
Fax 31 78 6576 474
E-Mail <orderdept@wkap.nl>

Electronic Services <http://www.wkap.nl>

Library of Congress Cataloging-in-Publication

Galántai, Aurél
Projectors and Projection Methods
ISBN 1-4020-7572-3

Printed on acid-free paper.

Printed in the United States of America

Contents

Preface

The projectors are considered as simple but important type of matrices and operators. Their basic theory can be found in many books, among which Halmos [177], [178] are of particular significance. The projectors or projections became an active research area in the last two decades due to ideas generated from linear algebra, statistics and various areas of algorithmic mathematics. There has also grown up a great and increasing number of projection methods for different purposes. The aim of this book is to give a unified survey on projectors and projection methods including the most recent results. The words projector, projection and idempotent are used as synonyms, although the word projection is more common.

We assume that the reader is familiar with linear algebra and mathematical analysis at a bachelor level. The first chapter includes supplements from linear algebra and matrix analysis that are not incorporated in the standard courses. The second and the last chapter include the theory of projectors. Four chapters are devoted to projection methods for solving linear and nonlinear systems of algebraic equations and convex optimization problems.

For unification purposes the material will be treated in finite dimensional Euclidean spaces except for the last chapter where a Hilbert space setting is given. For projection methods not included here we refer to the books of Krasnosel'skii et al [230], Saad [306], Brezinski [57] and McCormick [253] which might be interesting for the reader. Many of my colleagues discussed various aspects of this book with me and provided important suggestions. I am indebted to J. Abaffy, C. Broyden, E. Spedicato, C. Hegedûs, P. Rózsa, K. Agbeko, J. Mészáros, and S. Fegyverneki for their help and encouragement.

A. Galántai
Miskolc, Hungary

Chapter 1

SUPPLEMENTS FOR LINEAR ALGEBRA

The purpose of this chapter is to collect the notation, concepts and results of linear algebra that are usually not part of the standard linear algebra courses and which we need to develop our material. We use the following standard notation and concepts.

$\mathbb{R}$ denotes the set of real numbers, while $\mathbb{R}^+$ is the set of nonnegative real numbers. $\mathbb{C}$ denotes the set of complex numbers. $\mathbb{F}$ denotes a field (here $\mathbb{R}$ or $\mathbb{C}$).

$\mathbb{F}^n$ is the vector space of n-tuples of elements over $\mathbb{F}$. Similarly, $\mathbb{F}^{m\times n}$ is the vector space of $m \times n$ matrices over $\mathbb{F}$. The matrices and the subspaces will be denoted by capital letters. Subspaces are always linear subspaces. Vector $e_i \in \mathbb{R}^n$ is the ith unit vector ($i = 1, \ldots, n$). I is the unit matrix of appropriate order, I_n is the $n \times n$ unit matrix. Let $A \in \mathbb{F}^{m\times n}$, $\alpha = \{i_1, \ldots, i_k\} \subseteq \{1, \ldots, m\}$, $\beta = \{j_1, \ldots, j_k\} \subseteq \{1, \ldots, n\}$, $\alpha' = \{1, \ldots, m\} \setminus \alpha$ and $\beta' = \{1, \ldots, n\} \setminus \beta$. Then $A[\alpha, \beta]$ denotes the submatrix of A lying in the rows indicated by α and the columns indicated by β. Furthermore let $\Pi_\alpha = [e_{i_1}, \ldots, e_{i_k}]$ denote the partial permutation matrix. Thus $A[\alpha, \beta] = \Pi_\alpha^T A \Pi_\beta$.

A mapping $f : \mathbb{F}^n \to \mathbb{F}^m$ is a linear transformation if and only if there is a matrix $A \in \mathbb{F}^{m\times n}$ (called the representation of f) such that $f(x) = Ax$ ($x \in \mathbb{F}^n$). Hence we shall make no distinction between f and its representation A. Whenever we change from f to A or from A to f, it is to be understood in the standard bases of $\mathbb{F}^n$ and $\mathbb{F}^m$.

$\mathcal{R}(A) = \{Ax \mid x \in \mathbb{F}^n\} \subset \mathbb{F}^m$ denotes the range of matrix $A \in \mathbb{F}^{m\times n}$. Generally $\mathcal{R}(f)$ denotes the range of any mapping f. $\mathcal{D}(f)$ denotes the domain of mapping f. The null space of matrix $A \in \mathbb{F}^{m\times n}$ is given by $\mathcal{N}(A) = \{x \mid Ax = 0\} \subset \mathbb{F}^n$. The linear span of vectors $\{v_i\}_{i=1}^k \subset \mathbb{F}^n$ is denoted by $\mathcal{L}\left(\{v_i\}_{i=1}^k\right)$. If $A = [a_1, \ldots, a_n] \in \mathbb{F}^{m\times n}$, then $\mathcal{R}(A) = \mathcal{L}(a_1, \ldots, a_n)$.

A matrix $A \in \mathbb{C}^{m\times n}$ is Hermitian if its conjugate transpose $A^H = \overline{A}^T \in \mathbb{C}^{n\times m}$ is equal to A. A matrix $Q \in \mathbb{C}^{n\times n}$ is unitary , if $Q^H Q = I$. The real unitary matrices are called orthogonal . The trace of a square matrix $A = [a_{ij}]_{i,j=1}^n \in \mathbb{C}^{n\times n}$ is $\mathrm{tr}(A) = \sum_{i=1}^n a_{ii}$. It can be shown that $\mathrm{tr}(X^{-1}AX) = \mathrm{tr}(A)$ and $\mathrm{tr}(A) = \sum_{i=1}^n \lambda_i$, where λ_i is the ith eigenvalue of A.

Two vectors $x, y \in \mathbb{C}^n$ are orthogonal, written $x \perp y$, if $y^H x = 0$. A vector $x \in \mathbb{C}^n$ is orthogonal to a set $M \subset \mathbb{C}^n$, written $x \perp M$, if $x \perp y$ for all $y \in M$. Similarly we say that two sets $M, N \subset \mathbb{C}^n$ are orthogonal and write $M \perp N$, if $x \perp y$ for all $x \in M$ and $y \in N$. The orthogonal complement of $M \subset \mathbb{C}^n$ is defined by $M^\perp = \{x \in \mathbb{C}^n \mid x \perp y \text{ for all } y \in M\}$. Notice that $M^\perp$ is a linear subspace and $M^{\perp\perp} = \mathcal{L}(M)$.

Suppose that $U, V \subset \mathbb{F}^n$ are subspaces. The sum of the linear subspaces U and V is denoted by

$$U + V = \{u + v \mid u \in U, v \in V\}.$$

Similarly we define the operations

$$\lambda U = \{\lambda u \mid u \in U\}$$

and

$$c + U = \{c + u \mid u \in U\}.$$

The subspace $W \subseteq \mathbb{F}^n$ is a direct sum of the subspaces U and V if

$$U \cap V = \{0\}, \quad U + V = W.$$

This fact is denoted by $W = U \oplus V$. The subspaces $L, M \subset \mathbb{F}^n$ are complementary if $L \oplus M = \mathbb{F}^n$. We often use the following statements.

LEMMA 1.1 *Assume that $M, N \subset \mathbb{F}^n$ are linear subspaces. Then the following relations hold:*
(i) M and N are complementary, if and only if $M^\perp$ and $N^\perp$ are complementary;
(ii) $M \oplus M^\perp = \mathbb{F}^n$;
(iii) $M^{\perp\perp} = M$;
(iv) $(M + N)^\perp = M^\perp \cap N^\perp$;
(v) $(M \cap N)^\perp = M^\perp + N^\perp$.

COROLLARY 1.2 *Assume that $M_1, \ldots, M_k \subset \mathbb{F}^n$ are linear subspaces. Then*
(i) $\left(\sum_{i=1}^k M_i\right)^\perp = \cap_{i=1}^k M_i^\perp$;
(ii) $\left(\cap_{i=1}^k M_i\right)^\perp = \sum_{i=1}^k M_i^\perp$.

LEMMA 1.3 *Suppose that* $A \in \mathbb{C}^{m\times n}$. *Then* $\mathcal{N}(A) = \mathcal{R}^{\perp}(A^H)$.

Proof. By definition $\mathcal{N}(A) = \{x | Ax = 0\}$ and

$$\mathcal{R}(A^H) = \{y | \exists z : A^H z = y\}.$$

If $x \in \mathcal{N}(A)$, then $Ax = 0$ and $x^H A^H = 0$. Since $x^H(A^H z) = 0$ also holds for any z, we have that $x \perp y$ for all $y = A^H z \in \mathcal{R}(A^H)$ and hence $\mathcal{N}(A) \subset \mathcal{R}^{\perp}(A^H)$. If $x \in \mathcal{R}^{\perp}(A^H)$, then $x^H y = x^H A^H z = z^H(Ax) = 0$ for all z. The choice $z = Ax$ implies $Ax = 0$ and $\mathcal{R}^{\perp}(A^H) \subset \mathcal{N}(A)$. Therefore, $\mathcal{R}^{\perp}(A^H) = \mathcal{N}(A)$ or equivalently $\mathcal{R}(A) = \mathcal{N}^{\perp}(A^H)$. ■

The rank of a matrix A is denoted by rank(A). The rank satisfies the following properties:

$$\operatorname{rank}(A+B) \leq \operatorname{rank}(A) + \operatorname{rank}(B), \tag{1.1}$$

$$\operatorname{rank}(AB) \leq \min\{\operatorname{rank}(A), \operatorname{rank}(B)\}, \tag{1.2}$$

$$\operatorname{rank}(A) = n - \dim(\mathcal{N}(A)) \quad (A \in \mathbb{F}^{m\times n}), \tag{1.3}$$

$$\operatorname{rank}(AA^H) = \operatorname{rank}(A^H A) = \operatorname{rank}(A). \tag{1.4}$$

DEFINITION 1.4 *For any* $A \in \mathbb{F}^n$ *let*

$$|A| = [|a_{ij}|]_{i,j=1}^{m,n} \in \mathbb{R}^{m\times n}.$$

We define the natural partial ordering of real vectors and matrices as follows.

DEFINITION 1.5 *Let* $A, B \in \mathbb{R}^{m\times n}$. *Then* $A \leq B$ *if and only if* $a_{ij} \leq b_{ij}$ *for all* $i = 1, \ldots, m$ *and* $j = 1, \ldots, n$.

The absolute value $|A|$ of matrix A satisfies the following properties:
(i) $|A| \geq 0$ $(A \in \mathbb{F}^{m\times n})$, $|A| = 0 \Leftrightarrow A = 0$;
(ii) $|\lambda A| = |\lambda|\,|A|$ $(\lambda \in \mathbb{F})$;
(iii) $|A+B| \leq |A| + |B|$ $(A, B \in \mathbb{F}^{m\times n})$;
(iv) $|AB| \leq |A|\,|B|$ $(A \in \mathbb{F}^{m\times k},\ B \in \mathbb{F}^{k\times n})$.

DEFINITION 1.6 *A matrix* $A \in \mathbb{R}^{m\times m}$ *is said to be an* M*-matrix, if* $A = sI - B$, *where* $s > 0$, $B \geq 0$ *and* $s \geq \rho(B)$, *with* $\rho(B)$ *denoting the spectral radius of* B.

The M-matrix $A = sI - B$ is nonsingular if $s > \rho(B)$. An equivalent definition is given by

DEFINITION 1.7 *A matrix* $A \in \mathbb{R}^{n\times n}$ *is said to be a nonsingular* M*-matrix if* $a_{ij} \leq 0$ *for all* $i \neq j$ *and* $A^{-1} \geq 0$.

DEFINITION 1.8 *$A \in \mathbb{R}^{n\times n}$ is a Z-matrix, if $a_{ij} \leq 0$ $(i \neq j)$.*

The $n \times n$ type Z-matrices are denoted by $Z^{n\times n}$.

LEMMA 1.9 *Assume that $B \in Z^{n\times n}$ and A is an M-matrix with $A \leq B$. Then B is also an M-matrix and $0 \leq B^{-1} \leq A^{-1}$.*

THEOREM 1.10 *Let $A, B \in \mathbb{R}^{n\times n}$. If $|A| \leq B$, then $\rho(A) \leq \rho(|A|) \leq \rho(B)$.*

COROLLARY 1.11 *Let $A, B \in \mathbb{R}^{n\times n}$. If $0 \leq A \leq B$, then $\rho(A) \leq \rho(B)$.*

For more on M-matrices the reader is referred to [43], [314], [203], [361], [204].

1. Schur complements

The Schur complements are very useful in many areas (see, e.g., [76] and [272]). The following result of Guttman [175] plays a key role in this theory (see [272]).

LEMMA 1.12 *(Guttman). Let*

$$A = \begin{bmatrix} E & F \\ G & H \end{bmatrix} \in \mathbb{F}^{m\times n} \quad \left(E \in \mathbb{F}^{k\times k}\right). \tag{1.5}$$

If E is nonsingular, then

$$rank(A) = rank(E) + rank\left(H - GE^{-1}F\right). \tag{1.6}$$

If H is nonsingular, then

$$rank(A) = rank(H) + rank\left(E - FH^{-1}G\right). \tag{1.7}$$

Proof. Assume that E is nonsingular. The matrix A can be written in the form

$$A = \begin{bmatrix} I_k & 0 \\ GE^{-1} & I_{m-k} \end{bmatrix} \begin{bmatrix} E & 0 \\ 0 & H - GE^{-1}F \end{bmatrix} \begin{bmatrix} I_k & E^{-1}F \\ 0 & I_{n-k} \end{bmatrix}.$$

The lower and upper triangular matrices of the right side are nonsingular. Hence

$$\text{rank}(A) = \text{rank}\left(\begin{bmatrix} E & 0 \\ 0 & H - GE^{-1}F \end{bmatrix}\right) = \text{rank}(E) + \text{rank}\left(H - GE^{-1}F\right).$$

The proof of the other case is similar. ■

COROLLARY 1.13 *If $A \in \mathbb{F}^{n\times n}$ is nonsingular, then $S = H - GE^{-1}F$ is also nonsingular.*

DEFINITION 1.14 *The Schur complement of E in A is defined by*

$$S = (A/E) = H - GE^{-1}F. \tag{1.8}$$

Similarly, the Schur complement of H in A is given by

$$T = (A/H) = E - FH^{-1}G. \tag{1.9}$$

THEOREM 1.15 *(Banachiewicz, Frazer, Duncan, Collar). Let*

$$A = \begin{bmatrix} E & F \\ G & H \end{bmatrix} \in \mathbb{F}^{n\times n} \quad \left(E \in \mathbb{F}^{k\times k}\right) \tag{1.10}$$

and E be nonsingular. Then A^{-1} can be expressed in the partitioned form

$$A^{-1} = \begin{bmatrix} E^{-1} + E^{-1}FS^{-1}GE^{-1} & -E^{-1}FS^{-1} \\ -S^{-1}GE^{-1} & S^{-1} \end{bmatrix}. \tag{1.11}$$

Formula (1.11) can be written in the alternative form

$$\begin{bmatrix} E & F \\ G & H \end{bmatrix}^{-1} = \begin{bmatrix} E^{-1} & 0 \\ 0 & 0 \end{bmatrix} + \begin{bmatrix} E^{-1}F \\ -I \end{bmatrix} S^{-1} \left[GE^{-1}, -I\right]. \tag{1.12}$$

The bordered inversion method of Frazer, Duncan and Collar exploits formula (1.11) (see, e.g., [175], [272]). For $k = 1, \ldots, m$ let

$$E_{k+1} = \begin{bmatrix} E_k & F_k \\ G_k & H_k \end{bmatrix} \tag{1.13}$$

with

$$E_k \in \mathbb{F}^{n_k\times n_k}, \quad F_k, \quad G_k^T \in \mathbb{F}^{n_k\times p_k}, \quad H_k \in \mathbb{F}^{p_k\times p_k} \tag{1.14}$$

and let $E_{m+1} = E$. If E_k and $S_k = H_k - G_kE_k^{-1}F_k$ are nonsingular, then E_{k+1} is also nonsingular and by Theorem 1.15 we can calculate E_{k+1}^{-1}. Note that $n_{k+1} = n_k + p_k$.

The bordered inversion algorithm

Calculate E_1^{-1}.
for $k = 1, \ldots, m$

$S_k = H_k - G_kE_k^{-1}F_k$

$$E_{k+1}^{-1} = \begin{bmatrix} E_k^{-1} + E_k^{-1}F_kS_k^{-1}G_kE_k^{-1} & -E_k^{-1}F_kS_k^{-1} \\ -S_k^{-1}G_kE_k^{-1} & S_k^{-1} \end{bmatrix}$$

end

If S_k is nonsingular for $k = 1, \ldots, m$, then $E_{m+1}^{-1} = E^{-1}$. If S_k is singular for an index k, then the Guttman lemma implies that E_{k+1} is singular.

THEOREM 1.16 *(Guttman). Let E be nonsingular. The bordered inversion algorithm gives the inverse of E if and only if E_k is nonsingular for all $k = 1, \ldots, m$.*

2. Factorizations and decompositions

We use various multiplicative and additive decompositions (factorizations) of matrices and the CS decomposition of orthogonal matrices.

DEFINITION 1.17 *Assume that $A \in \mathbb{F}^{m\times n}$ has rank $r \geq 1$. Decomposition $A = XY$ is called full rank factorization, if $X \in \mathbb{F}^{m\times r}$, $Y \in \mathbb{F}^{r\times n}$ and $rank(X) = rank(Y) = r$.*

Every nonzero matrix has a full rank factorization. The full rank factorization is not unique for if $A = XY$ is a full rank factorization, then $A = (XM)\left(M^{-1}Y\right)$ is also a full rank factorization of A whenever $M \in \mathbb{F}^{r\times r}$ is nonsingular.

PROPOSITION 1.18 *Let $A \in \mathbb{F}^{m\times n}$ have rank $r \geq 1$. Assume that A has the full rank factorizations $A = XY$ and $A = X_1Y_1$. Then there exists a nonsingular matrix M such that $X_1 = XM^{-1}$ and $Y_1 = MY$.*

Proof. The equality $XY = X_1Y_1$ implies $X_1^H XY = X_1^H X_1 Y_1$ and

$$\left[\left(X_1^H X_1\right)^{-1} X_1^H X\right] Y = Y_1.$$

Also we have $XYY_1^H = X_1Y_1Y_1^H$ and $X\left[YY_1^H\left(Y_1Y_1^H\right)^{-1}\right] = X_1$. Let $M = \left(X_1^H X_1\right)^{-1} X_1^H X$ and $N = YY_1^H\left(Y_1Y_1^H\right)^{-1}$. The next calculation shows that $N = M^{-1}$:

$$\begin{aligned} MN &= \left[\left(X_1^H X_1\right)^{-1} X_1^H X\right]\left[YY_1^H\left(Y_1Y_1^H\right)^{-1}\right] \\ &= \left(X_1^H X_1\right)^{-1} X_1^H A Y_1^H \left(Y_1Y_1^H\right)^{-1} \\ &= \left(X_1^H X_1\right)^{-1} X_1^H \left(X_1Y_1\right) Y_1^H \left(Y_1Y_1^H\right)^{-1} = I. \end{aligned}$$

■

Full rank factorizations are frequently given in the form $A = XZY$, where $Z \in \mathbb{F}^{r\times r}$ is nonsingular. Other notations of full rank factorizations are $A = XY^H$ ($X \in \mathbb{F}^{m\times r}$, $Y \in \mathbb{F}^{n\times r}$) and $A = XZY^H$, respectively.

PROPOSITION 1.19 *Let $A = XY$ ($X \in \mathbb{F}^{m\times r}$, $Y \in \mathbb{F}^{r\times n}$) be a full rank factorization. Then*

$$\mathcal{R}(A) = \mathcal{R}(X), \quad \mathcal{N}(A) = \mathcal{N}(Y). \tag{1.15}$$

Proof. $z \in \mathcal{N}(A) \Leftrightarrow Az = X(Yz) = 0 \Leftrightarrow Yz = 0 \Leftrightarrow z \in \mathcal{N}(Y)$. Similarly, $\mathcal{N}\left(A^H\right) = \mathcal{N}\left(X^H\right)$. Lemma 1.3 implies that $\mathcal{R}(A) = \mathcal{N}^{\perp}\left(A^H\right) = \mathcal{N}^{\perp}\left(X^H\right) = \mathcal{R}(X)$. ■

Let $A = XY^H$, where $X = [x_1, \ldots, x_r]$ and $Y = [y_1, \ldots, y_r]$. We can write A as the sum of outer products:

$$A = \sum_{i=1}^{r} x_i y_i^H. \tag{1.16}$$

This representation is minimal in the sense that A cannot be decomposed as the sum of less than r outer products. Hence the number of outer products in the above minimal representation defines the rank of matrix A.

Any partition of the full rank factorization

$$A = XY^H = [X_1, \ldots, X_k]\,[Y_1, \ldots, Y_k]^H \quad \left(X_j \in \mathbb{F}^{m \times l_j},\ Y_j \in \mathbb{F}^{n \times l_j}\right)$$

gives the rank additive decomposition

$$A = \sum_{j=1}^{k} X_j Y_j^H \tag{1.17}$$

with

$$\operatorname{rank}(A) = \sum_{j=1}^{k} \operatorname{rank}\left(X_j Y_j^H\right) = \sum_{j=1}^{k} l_j. \tag{1.18}$$

The rank additive decompositions have special properties (see Marsaglia and Styan [244], Carlson [76]).

DEFINITION 1.20 *The matrix $A \in \mathbb{F}^{n \times n}$ is said to be lower triangular if $a_{ij} = 0$ for all $i < j$. If $a_{ij} = 0$ for all $i \leq j$, then A is strictly lower triangular. If $a_{ij} = 0$ for $i < j$ and $a_{ii} = 1$ $(i = 1, \ldots, n)$, then A is unit lower triangular.*

DEFINITION 1.21 *The matrix $A \in \mathbb{F}^{n \times n}$ is said to be upper triangular if $a_{ij} = 0$ for all $i > j$. If $a_{ij} = 0$ for all $i \geq j$, then A is strictly upper triangular. If $a_{ij} = 0$ for $i > j$ and $a_{ii} = 1$ $(i = 1, \ldots, n)$, then A is unit upper triangular.*

The LU decomposition of a matrix $A \in \mathbb{F}^{n \times n}$ is defined by $A = LU$, where L is lower triangular and U is upper triangular. The LU decomposition, if it exists, is not unique. For any nonsingular diagonal matrix D, decomposition $A = (LD)\left(D^{-1}U\right)$ is also an LU factorization of the matrix. The sufficient part of the following result was proved by Turing (see, e.g., [198]).

THEOREM 1.22 *Let $A \in \mathbb{F}^{n \times n}$ be a nonsingular matrix. The matrix A has a unique LU decomposition*

$$A = LU, \tag{1.19}$$

where L is unit lower triangular and U is upper triangular, if and only if the leading principal submatrices are all nonsingular.

DEFINITION 1.23 *The block LU decomposition of the partitioned matrix*

$$A = [A_{ij}]_{i,j=1}^{k} \in \mathbb{F}^{n\times n} \quad (A_{ij} \in \mathbb{F}^{l_i\times l_j})$$

is defined by $A = LU$, where L and U are block triangular matrices of the form

$$L = [L_{ij}]_{i,j=1}^{k} \quad (L_{ij} \in \mathbb{F}^{l_i\times l_j};\ L_{ij} = 0,\ i < j)$$

and

$$U = [U_{ij}]_{i,j=1}^{k} \quad (U_{ij} \in \mathbb{F}^{l_i\times l_j};\ U_{ij} = 0,\ i > j).$$

THEOREM 1.24 *The matrix A has a block LU decomposition if and only if the first $k-1$ leading principal block submatrices of A are nonsingular.*

For proofs of the above theorems, see e.g., [150], [203], [198].

A block triangular matrix is unit block triangular if its diagonal elements are unit matrices. If L or U is unit block triangular, then the block LU decomposition is unique.

DEFINITION 1.25 *A partitioned nonsingular matrix $A \in \mathbb{F}^{n\times n}$ is said to be (block) strongly nonsingular, if A has a (block) LU decomposition.*

The conditions of strong nonsingularity are clearly the most restrictive in the case when all blocks are scalars. Such a case implies block strong nonsingularity for any allowed partition. A geometric characterization of the strong nonsingularity will be given in Section 1.8.

Occasionally we denote the unique LU factorization of A by $A = L_1U$, where L_1 stands for the unit lower triangular component. The unique LDU factorization of A is defined by $A = L_1DU_1$, where L_1 is unit lower triangular, D is diagonal and U_1 is unit upper triangular ($U = DU_1$). For technical reasons we also use the unique LU factorization $A = LU_1$, where $L = L_1D$ is lower triangular and U_1 is unit upper triangular. A special case of the LU factorizations is the Cholesky factorization.

THEOREM 1.26 *(Cholesky decomposition). Let $A \in \mathbb{C}^{n\times n}$ be Hermitian and positive definite. Then A can be written in the form*

$$A = LL^H, \tag{1.20}$$

where L is lower triangular with positive diagonal entries. If A is real, L may be taken to be real.

The next important decomposition is the QR factorization .

THEOREM 1.27 *(QR factorization). If $A \in \mathbb{C}^{m\times n}$ and $m \geq n$, then there is a matrix $Q \in \mathbb{F}^{m\times n}$ with orthonormal columns and an upper triangular matrix $R \in \mathbb{F}^{n\times n}$ such that*

$$A = QR. \tag{1.21}$$

If $m = n$, Q is unitary; if in addition A is nonsingular, then R may be chosen so that all its diagonal entries are positive, and in this case, the factors Q and R are both unique. If $A \in \mathbb{R}^{m\times n}$, then both Q and R may be real matrices.

If A has rank n, then the QR factorization is also a full rank factorization. The QR factorization can be written in the form

$$A = Q\begin{bmatrix} R_1 \\ 0 \end{bmatrix} \quad \left(Q \in \mathbb{F}^{m\times m},\ R_1 \in \mathbb{F}^{n\times n}\right), \tag{1.22}$$

where Q is unitary and R_1 is nonsingular upper triangular.

THEOREM 1.28 *(Schur). Given any $A \in \mathbb{C}^{n\times n}$ with eigenvalues $\lambda_1, \dots, \lambda_n$ in any prescribed order, there is a unitary matrix $Q \in \mathbb{C}^{n\times n}$ such that*

$$Q^H A Q = R = [r_{ij}]_{i,j=1}^{n} \tag{1.23}$$

is upper triangular, with diagonal entries $r_{ii} = \lambda_i$, $i = 1, \dots, n$. Furthermore, if $A \in \mathbb{R}^{n\times n}$ and if all the eigenvalues of A are real, then Q may be chosen to be real and orthogonal.

DEFINITION 1.29 *A Jordan block $J_k(\lambda) \in \mathbb{C}^{k\times k}$ is a $k \times k$ upper triangular matrix of the form*

$$J_k(\lambda) = \begin{bmatrix} \lambda & 1 & 0 & \dots & 0 \\ 0 & \lambda & 1 & \ddots & \vdots \\ \vdots & \ddots & \ddots & \ddots & 0 \\ \vdots & & \ddots & \ddots & 1 \\ 0 & \dots & \dots & 0 & \lambda \end{bmatrix}. \tag{1.24}$$

A Jordan matrix $J \in \mathbb{C}^{n\times n}$ has the form

$$J = \begin{bmatrix} J_{n_1}(\lambda_1) & & \\ & J_{n_2}(\lambda_2) & \\ & & \ddots \\ & & & J_{n_k}(\lambda_k) \end{bmatrix}, \quad \left(\sum_{i=1}^{k} n_i = n\right) \tag{1.25}$$

in which the orders n_i may not be distinct and the values λ_i need not be distinct either.

Notice that if each Jordan block is one dimensional, that is $J_{n_i}(\lambda_i) = [\lambda_i]$, then the Jordan matrix J is diagonal.

THEOREM 1.30 *(Jordan). Let $A \in \mathbb{C}^{n\times n}$. There is a nonsingular matrix $X \in \mathbb{C}^{n\times n}$ such that*

$$A = X \begin{bmatrix} J_{n_1}(\lambda_1) & & \\ & J_{n_2}(\lambda_2) & \\ & & \\ & & J_{n_k}(\lambda_k) \end{bmatrix} X^{-1} = XJX^{-1} \tag{1.26}$$

and $\sum_{i=1}^{k} n_i = n$. The Jordan matrix J of A is unique up to permutations of the diagonal blocks. The eigenvalues λ_i $(i = 1, \ldots, k)$ are not necessarily distinct. If A is a real matrix with only real eigenvalues, then the similarity matrix X can be taken to be real.

It is important to know whether or not a matrix can be transformed to diagonal form by similarity transformations. Let $p(t) = a_k t^k + a_{k-1} t^{k-1} + \cdots + a_1 t + a_0$ $(a_k \neq 0)$ be a given polynomial. If $a_k = 1$, then $p(t)$ is said to be monic . For any $A \in \mathbb{C}^{n\times n}$ we define

$$p(A) = a_k A^k + a_{k-1} A^{k-1} + \cdots + a_1 A + a_0 I. \tag{1.27}$$

THEOREM 1.31 *Let $A \in \mathbb{C}^{n\times n}$. There exists a unique monic polynomial $q_A(t)$ of minimum degree such that $q_A(A) = 0$. The degree of $q_A(t)$ is at most n. If $p(t)$ is any polynomial such that $p(A) = 0$, then $q_A(t)$ divides $p(t)$.*

The unique polynomial $q_A(t)$ is called the minimal polynomial of A.

THEOREM 1.32 *Let $A \in \mathbb{C}^{n\times n}$. Then A is diagonalizable if and only if every root of $q_A(t)$ has multiplicity* 1.

The singular value decomposition (SVD) is a very important tool in many fields. The early history and applications of the SVD can be found in Stewart [329].

THEOREM 1.33 *(Singular value decomposition). If $A \in \mathbb{C}^{m\times n}$ has rank k, then it may be written in the form*

$$A = U\Sigma V^H, \tag{1.28}$$

where $U \in \mathbb{C}^{m\times m}$ and $V \in \mathbb{C}^{n\times n}$ are unitary. The matrix $\Sigma = [\sigma_{ij}]_{i,j=1}^{m,n}$ is such that $\sigma_{ij} = 0$ for all $i \neq j$, and

$$\sigma_{11} \geq \sigma_{22} \geq \cdots \geq \sigma_{kk} > \sigma_{k+1,k+1} = \cdots = \sigma_{qq} = 0, \tag{1.29}$$

where $q = \min\{m, n\}$. *The numbers* $\{\sigma_{ij}\} = \{\sigma_i\}$ *are the nonnegative square roots of the eigenvalues of* AA^H *and the columns of* V *are eigenvectors of* $A^H A$ *(arranged in the same order as the corresponding eigenvalues* σ_i^2*).*

The "diagonal entries" $\sigma_i = \sigma_{ii}$ $(i = 1, \ldots, \min\{m, n\})$ of Σ are called the singular values of A, and the columns of U and the columns of V are referred to as the (respectively, left and right) singular vectors of A. A variational characterization of the singular values will be given in Section 1.6

If $A \in \mathbb{C}^{m\times n}$ has rank k, then its singular value decomposition can be written in the partitioned form

$$A = [U_1, U_2] \begin{bmatrix} \Sigma_1 & 0 \\ 0 & 0 \end{bmatrix} \begin{bmatrix} V_1^H \\ V_2^H \end{bmatrix},$$

where $U_1 \in \mathbb{C}^{m\times k}$, $\Sigma_1 = \operatorname{diag}(\sigma_i) \in \mathbb{R}^{k\times k}$ and $V_1^H \in \mathbb{C}^{k\times n}$. Hence

$$A = U_1 \Sigma_1 V_1^H \tag{1.30}$$

is another form of the singular value decomposition. This form of the singular value decomposition is also a full rank factorization of the matrix A. Let $U_1 = [u_1, \ldots, u_k]$ and $V_1 = [v_1, \ldots, v_k]$. Then

$$A = \sum_{i=1}^{k} \sigma_i u_i v_i^H. \tag{1.31}$$

If A is Hermitian, then $u_i = v_i$ $(i = 1, \ldots, k)$ and the above expression is the spectral decomposition of A (see, e.g., [178]).

The singular value decomposition is essentially unique (cf. [330]).

THEOREM 1.34 *Let* $A \in \mathbb{C}^{m\times n}$ $(m \geq n)$ *have the singular value decomposition*

$$U^H A V = \begin{bmatrix} \Sigma \\ 0 \end{bmatrix}.$$

Let

$$\widehat{U}^H A \widehat{V} = \begin{bmatrix} \widehat{\Sigma} \\ 0 \end{bmatrix}$$

be another singular value decomposition. Then $\Sigma = \widehat{\Sigma}$. *Moreover,* $\widehat{V} = VQ$ *where* $Q = V^H \widehat{V}$ *is unitary and*

$$\sigma_i \neq \sigma_j \Rightarrow q_{ij} = 0.$$

A similar statement is true for U *provided we regard the singular vectors* $u_{p+1}, \ldots, u_n$ *as corresponding to zero singular values.*

The CS decomposition was discovered by Davis and Kahan [93], [94] concerning the analysis of invariant subspace perturbation problems. The greater generality of the CS decomposition was discovered by Stewart [326], who also gave an algorithm for computing it ([327]). Paige and Wei [274] give a review of the development and applications of the CS decomposition. Here we quote the most general form of the CS decomposition due to Paige and Saunders [273] (see also [274]).

THEOREM 1.35 *(CS decomposition). Let*

$$Q = \begin{bmatrix} Q_{11} & Q_{12} \\ Q_{21} & Q_{22} \end{bmatrix} \begin{matrix} m \\ p \end{matrix} \qquad \left(Q_{11} \in \mathbb{C}^{m\times k},\ Q_{22} \in \mathbb{C}^{p\times q}\right)$$
$$\qquad\quad k \qquad\ \ q$$

be a unitary matrix of dimension $m + p = k + q$. *Then there exist unitary matrices* U_1, U_2, V_1, V_2 *of dimensions* m, p, k, q *respectively, so that*

$$U^H Q V = \underset{m \qquad p}{\begin{bmatrix} U_1^H & 0 \\ 0 & U_2^H \end{bmatrix}} \begin{bmatrix} Q_{11} & Q_{12} \\ Q_{21} & Q_{22} \end{bmatrix} \underset{k \qquad q}{\begin{bmatrix} V_1 & 0 \\ 0 & V_2 \end{bmatrix}} = D,$$

$$D = \begin{bmatrix} D_{11} & D_{12} \\ D_{21} & D_{22} \end{bmatrix} = \left[\begin{array}{ccc|ccc} I & & & O_S^H & & \\ & C & & & S & \\ & & O_C & & & I \\ \hline O_S & & & I & & \\ & S & & & -C & \\ & & I & & & O_C^H \end{array}\right] \begin{matrix} r \\ s \\ m-r-s \\ p-k+r \\ s \\ k-r-r \end{matrix}$$
$$r \quad s \quad k{-}r{-}s \quad p{-}k{+}r \quad s \quad m{-}r{-}s$$

where

$$C = diag\,(\alpha_{r+1}, \ldots, \alpha_{r+s})\,, \quad 1 > \alpha_{r+1} \geq \cdots \geq \alpha_{r+s} > 0,$$
$$S = diag\,(\beta_{r+1}, \ldots, \beta_{r+s})\,, \quad 0 < \beta_{r+1} \leq \cdots \leq \beta_{r+s} < 1,$$
$$C^2 + S^2 = I.$$

The O_S *and* O_C *are matrices of zeros and, depending on* Q *and the partition, may have no rows and or no columns. Some of the unit matrices may be nonexistent, and no two of them need be equal. The four* C *and* S *matrices are square with the same dimensions, and may be nonexistent.*

The four unitary matrices U_1, U_2, V_1 and V_2 essentially give the singular value decompositions of all four subblocks

$$Q_{ij} = U_i D_{ij} V_j^H \quad (i, j = 1, 2)\,,$$

although eight different unitary matrices would be required if Q were a general matrix.

3. Norms

There are several important results on matrix and vector norms. For the interested reader we recommend Horn and Johnson [203], and Belitskii and Lyubich [36].

DEFINITION 1.36 *Let V be a vector space over a field $\mathbb{F}$ ($\mathbb{R}$ or $\mathbb{C}$). A function $\|\cdot\| : V \to \mathbb{R}$ is a vector norm if for all $x, y \in V$,*
(i) $\|x\| \geq 0$, $\|x\| = 0 \Leftrightarrow x = 0$;
(ii) $\|\lambda x\| = |\lambda| \, \|x\|$ $(\lambda \in \mathbb{F})$;
(iii) $\|x + y\| \leq \|x\| + \|y\|$.

The most common vector norms on $\mathbb{C}^n$ are the sum (or l_1) norm

$$\|x\|_1 = \sum_{i=1}^{n} |x_i| \,, \tag{1.32}$$

the Euclidean (or l_2) norm

$$\|x\|_2 = \left(\sum_{i=1}^{n} |x_i|^2 \right)^{1/2} , \tag{1.33}$$

the max or (l_∞) norm

$$\|x\|_\infty = \max_{1 \leq i \leq n} |x_i| \tag{1.34}$$

and the Hölder p-norm (l_p) norm ($p \geq 1$)

$$\|x\|_p = \left(\sum_{i=1}^{n} |x_i|^p \right)^{1/p} . \tag{1.35}$$

Another important example is the weighted Euclidean norm

$$\|x\|_W = \left(x^H W x \right)^{1/2} , \tag{1.36}$$

where W is some given Hermitian positive definite matrix.

THEOREM 1.37 *(Equivalence of norms). Let $\|\cdot\|_\alpha$ and $\|\cdot\|_\beta$ be any two vector norms on $\mathbb{F}^n$. Then there exist finite positive constants C_m and C_M such that*

$$C_m \|x\|_\alpha \leq \|x\|_\beta \leq C_M \|x\|_\alpha \quad (x \in \mathbb{F}^n) . \tag{1.37}$$

The constants C_m and C_M depend only on α and β. For some α and β they can be found in [203] or [198]. It follows that for any sequence $\left\{ x^{(k)} \right\} \subset \mathbb{F}^n$, $\left\| x^{(k)} - x \right\|_\alpha \to 0$ holds if and only if $\left\| x^{(k)} - x \right\|_\beta \to 0$ holds in any other norm.

DEFINITION 1.38 *A norm* $\|\cdot\|$ *on* $\mathbb{C}^n$ *is monotone if* $|x| \leq |y| \Rightarrow \|x\| \leq \|y\|$ *for all* $x, y \in \mathbb{C}^n$, *and absolute if* $\||x|\| = \|x\|$ *for all* $x \in \mathbb{C}^n$.

The above two properties are equivalent. The p-norms are monotone and absolute.

THEOREM 1.39 *(Bauer, Stoer, Witzgall). A norm on* $\mathbb{C}^n$ *is monotone if and only if it is absolute.*

DEFINITION 1.40 *Let* $\mathbb{F}$ *be a field. A function* $\langle \cdot, \cdot \rangle : \mathbb{F}^n \times \mathbb{F}^n \to \mathbb{F}$ *is an inner product if for all* $x, y, z \in \mathbb{F}^n$,
(i) $\langle x, x \rangle \geq 0$; $\langle x, x \rangle = 0 \Leftrightarrow x = 0$;
(ii) $\langle x + y, z \rangle = \langle x, z \rangle + \langle y, z \rangle$;
(iii) $\langle \lambda x, y \rangle = \lambda \langle x, y \rangle$ $(\lambda \in \mathbb{F})$;
(iv) $\langle x, y \rangle = \overline{\langle y, x \rangle}$.

The natural or Euclidean inner product on $\mathbb{R}^n$ is $\langle x, y \rangle = y^T x$. Similarly, $\langle x, y \rangle = y^H x$ is the Euclidean inner product on $\mathbb{C}^n$. For any Hermitian and positive definite $W \in \mathbb{C}^{n \times n}$, the function $\langle x, y \rangle_W = y^H W x$ is also an inner product.

The inner product extends the concept of orthogonality. Two vectors $x, y \in \mathbb{F}^n$ are orthogonal $(x \perp y)$ with respect to the inner product $\langle \cdot, \cdot \rangle : \mathbb{F}^n \times \mathbb{F}^n \to \mathbb{F}$, if $\langle x, y \rangle = 0$. Similarly, two sets of vectors X and Y are orthogonal $(X \perp Y)$, if $\langle x, y \rangle = 0$ for all $x \in X$ and $y \in Y$.

THEOREM 1.41 *(Cauchy-Schwarz inequality). If* $\langle \cdot, \cdot \rangle : \mathbb{F}^n \times \mathbb{F}^n \to \mathbb{F}$ *is an inner product, then*

$$|\langle x, y \rangle|^2 \leq \langle x, x \rangle \langle y, y \rangle \quad (x, y \in \mathbb{F}^n). \tag{1.38}$$

Equality occurs if and only if $x = \alpha y$ *for some* $\alpha \in \mathbb{F}$.

COROLLARY 1.42 *If* $\langle \cdot, \cdot \rangle : \mathbb{F}^n \times \mathbb{F}^n \to \mathbb{F}$ *is an inner product, then* $\|x\| = \langle x, x \rangle^{1/2}$ *is a norm on* $\mathbb{F}^n$.

The weighted Euclidean norm is easily derived from the inner product $\langle x, y \rangle_W$ as follows: $\|x\|_W = \langle x, x \rangle_W^{1/2}$.

DEFINITION 1.43 *We call a function* $\|\cdot\| : \mathbb{F}^{n \times n} \to \mathbb{R}$ *a matrix norm if for all* $A, B \in \mathbb{F}^{n \times n}$ *it satisfies the following three axioms:*
(i) $\|A\| \geq 0$, $\|A\| = 0 \Leftrightarrow A = 0$;
(ii) $\|\lambda A\| = |\lambda| \, \|A\|$ $(\lambda \in \mathbb{F})$;
(iii) $\|A + B\| \leq \|A\| + \|B\|$.

An important example of matrix norms is the Frobenius norm

$$\|A\|_F = \left(\sum_{i=1}^{n} \sum_{j=1}^{n} |a_{ij}|^2 \right)^{1/2}. \tag{1.39}$$

DEFINITION 1.44 *Given a vector norm on $\mathbb{C}^n$, the corresponding subordinate or induced matrix norm on $\mathbb{C}^{n\times n}$ is defined by*

$$\|A\| = \max_{x\neq 0} \frac{\|Ax\|}{\|x\|} = \max_{\|x\|=1} \|Ax\| . \tag{1.40}$$

The subordinate norm is sometimes called the operator norm or *lub* (least upper bound) norm. For the l_1-, l_∞-, and l_2-vector norms it can be shown that the corresponding induced matrix norms are the following:

$$\|A\|_1 = \max_{1\leq j\leq n} \sum_{i=1}^{n} |a_{ij}| , \tag{1.41}$$

$$\|A\|_\infty = \max_{1\leq i\leq n} \sum_{j=1}^{n} |a_{ij}| \tag{1.42}$$

and

$$\|A\|_2 = \left(\rho\left(A^H A\right)\right)^{1/2} = \sigma_{\max}(A) , \tag{1.43}$$

where the spectral radius

$$\rho(B) = \max\{|\lambda| \mid \det(B - \lambda I) = 0\} ,$$

and where $\sigma_{\max}(A)$ denotes the largest singular value of A. $\|A\|_2$ is called the spectral norm of A.

It is an often used fact that in any induced matrix norm

$$\max_i |\lambda_i| \leq \|A\| , \tag{1.44}$$

where $A \in \mathbb{C}^{n\times n}$ and λ_i denotes the ith eigenvalue of A. This inequality simply follows from the relation

$$|\lambda_i| \|x\| = \|\lambda_i x\| = \|Ax\| \leq \|A\| \|x\| .$$

DEFINITION 1.45 *A matrix norm $\|\cdot\|$ is submultiplicative or consistent if it satisfies the inequality $\|AB\| \leq \|A\| \|B\|$.*

The induced matrix norms are submultiplicative.

DEFINITION 1.46 *A matrix norm $\|\cdot\|_M$ is compatible with a vector norm $\|\cdot\|_v$ if $\|Ax\|_v \leq \|A\|_M \|x\|_v$.*

All vector norms are compatible with the induced matrix norms. The Frobenius norm $\|\cdot\|_F$ is compatible with the Euclidean vector norm $\|\cdot\|_2$.

DEFINITION 1.47 *Assume that $\|\cdot\| : \mathbb{F}^{n\times n} \to \mathbb{R}^+$ is a submultiplicative matrix norm. The quantity*

$$cond(A) = \|A\| \left\|A^{-1}\right\| \tag{1.45}$$

is called the condition number of A.

Note that $\mathrm{cond}(A) \geq 1$.

DEFINITION 1.48 *A matrix norm $\|.\| : \mathbb{F}^{n\times n} \to \mathbb{R}^+$ is called unitarily invariant, if $\|A\| = \|UAV\|$ holds for all unitary matrices U and V.*

Let $A = U\Sigma V^H$ be the singular value decomposition of $A \in \mathbb{F}^{n\times n}$, where $U, V \in \mathbb{F}^{n\times n}$ are unitary matrices and

$$\Sigma = \mathrm{diag}\,(\sigma_1, \ldots, \sigma_n) \in \mathbb{R}^{n\times n}$$

with $\sigma_1 \geq \cdots \geq \sigma_n \geq 0$. Then $\|A\| = \|\Sigma\|$ holds for all $A \in \mathbb{F}^{n\times n}$ in any unitarily invariant matrix norm. Hence $\|A\| = \left\|A^H\right\|$ also holds. Let us denote by $\sigma(A) = [\sigma_1, \ldots, \sigma_n]$ the decreasingly ordered vector of singular values .

DEFINITION 1.49 *A function $\phi : \mathbb{R}^n \to \mathbb{R}$ is called a symmetric gauge function, if*
(i) $\phi(u) > 0$ for all $0 \neq u \in \mathbb{R}^n$;
(ii) $\phi(\gamma u) = |\gamma|\,\phi(u)$ for all $\gamma \in \mathbb{R}$ and $u \in \mathbb{R}^n$;
(iii) $\phi(u+v) \leq \phi(u) + \phi(v)$ for all $u, v \in \mathbb{R}^n$;
(iv) $\phi(u) = \phi(\mathrm{diag}\,(\epsilon_1, \ldots, \epsilon_n)\,\Pi u)$, whenever Π is a permutation matrix and $\epsilon_i = \pm 1$ $(i = 1, \ldots, n)$.

Von Neumann [265] proved the following result (see also Schatten [311], or Horn and Johnson [203]).

THEOREM 1.50 *(von Neumann). If ϕ is a symmetric gauge function and $A \in \mathbb{F}^{n\times n}$, then the matrix function defined by $\phi(\sigma_1(A), \ldots, \sigma_n(A))$ is a unitarily invariant matrix norm. Conversely, every unitarily invariant matrix norm $\|.\|$ has a representation of the form*

$$\|A\| = \phi(\sigma_1(A), \ldots, \sigma_n(A)), \tag{1.46}$$

where ϕ is a symmetric gauge function.

The following examples of symmetric gauge functions are well known. Function

$$\phi(u) = \max_{1\leq i\leq n} |u_i| \tag{1.47}$$

corresponds to the spectral norm. Function

$$\phi(u) = \left(\sum_{i=1}^{n} |u_i|^p\right)^{1/p}, \quad p \geq 1 \tag{1.48}$$

generates Schatten's p-norm. For $p = 1$ and $p = 2$ it corresponds to the trace and the Frobenius norms, respectively. Finally, function

$$\phi_k(u) = \max_{i_1 < i_2 < \cdots < i_k} \left(|u_{i_1}| + \cdots + |u_{i_k}| \right). \tag{1.49}$$

generates the so-called Ky Fan k-norm. Function (1.47) is Ky Fan's k-norm for $k = 1$.

The proof of the following result can be found in Schatten [311].

THEOREM 1.51 *(Schatten). For symmetric gauge functions ϕ, the relation $0 \le u \le v$ $(u, v \in \mathbb{R}^n)$ implies*

$$\phi(u_1, \ldots, u_n) \le \phi(v_1, \ldots, v_n). \tag{1.50}$$

If the symmetric gauge function ϕ is normalized such that

$$\phi(1, 0, \ldots, 0) = 1, \tag{1.51}$$

then

$$\max_{1 \le i \le n} |u_i| \le \phi(u) \le \sum_{i=1}^{n} |u_i| \quad (u \in \mathbb{R}^n) \tag{1.52}$$

holds.

A unitarily invariant matrix norm is submultiplicative if and only if $\|A\| \ge \sigma_1(A)$ for all $A \in \mathbb{F}^{n \times n}$. Particularly, the norms generated by the symmetric gauge functions (1.47)-(1.49) are all submultiplicative (see, Horn, Johnson [203], p. 450).

We use the following generalization of the unitarily invariant norms to rectangular matrices.

Let $A \in \mathbb{F}^{r \times s}$ be an arbitrary rectangular matrix and $1 \le r, s \le n$. Let A be augmented by zeros such that

$$\hat{A} = \begin{bmatrix} A & 0 \\ 0 & 0 \end{bmatrix} \in \mathbb{F}^{n \times n}. \tag{1.53}$$

If $\|.\|$ denotes any unitarily invariant matrix norm on $\mathbb{F}^{n \times n}$, then the quantity

$$\|A\| = \left\| \hat{A} \right\| \tag{1.54}$$

defines the unitarily invariant matrix norm of the rectangular matrix A (see, e.g., [245], [246] or [203]). The corresponding condition number of $A \in \mathbb{F}^{r \times s}$ $(1 \le r, s \le n)$ is then defined as

$$\operatorname{cond}(A) = \|A\| \left\| A^+ \right\|, \tag{1.55}$$

where $A^+ \in \mathbb{F}^{s \times r}$ is the Moore-Penrose inverse of matrix A (see [245], [246] and the next section). Assuming that ϕ is given for any $n \ge 1$, Stewart [326] defines the unitarily invariant norm of $A \in \mathbb{F}^{r \times s}$ $(1 \le r, s \le n)$ by

$$\|A\| = \phi(\sigma_1(A), \ldots, \sigma_p(A)) \quad (p = \min\{r, s\}). \tag{1.56}$$

4. Generalized inverses

Generalized inverses, which enjoy many important properties of the classic inverse of square matrices, play important roles in theory and applications. There are several different types of generalized inverses. We only quote those concepts and results we need later. The interested reader may consult with any of the books [52], [300], [10], [40], [263], [243] and [75].

DEFINITION 1.52 *Let $A \in \mathbb{C}^{m\times n}$. The matrix $A_R^{-1} \in \mathbb{C}^{n\times m}$ is a right inverse of A, if*

$$AA_R^{-1} = I. \tag{1.57}$$

The matrix $A_L^{-1} \in \mathbb{C}^{n\times m}$ is a left inverse of A, if

$$A_L^{-1}A = I. \tag{1.58}$$

The following theorem is easy to prove [300].

THEOREM 1.53 *(i) The matrix $A \in \mathbb{C}^{m\times n}$ has a right inverse if and only if its rank is equal to m. In that case*

$$A_R^{-1} = VA^H\left(AVA^H\right)^{-1},$$

where $V \in \mathbb{C}^{n\times n}$ is arbitrary such that $\text{rank}\left(AVA^H\right) = \text{rank}(A)$.
(ii) The matrix $A \in \mathbb{C}^{m\times n}$ has a left inverse if and only if its rank is equal to n. In that case

$$A_L^{-1} = \left(A^HVA\right)^{-1}A^HV,$$

where $V \in \mathbb{C}^{m\times m}$ is arbitrary such that $\text{rank}\left(A^HVA\right) = \text{rank}(A)$.

The most common generalized inverse is due to Moore and Penrose.

DEFINITION 1.54 *(Penrose). Let $A \in \mathbb{C}^{m\times n}$ be an arbitrary matrix. Then $A^+ \in \mathbb{C}^{n\times m}$ is the unique matrix such that*
(1) $AA^+A = A$;
(2) $A^+AA^+ = A^+$;
(3) $(AA^+)^H = AA^+$;
(4) $(A^+A)^H = A^+A$.

THEOREM 1.55 *Let $A \in \mathbb{C}^{m\times n}$ have the full rank factorization $A = XY$, where $X \in \mathbb{C}^{m\times r}$, $Y \in \mathbb{C}^{r\times n}$ and $\text{rank}(X) = \text{rank}(Y) = \text{rank}(A) = r$. Then*

$$A^+ = Y^H\left(YY^H\right)^{-1}\left(X^HX\right)^{-1}X^H. \tag{1.59}$$

COROLLARY 1.56 *If $A \in \mathbb{F}^{m\times n}$ has rank m, then $A^+ = A^H\left(AA^H\right)^{-1}$. If $\text{rank}(A) = n$, then $A^+ = \left(A^HA\right)^{-1}A^H$.*

REMARK 1.57 *If $A \in \mathbb{F}^{m\times n}$ has rank n and the QR factorization*

$$A = Q\begin{bmatrix} R_1 \\ 0 \end{bmatrix} \quad (R_1 \in \mathbb{F}^{n\times n}),$$

then $A^+ = \left[R_1^{-1}, 0\right] Q^H$ (see e.g., [157]).

THEOREM 1.58 *Let $A \in \mathbb{C}^{m\times n}$ have the singular value decomposition*

$$A = U\begin{bmatrix} \Sigma_+ & 0 \\ 0 & 0 \end{bmatrix} V^H,$$

where $\Sigma_+ = diag(\sigma_i) \in \mathbb{R}^{k\times k}$ (k =rank(A)) is nonsingular. Then

$$A^+ = V\begin{bmatrix} \Sigma_+^{-1} & 0 \\ 0 & 0 \end{bmatrix} U^H.$$

The following theorems provide useful properties of the Moore-Penrose inverse. For other properties, see e.g., [300].

THEOREM 1.59 *Suppose that $A \in \mathbb{C}^{m\times n}$. Then*
(i) $(A^+)^+ = A$;
(ii) $(A^+)^H = \left(A^H\right)^+$;
(iii) If $\lambda \in \mathbb{C}$, $(\lambda A)^+ = \lambda^+ A^+$ where $\lambda^+ = \frac{1}{\lambda}$ if $\lambda \neq 0$ and $\lambda^+ = 0$ if $\lambda = 0$;
(iv) $A^H = A^H A A^+ = A^+ A A^H$;
(v) $\left(A^H A\right)^+ = A^+ \left(A^H\right)^+$;
(vi) $A^+ = \left(A^H A\right)^+ A^H = A^H \left(A A^H\right)^+$;
(vii) $(UAV)^+ = V^H A^+ U^H$ where U, V are unitary matrices.

THEOREM 1.60 *If $A \in \mathbb{C}^{m\times n}$, then*
(i) $\mathcal{R}(A) = \mathcal{R}(AA^+) = \mathcal{R}\left(AA^H\right)$;
(ii) $\mathcal{R}(A^+) = \mathcal{R}\left(A^H\right) = \mathcal{R}(A^+A) = \mathcal{R}\left(A^H A\right)$;
(iii) $\mathcal{R}(I - AA^+) = \mathcal{N}(AA^+) = \mathcal{N}\left(A^H\right) = \mathcal{N}(A^+) = \mathcal{R}^\perp(A)$;
(iv) $\mathcal{R}(I - A^+A) = \mathcal{N}(A^+A) = \mathcal{N}(A) = \mathcal{R}^\perp\left(A^H\right)$.

DEFINITION 1.61 *The matrix $G \in \mathbb{C}^{n\times m}$ is called an $(i_1, \ldots, i_k)$-inverse for $A \in \mathbb{C}^{m\times n}$ if G satisfies the Penrose equations $(i_1), \ldots, (i_k)$ from among the following equations:*
(1) $AGA = A$;
(2) $GAG = G$;
(3) $(AG)^H = AG$;
(4) $(GA)^H = GA$.
Furthermore, the set of all $(i_1, \ldots, i_k)$-inverses for A will be denoted by $A^{\{i_1,\ldots,i_k\}}$.

A particular $(i_1, \ldots, i_k)$-inverse will be denoted by $A^{(i_1,\ldots,i_k)}$. The (1)-inverse is often called g-inverse or simply generalized inverse. The usual notation of g-inverses is A^-. If A^- is a g-inverse of A, then rank$(A^-) \geq$rank(A). The $(1,2)$-inverses are called reflexive inverses . A g-inverse A^- is a reflexive inverse if and only if rank$(A^-) =$rank(A) .

THEOREM 1.62 *(Urquhart).* $A^+ = A^{(1,4)}AA^{(1,3)}$.

For the proof, see [40]. We now give the following characterizations of g-inverses and reflexive inverses.

THEOREM 1.63 *Let* $A \in \mathbb{C}^{m\times n}$ *be of rank* r, *and let* $E \in \mathbb{C}^{m\times m}$ *and* $P \in \mathbb{C}^{n\times n}$ *be nonsingular matrices such that*

$$EAP = \begin{bmatrix} I_r & K \\ 0 & 0 \end{bmatrix}. \tag{1.60}$$

Then for any $L \in \mathbb{C}^{(n-r)\times(m-r)}$, *the* $n \times m$ *matrix*

$$G = P \begin{bmatrix} I_r & 0 \\ 0 & L \end{bmatrix} E \tag{1.61}$$

is a g*-inverse of* A.

Since rank$(X) = r+$rank(L), it follows that a g-inverse of A exists having any specified rank between r and $\min\{m,n\}$. The form of EAP can be computed by the Gaussian elimination.

THEOREM 1.64 *Let* $A = XY$ *be a full rank factorization. Then* $G \in A^{\{1,2\}}$ *if and only if it can be expressed as*

$$G = Y_R^{-1}X_L^{-1}, \tag{1.62}$$

where X_L^{-1} *is the left inverse of* X *and* Y_R^{-1} *is the right inverse of* Y.

THEOREM 1.65 *A* g*-inverse* A^- *is a reflexive inverse of* A *if and only if* A^- *can be expressed as*

$$A^- = XAY, \tag{1.63}$$

where $X, Y \in A^{\{1\}}$.

A particular $\{1,2\}$-inverse is given by

$$G = P \begin{bmatrix} I_r & 0 \\ 0 & 0 \end{bmatrix} E.$$

Other representations of $(i_1, \ldots, i_k)$-inverses can be found in [52], [300], [10], [263], [75] and especially in [40].

There are generalized inverses that have some of the spectral properties of the inverse of a nonsingular matrix.

DEFINITION 1.66 *Let $A \in \mathbb{C}^{n\times n}$. The smallest nonnegative integer k such that $rank(A^k)$ $=rank(A^{k+1})$, is called the index of A and is denoted by $Ind(A)$.*

If A is nonsingular, then $\text{Ind}(A) = 0$.

DEFINITION 1.67 *(Drazin inverse). If $A \in \mathbb{C}^{n\times n}$ with $Ind(A) = k$ and if $A^D \in \mathbb{C}^{n\times n}$ is such that*
$A^D A A^D = A^D$,
$AA^D = A^D A$,
$A^{k+1}A^D = A^k$,
then A^D is called the Drazin inverse of A.

The Drazin inverse exists and is unique for all $A \in \mathbb{C}^{n\times n}$. The following result gives a canonical form for A and A^D.

THEOREM 1.68 *If $A \in \mathbb{C}^{n\times n}$ is such that $Ind(A) = k > 0$, then there exists a nonsingular matrix P such that*

$$A = P\begin{bmatrix} C & 0 \\ 0 & N \end{bmatrix}P^{-1} \tag{1.64}$$

where C is any nonsingular matrix and N is nilpotent of index k. Furthermore, if P, C and N are any matrices satisfying the above conditions, then

$$A^D = P\begin{bmatrix} C^{-1} & 0 \\ 0 & 0 \end{bmatrix}P^{-1}. \tag{1.65}$$

In special cases the Drazin inverse is a g-inverse of $A \in \mathbb{C}^{n\times n}$.

THEOREM 1.69 *For $A \in \mathbb{C}^{n\times n}$, $AA^D A = A$ if and only if $Ind(A) \leq 1$.*

Proof. If $\text{Ind}(A) = 0$, then $A^D = A^{-1}$ and $AA^D A = A$. Suppose that $\text{Ind}(A) \geq 1$. Then by the preceding theorem we have $AA^D A = A$ if and only if $N = 0$. But $N = 0$ if and only if $\text{Ind}(A) = 1$. ■

DEFINITION 1.70 *If $A \in \mathbb{C}^{n\times n}$ is such that $Ind(A) \leq 1$, then the Drazin inverse of A is called the group inverse of A and is denoted by $A^{\#}$. When it exists, $A^{\#}$ is characterized as the unique matrix satisfying the three equations*

$$AA^{\#}A = A, \quad A^{\#}AA^{\#} = A^{\#}, \quad AA^{\#} = A^{\#}A. \tag{1.66}$$

The following existence result is a special case of Theorem 1.68.

THEOREM 1.71 *For $A \in \mathbb{C}^{n\times n}$, $A^{\#}$ exists if and only if there exist nonsingular matrices P and C such that*

$$A = P\begin{bmatrix} C & 0 \\ 0 & 0 \end{bmatrix} P^{-1}. \tag{1.67}$$

THEOREM 1.72 *(Cline). Let $A \in \mathbb{C}^{n\times n}$ have the full rank factorization*

$$A = FG. \tag{1.68}$$

Then A has a group inverse if and only if GF is nonsingular, in which case

$$A^{\#} = F\,(GF)^{-2}\,G. \tag{1.69}$$

Proof. Let r =rank(A). Then $GF \in \mathbb{C}^{r\times r}$ and $A^2 = FGFG$. Since rank(A^2) $\leq$rank(GF) $\leq$rank(A), the condition rank(A^2) =rank(A) holds if and only if rank(A) =rank$(GF) = r$. This is equivalent to the nonsingularity of GF. Hence $A^{\#}$ exists. It is easy to verify that the above expression of $A^{\#}$ satisfies the three equations of (1.66). ■

The following results show how the generalized inverse can characterize the solution of linear equations.

THEOREM 1.73 *Assume that $A \in \mathbb{C}^{m\times n}$ and let A^- be an arbitrary g-inverse of A. The equation $Ax = b$ is consistent if and only if*

$$AA^-b = b$$

in which case the general solution is

$$x = A^-b + \left(I - A^-A\right)z \quad (z \in \mathbb{C}^n).$$

The g-inverses can be characterized as equation solving inverses.

THEOREM 1.74 *(Bose). Let $A \in \mathbb{C}^{m\times n}$, $X \in \mathbb{C}^{n\times m}$. Then $X \in A^{\{1\}}$ if and only if $x = Xb$ is a solution of $Ax = b$ for every $b \in \mathcal{R}(A)$.*

THEOREM 1.75 *Let $A \in \mathbb{C}^{m\times n}$, $b \in \mathbb{C}^m$. $X \in A^{\{1,4\}}$ if and only if $x = Xb$ is the minimum norm solution of $Ax = b$ for every $b \in \mathcal{R}(A)$.*

We close the section with a result on the Moore-Penrose inverse of a partitioned matrix (see e.g., [10] or [75], p. 54).

THEOREM 1.76 *(Greville). For $B \in \mathbb{C}^{m\times n}$, $c \in \mathbb{C}^m$, and $P = [B, c]$, let $k = B^+c$ and $u = (I - BB^+)\,c = c - Bk$. The generalized inverse of P is given by*

$$P^+ = \begin{bmatrix} B^+ - ky \\ y \end{bmatrix}, \tag{1.70}$$

where

$$y = \begin{cases} u^+, & u \neq 0, \\ \left(1 + k^Hk\right)^{-1} k^H B^+, & u = 0. \end{cases} \tag{1.71}$$

5. Linear least squares problems

The linear least squares problem is the solution of the minimization problem

$$\|Au-b\|_2^2 \to \min, \tag{1.72}$$

where $A \in \mathbb{C}^{m\times n}$ and $b \in \mathbb{C}^m$.

DEFINITION 1.77 *Suppose that $A \in \mathbb{C}^{m\times n}$ and $b \in \mathbb{C}^m$. Then a vector $u \in \mathbb{C}^n$ is called a least squares solution to $Ax=b$, if $\|Au-b\|_2 \le \|Av-b\|_2$ for all $v \in \mathbb{C}^n$. A vector u is called the minimal (norm) least squares solution to $Ax=b$ if u is a least squares solution to $Ax=b$ and $\|u\|_2 < \|w\|_2$ for all other least squares solutions w.*

THEOREM 1.78 *Suppose that $A \in \mathbb{C}^{m\times n}$ and $b \in \mathbb{C}^m$. Then A^+b is the minimal least squares solution to $Ax=b$.*

THEOREM 1.79 *Suppose that $A \in \mathbb{C}^{m\times n}$ and $b \in \mathbb{C}^m$. Then the following statements are equivalent:*
(i) u is a least squares solution of $Ax=b$;
(ii) u is a solution of $Ax=AA^+b$;
(iii) u is a solution of $A^HAx=A^Hb$;
(iv) u is of the form A^+b+h, where $h \in \mathcal{N}(A)$.

LEMMA 1.80 *$x=A^+b$ is the unique vector which satisfies $x \in \mathcal{R}\left(A^T\right)$ and $(Ax-b)\perp\mathcal{R}(A)$, or equivalently $x\perp\mathcal{N}(A)$ and $(Ax-b) \in \mathcal{N}\left(A^T\right)$.*

The weighted least squares problem is defined by

$$\|Ax-b\|_W^2 \to \min .$$

It is known [40], [191] that x minimizes $\|Ax-b\|_W$ if and only if x is a solution of

$$A^TGAx = A^TGb. \tag{1.73}$$

6. Variational characterizations of eigenvalues

We exploit the following results.

THEOREM 1.81 *(Rayleigh-Ritz). Let $A \in \mathbb{C}^{n\times n}$ be Hermitian, and let the eigenvalues of A be ordered as*

$$\lambda_{\min} = \lambda_1 \le \lambda_2 \le \cdots \le \lambda_{n-1} \le \lambda_n = \lambda_{\max}. \tag{1.74}$$

Then

$$\lambda_1 x^Hx \le x^HAx \le \lambda_n x^Hx \quad \left(x \in \mathbb{C}^n\right), \tag{1.75}$$

$$\lambda_{\max} = \lambda_n = \max_{x\neq 0}\frac{x^HAx}{x^Hx} = \max_{x^Hx=1} x^HAx, \tag{1.76}$$

$$\lambda_{\min} = \lambda_1 = \min_{x \neq 0} \frac{x^H A x}{x^H x} = \min_{x^H x} x^H A x. \tag{1.77}$$

THEOREM 1.82 *(Courant-Fischer). Let $A \in \mathbb{C}^{n \times n}$ be a Hermitian matrix with eigenvalues $\lambda_1 \leq \lambda_2 \leq \cdots \leq \lambda_n$, and let k be a given integer with $1 \leq k \leq n$. Then*

$$\min_{w_1,\ldots,w_{n-k} \in \mathbb{C}^n} \max_{\substack{x \neq 0,\ x \in \mathbb{C}^n \\ x \perp w_1,\ldots,w_{n-k}}} \frac{x^H A x}{x^H x} = \lambda_k \tag{1.78}$$

and

$$\max_{w_1,\ldots,w_{k-1} \in \mathbb{C}^n} \min_{\substack{x \neq 0,\ x \in \mathbb{C}^n \\ x \perp w_1,\ldots,w_{k-1}}} \frac{x^H A x}{x^H x} = \lambda_k. \tag{1.79}$$

An application of the Courant-Fischer theorem is the following.

THEOREM 1.83 *Let $A \in \mathbb{C}^{m \times n}$, let $q = \min\{m, n\}$, let $\sigma_1 \geq \sigma_2 \geq \cdots \geq \sigma_q$ be the ordered singular values of A, and let k be a given integer with $1 \leq k \leq q$. Then*

$$\min_{w_1,\ldots,w_{k-1} \in \mathbb{C}^n} \max_{\substack{x \neq 0,\ x \in \mathbb{C}^n \\ x \perp w_1,\ldots,w_{k-1}}} \frac{\|Ax\|_2}{\|x\|_2} = \sigma_k \tag{1.80}$$

and

$$\max_{w_1,\ldots,w_{n-k} \in \mathbb{C}^n} \min_{\substack{x \neq 0,\ x \in \mathbb{C}^n \\ x \perp w_1,\ldots,w_{n-k}}} \frac{\|Ax\|_2}{\|x\|_2} = \sigma_k. \tag{1.81}$$

For the proofs we refer to [203].

THEOREM 1.84 *(Inclusion principle). Let $A \in \mathbb{C}^{n \times n}$ be a Hermitian matrix, let r be an integer with $1 \leq r \leq n$, and let A_r denote any principal submatrix of A (obtained by deleting $n-r$ columns and the corresponding columns from A). If the eigenvalues of A and A_r are arranged in increasing order (1.74), we have*

$$\lambda_k(A) \leq \lambda_k(A_r) \leq \lambda_{k+n-r}(A) \quad (1 \leq k \leq r). \tag{1.82}$$

THEOREM 1.85 *(Poincaré separation theorem). Let $A \in \mathbb{C}^{n \times n}$ be Hermitian, let r be a given integer with $1 \leq r \leq n$, and let $u_1, \ldots, u_r \in \mathbb{C}^n$ be r given orthonormal vectors. Let $B_r = \left[u_i^H A u_j\right]_{i,j=1}^{r} \in \mathbb{C}^{r \times r}$. If the eigenvalues of A and B_r are arranged in increasing order (1.74), we have*

$$\lambda_k(A) \leq \lambda_k(B_r) \leq \lambda_{k+n-r}(A) \quad (k = 1, 2, \ldots, r). \tag{1.83}$$

THEOREM 1.86 *Let $A, B \in \mathbb{C}^{m \times n}$ be given and let $q = \min\{m, n\}$. The following inequalities hold for the decreasingly ordered singular values of A,*

B and $A+B$:
(i) $\sigma_{i+j-1}(A+B) \leq \sigma_i(A) + \sigma_j(B)$*;*
(ii) $\sigma_{i+j-1}(AB^H) \leq \sigma_i(A)\sigma_j(B)$*;*
for $1 \leq i, j \leq q$ *and* $i+j \leq q+1$*. In particular,*
(iii) $|\sigma_i(A+B) - \sigma_i(A)| \leq \sigma_1(B)$ *for* $i = 1, \ldots, q$
and
(iv) $\sigma_i(AB^H) \leq \sigma_i(A)\sigma_1(B)$ *for* $i = 1, \ldots, q$*.*

7. Angles between subspaces and the CS decomposition

The concept of principal angles between subspaces was first introduced by Jordan. Since then it was rediscovered and developed several times, sometimes independently (see [8], [94], [274], [180]). The principal angles are means to describe the relative orientation of two subspaces.

Let $M, N \subseteq \mathbb{R}^n$ be two subspaces such that $p = \dim(M) \geq \dim(N) = q$. The *principal angles* $\theta_1, \ldots, \theta_q \in [0, \pi/2]$ between M and N are defined recursively. The first principal angle is the smallest angle that satisfies

$$\cos\theta_1 = \max\left\{u^T v \mid u \in M, \|u\|_2 = 1, v \in N, \|v\|_2 = 1\right\}. \tag{1.84}$$

Clearly $0 \leq \cos(\theta_1) \leq 1$, $\theta_1 = \pi/2$ if and only if $M \perp N$, and $\theta_1 = 0$ if and only if $M \cap N \neq \{0\}$. Suppose the maximum is attained when $u = u_1$ and $v = v_1$. The second principal angle θ_2 is defined by

$$\begin{aligned} \cos\theta_2 = \max\{u^T v \mid\ & u \in M \cap \mathcal{R}^{\perp}(u_1), \|u\|_2 = 1, \\ & v \in N \cap \mathcal{R}^{\perp}(v_1), \|v\|_2 = 1\}. \end{aligned}$$

Let u_2, v_2 be the pair for which the maximum is attained. In general the principal angles are defined recursively for $i = 1, \ldots, q$ by

$$\cos\theta_i = \max\left\{u^T v \mid u \in M_i, \|u\|_2 = 1, v \in N_i, \|v\|_2 = 1\right\}, \tag{1.85}$$

where

$$M_i = M \cap \mathcal{R}^{\perp}(u_1, \ldots, u_{i-1}), \quad N_i = N \cap \mathcal{R}^{\perp}(v_1, \ldots, v_{i-1}). \tag{1.86}$$

The vectors $u_i \in M_i$ and $v_i \in N_i$ are chosen to be a pair for which the maximum is attained. It is clear that

$$0 \leq \theta_1 \leq \theta_2 \leq \cdots \leq \theta_q \leq \pi/2. \tag{1.87}$$

The orthonormal vectors $u_1, \ldots, u_q \in M$ and $v_1, \ldots, v_q \in N$ are called *principal vectors* of the pair (M, N). If $i \neq j$, then $u_i^T v_j = 0$. The principal angles are uniquely determined, while the principal vectors are not. If

$\dim(M \cap N) = i$, then $\theta_1 = \theta_2 = \cdots = \theta_i = 0$ and $\theta_{i+1} \neq 0$ (if $i < q$). The first principal angle is also called the minimal angle between M and N. We also use the notations $\theta_{\min}$ and $\theta_{\max}$ for the smallest and the largest principal angles, respectively.

The cosines of the principal angles can be computed using the following result of Björck and Golub [48] (see also [165]).

THEOREM 1.87 *(Björck, Golub). Let the columns of $Q_M \in \mathbb{R}^{n\times p}$ and $Q_N \in \mathbb{R}^{n\times q}$ be orthonormal bases for M and N, respectively. Put $C = Q_M^T Q_N$, and let the singular value decomposition of this $p \times q$ matrix be*

$$C = YCZ^T, \quad C = diag(\sigma_1, \ldots, \sigma_q), \ Y \in \mathbb{R}^{p\times q}, \ Z \in \mathbb{R}^{q\times q}, \tag{1.88}$$

where $Y^T Y = Z^T Z = ZZ^T = I_q$. Then the principal angles and principal vectors associated with this pair of subspaces are given by

$$\cos(\theta_i) = \sigma_i(C) \quad (i = 1, \ldots, q), \quad U = Q_M Y, \quad V = Q_N Z. \tag{1.89}$$

As $\sigma_i(A) = \sigma_i(A^T)$, the singular value decomposition of $Q_N^T Q_M \in \mathbb{R}^{q\times p}$ also gives the cosines of the principal angles θ_i $(i = 1, \ldots, q)$.

Sommerville [323] presents Jordan's classic work on the principal angles. Modernized versions of this material can be found in Gluck [158] and Csikós [92]. A neat elementary introduction of the principal angles can be found in Watkins [354] for $p = q$. Bhatia [46] gives a different setting using the CS decomposition, also in the case $p = q$. Bhatia's presentation is related to the work of Davis and Kahan [93], [94]. A third approach, which is based on the simultaneous block diagonalization of projections with orthogonal similarity is due to Wedin [355].

We need a special, easy to prove, consequence of the CS decomposition [143].

THEOREM 1.88 *Let $W \in \mathbb{R}^{n\times n}$ be orthogonal and partitioned in the form*

$$W = \begin{bmatrix} A & B \\ C & D \end{bmatrix} \quad \left(A \in \mathbb{R}^{p\times q}, \ p \geq q\right). \tag{1.90}$$

If $p + q \leq n$, then the submatrices have the following singular value sets

$$\sigma(A) = \{\gamma_1, \ldots, \gamma_q\}, \quad \sigma(B) = \{\mu_1, \ldots, \mu_q, \underbrace{1, \ldots, 1}_{p-q}\}, \tag{1.91}$$

$$\sigma(C) = \{\mu_1, \ldots, \mu_q\}, \quad \sigma(D) = \{\gamma_1, \ldots, \gamma_q, \underbrace{1, \ldots, 1}_{n-p-q}\}, \tag{1.92}$$

where $\gamma_i^2 + \mu_i^2 = 1$ $(i = 1, \ldots, q)$. If $p + q > n$, then

$$\sigma(A) = \{\gamma_1, \ldots, \gamma_{n-p}, \underbrace{1, \ldots, 1,}_{p+q-n}\}, \quad \sigma(B) = \{\mu_1, \ldots, \mu_{n-p}, \underbrace{1, \ldots, 1}_{p-q}\}, \tag{1.93}$$

$$\sigma(C) = \{\mu_1, \ldots, \mu_{n-p}\}, \quad \sigma(D) = \{\gamma_1, \ldots, \gamma_{n-p}\}, \tag{1.94}$$

where $\gamma_i^2 + \mu_i^2 = 1$ $(i = 1, \ldots, n-p)$.

Proof. Let $r = n - p$ and $s = n - q$. Since the columns of A are orthonormal, $A^T A + C^T C = I_q$. Matrices $A^T A$ and $C^T C$ have common eigenvectors. If $A^T A x = \lambda x$, then $C^T C x = \mu x$, where $\lambda + \mu = 1$ and vice versa. Hence

$$\lambda_i\left(A^T A\right) + \lambda_i\left(C^T C\right) = 1 \quad (i = 1, \ldots, q). \tag{1.95}$$

Using similar arguments we obtain the relations

$$\lambda_i\left(B^T B\right) + \lambda_i\left(D^T D\right) = 1 \quad (i = 1, \ldots, s) \tag{1.96}$$

and

$$\lambda_i\left(CC^T\right) + \lambda_i\left(DD^T\right) = 1 \quad (i = 1, \ldots, r). \tag{1.97}$$

We exploit the relation between the singular values of A and the eigenvalues of $A^T A$ or AA^T. Since $p \geq q$, the singular value decomposition of A has the form

$$A = U \begin{bmatrix} \Sigma_1 \\ 0 \end{bmatrix} V^T \quad \left(\Sigma_1 = \operatorname{diag}\left(\sigma_i(A)\right) \in \mathbb{R}^{q \times q}\right).$$

Thus $A^T A = V \Sigma_1^2 V^T \in \mathbb{R}^{q \times q}$ and $AA^T = U \begin{bmatrix} \Sigma_1^2 & 0 \\ 0 & 0 \end{bmatrix} U^T \in \mathbb{R}^{p \times p}$. Hence we can write $\lambda_i\left(A^T A\right) = \sigma_i^2(A)$ for $i = 1, \ldots, q$, and

$$\lambda_i\left(AA^T\right) = \begin{cases} \sigma_i^2(A), & i = 1, \ldots, q, \\ 0, & i = q+1, \ldots, p. \end{cases} \tag{1.98}$$

We determine the possible block sizes in W. The assumption $p \geq q$ implies that $r \leq s$. If $r > s$ would hold, then $n = p + r > q + s = n$, which is a contradiction. Now there are two cases: $r \geq q$ $(n \geq p+q)$ or $r < q$ $(n < p+q)$. If $r \geq q$, then $p \leq s$ $(p > s$ would imply again that $n = p + r > s + q = n)$ and $q \leq n/2$. If $r < q$, then $p > s$ and $p \geq n/2$.

We first assume that $r \geq q$. Let $\{\gamma_i\}_{i=1}^q$ denote the singular values of A in any order. Similarly, let $\{\mu_i\}_{i=1}^q$ be the singular values of C. Then relation (1.95) implies that $\gamma_i^2 + \mu_i^2 = 1$ for $i = 1, \ldots, q$. Since $q \leq r \leq s$, a similar argument with (1.96) yields that $\sigma_i(D) = \sigma_i(A)$ for $i = 1, \ldots, q$ and $\sigma_i(D) = 1$ for $i = q+1, \ldots, r$. Note that $r - q = n - p - q$. The inequality $q \leq p \leq s$ and relation (1.97) together entail that $\sigma_i(B) = \mu_i$ for $i = 1, \ldots, q$ and $\sigma_i(B) = 1$ for $i = q+1, \ldots, p$.

If $r < q$, then C has $r = n - p$ singular values μ_i. Thus relation (1.95) yields that $\sigma_i(A) = \gamma_i$ for $i = 1, \ldots, r$ and $\sigma_i(A) = 1$ for $i = r+1, \ldots, q$. Since $r \leq s$, relation (1.96) gives $\sigma_i(D) = \gamma_i$ for $i = 1, \ldots, r$. Finally, it follows from $p \geq s \geq r$ and relation (1.97) that $\sigma_i(D) = \mu_i$ for $i = 1, \ldots, r$ and $\sigma_i(D) = 1$ for $i = r+1, \ldots, s$. ■

The sets of singular values given in the theorem are not ordered. In the sequel we assume that the singular values are in decreasing order as in Theorem 1.33.

Let $U = [u_1, \ldots, u_n] \in \mathbb{R}^{n \times n}$ be orthogonal such that $U_1 = [u_1, \ldots, u_p]$ is a basis for M, and $U_2 = [u_{p+1}, \ldots, u_n]$ is a basis for $M^\perp$. Similarly, let $V = [v_1, \ldots, v_n] \in \mathbb{R}^{n \times n}$ be an orthogonal matrix such that $V_1 = [v_1, \ldots, v_q]$ is a basis for N, and $V_2 = [v_{q+1}, \ldots, v_n]$ is a basis for $N^\perp$. Later we use these four bases for the representation of projections. Therefore we consider the partitioned orthogonal matrix

$$U^T V = \begin{bmatrix} U_1^T V_1 & U_1^T V_2 \\ U_2^T V_1 & U_2^T V_2 \end{bmatrix} \quad \left(U_1^T V_1 \in \mathbb{R}^{p \times q}\right). \tag{1.99}$$

Let $\theta_i = \theta_i(M, N)$ be the ith principal angle between the subspaces M and N and let

$$\cos(\theta(M, N)) = [\cos(\theta_1(M, N)), \ldots, \cos(\theta_q(M, N))]. \tag{1.100}$$

It follows from the theorem of Björck and Golub that

$$\sigma\left(U_1^T V_1\right) = \cos(\theta(M, N)), \quad \sigma\left(U_2^T V_1\right) = \cos\left(\theta\left(M^\perp, N\right)\right), \tag{1.101}$$

$$\sigma\left(U_1^T V_2\right) = \cos\left(\theta\left(M, N^\perp\right)\right), \quad \sigma\left(U_2^T V_2\right) = \cos\left(\theta\left(M^\perp, N^\perp\right)\right). \tag{1.102}$$

For $p + q \leq n$, Theorem 1.88 implies that $\gamma_i = \cos(\theta_i)$, $\mu_i = \sin(\theta_i)$ and $\theta_i\left(M^\perp, N\right) = \frac{\pi}{2} - \theta_{q+1-i}(M, N)$ for all $i = 1, \ldots, q$. Similar facts can be said about the other pairs of subspaces and also in the case $p + q > n$. Note that the ones among the singular values of $U_1^T V_2$ and $U_2^T V_2$ correspond to 0 principal angles. We can also observe the following.

PROPOSITION 1.89 *$\theta_i(M, N) = \theta_i\left(M^\perp, N^\perp\right)$ for all $i = 1, \ldots, q$, if $p + q = n$.*

The result is due to Jordan (see e.g. [274]). For the first principal angle and complementary subspaces it was also proved by Krein, Krasnosel'skii and Milman in abstract spaces (see, Gohberg and Krein [160]).

The Björck, Golub theorem provides the possibility to compare other definitions of the principal angles with the one given here. Zassenhaus [368] defines the angles (of inclination) between M and N as follows. Let the columns of $U \in \mathbb{R}^{n \times p}$ and $V \in \mathbb{R}^{n \times q}$ $(p \geq q)$ be arbitrary bases for M and N, respectively. Then the eigenvalues of the matrix

$$f(M, N) = V^T U \left(U^T U\right)^{-1} U^T V \left(V^T V\right)^{-1} \tag{1.103}$$

are real nonnegative and ≤ 1. The eigenvalues are the squares of the cosines of the angles between M and N. The eigenvalues are, of course, independent of the choice of the bases. Substitute U by Q_M and V by Q_N. Then

$$f(M,N) = Q_N^T Q_M Q_M^T Q_N = C^T C.$$

Since $\lambda_i\left(C^T C\right) = \sigma_i^2(C)$, the two definitions of the principal angles agree. Zassenhaus' definition has the following symmetric form [39] without any restriction on p and q.

THEOREM 1.90 *(Ben-Israel). Let M and N be two arbitrary subspaces in $\mathbb{R}^n$, and let U and V be two matrices whose columns span M and N, respectively. Then the nonzero eigenvalues of the matrix*

$$U^+ V V^+ U \tag{1.104}$$

are the squares of the cosines of the principal angles between the subspaces M and N.

8. Properties of triangular factorizations

The triangular factorizations will play a key role in several sections of the book. Here we give a geometric characterization of the LU decomposability [144] and investigate the stability of LU factorizations [146], [148].

We recall that for any nonsingular matrix $A \in \mathbb{R}^{n\times n}$ there is a unique QR factorization $A = Q_A R_A$ such that Q_A is orthogonal and the main diagonal of R_A contains only positive entries. If the Cholesky decomposition $A^T A = R^T R$ is known, then $Q_A = AR^{-1}$ and $R_A = R$.

We also need the Householder notation. Given any matrix A, the matrices $A^{\overline{k}}$, $A^{|k}$, $A^{\underline{k}}$ and $A^{k|}$ will denote the submatrices consisting of the first k rows, the first k columns, the last k rows and the last k columns, respectively. We will use the Householder notation for block matrices as well.

PROPOSITION 1.91 *A nonsingular matrix $A \in \mathbb{R}^{n\times n}$ has an LU factorization if and only if its orthogonal factor Q_A has an LU factorization.*

Proof. If the LU factorization $A = L_A U_A$ exists, then $L_A U_A = Q_A R_A$, from which $L_A\left(U_A R_A^{-1}\right) = Q_A$ follows. As $U_A R_A^{-1}$ is upper triangular, this proves the only if part. Conversely, if $Q_A = LU$, then $A = Q_A R_A = L\left(UR_A\right)$ proving the LU decomposability of matrix A. ■

Let us consider the orthogonal factor Q_A in 2×2 partitioned form:

$$Q_A = \begin{bmatrix} Q_{11}^{(k)} & Q_{12}^{(k)} \\ Q_{21}^{(k)} & Q_{22}^{(k)} \end{bmatrix} \quad \left(Q_{11}^{(k)} \in \mathbb{R}^{k\times k}\right). \tag{1.105}$$

The matrix Q_A has an LU factorization, if and only if $Q_{11}^{(k)}$ is nonsingular for all $k = 1, \ldots, n-1$. The matrix $Q_{11}^{(k)}$ is nonsingular, if and only if its singular values are positive.

Substituting $U = Q_A^T$, $V = I$, $U_1 = \left(Q_A^T\right)^{|k}$, $U_2 = \left(Q_A^T\right)^{n-k|}$, $V_1 = I^{|k}$ and $V_2 = I^{n-k|}$ in partition (1.99) we obtain partition (1.105). Thus

$$\sigma_i\left(Q_{11}^{(k)}\right) = \cos\left(\theta_i\left(\mathcal{R}\left(\left(Q_A^T\right)^{|k}\right), \mathcal{R}\left(I^{|k}\right)\right)\right), \quad i = 1, \ldots, k. \qquad (1.106)$$

Hence the matrix $Q_{11}^{(k)}$ is nonsingular, if $\cos(\theta_i) > 0$ holds for all i. This happens, if and only if $\theta_i < \pi/2$ for all $i = 1, \ldots, k$. In other words, the angles between the subspaces $\mathcal{R}\left(Q_A^{\bar{k}}\right)$ (row space) and $\mathcal{R}\left(I^{|k}\right)$ must be less than $\pi/2$. Thus we have proved

PROPOSITION 1.92 *Let $A \in \mathbb{R}^{n\times n}$ be nonsingular and let $A = Q_A R_A$ be its unique QR factorization with positive diagonal entries in R_A. Then A has an LU factorization if and only if*

$$\theta_k\left(\mathcal{R}\left(\left(Q_A^T\right)^{|k}\right), \mathcal{R}\left(I^{|k}\right)\right) < \pi/2 \quad (k = 1, \ldots, n-1). \qquad (1.107)$$

If a nonsingular matrix A has no LU factorization, then a permutation matrix P exists such that PA has an LU factorization. As $PA = (PQ_A)R_A$ is also a QR factorization, the multiplication of A by P (change of rows) improves on the orthogonal part Q_A of matrix. In other words, the change of row vectors keeps out $\pi/2$ as a principal angle.

8.1 Perturbations of triangular factorizations

We use the following notation. Let $A = [a_{ij}]_{i,j=1}^n$. Then

$$\operatorname{diag}(A) = \operatorname{diag}(a_{11}, a_{22}, \ldots, a_{nn}), \qquad (1.108)$$

$\operatorname{tril}(A, l) = [\alpha_{ij}]_{i,j=1}^n$ and $\operatorname{triu}(A, l) = [\beta_{ij}]_{i,j=1}^n$, where $0 \le |l| < n$ and

$$\alpha_{ij} = \begin{cases} a_{ij}, & i \ge j - l \\ 0, & i < j - l \end{cases}, \qquad \beta_{ij} = \begin{cases} a_{ij}, & i \le j - l \\ 0, & i > j - l \end{cases}.$$

Matrices $\operatorname{tril}(A, l)$ and $\operatorname{triu}(A, l)$ are called the lower and upper triangular parts of A, respectively. Related special notations are $\operatorname{tril}(A) = \operatorname{tril}(A, 0)$, $\operatorname{tril}^*(A) = \operatorname{tril}(A, -1)$, $\operatorname{triu}(A) = \operatorname{triu}(A, 0)$ and $\operatorname{triu}^*(A) = \operatorname{triu}(A, 1)$. Let $e_i \in \mathbb{R}^n$ be the ith unit vector and let $I^{(k)} = \sum_{i=1}^k e_i e_i^T$ for $1 \le k \le n$ and $I^{(k)} = 0$ for $k \le 0$. Similarly, let $I_{(k)} = \sum_{i=k+1}^n e_i e_i^T$ for $0 \le k < n$ and $I_{(k)} \doteq 0$ for $k \ge n$. Note that $I^{(n)} = I = I_{(0)}$ and $I_{(k)} + I^{(k)} = I$.

We first derive the exact perturbation terms for the LU and LDU factorizations, respectively.

LEMMA 1.93 *Let $C, B \in \mathbb{R}^{n\times n}$ and assume that $I - B$ is nonsingular and has LU factorization. Then the solution of the equation*

$$W = C + B\,triu\,(W, l) \quad (l \geq 0) \tag{1.109}$$

is given by

$$We_k = \left(I - BI^{(k-l)}\right)^{-1} Ce_k \quad (k = 1, \ldots, n). \tag{1.110}$$

Proof. For $k = 1, \ldots, n$ relations $We_k = Ce_k + B\text{triu}\,(W, l)\, e_k$ and $\text{triu}\,(W, l)\, e_k = I^{(k-l)} We_k$ imply $\left(I - BI^{(k-l)}\right) We_k = Ce_k$ which gives the result. The nonsingularity of $I - BI^{(k-l)}$ follows from the assumptions that $I - B$ is nonsingular and has LU factorization. ■

Note that $I^{(j)} = 0$ $(j \leq 0)$ implies that $We_k = Ce_k$ for $k \leq l$. Equation (1.109) can be transformed into one single system by using the *vec* operation:

$$\text{diag}\left(I - BI^{(1-l)}, \ldots, I - BI^{(n-l)}\right) vec\,(W) = vec\,(C). \tag{1.111}$$

COROLLARY 1.94 *Let $C, B \in \mathbb{R}^{n\times n}$ and assume that $I - B$ is nonsingular and has LU factorization. Then the solution of the equation*

$$W = C + tril\,(W, -l)\, B \quad (l \geq 0) \tag{1.112}$$

is given by $e_k^T W = e_k^T C \left(I - I^{(k-l)} B\right)^{-1}$ $(k = 1, \ldots, n)$.

Proof. By transposition

$$W^T = C^T + B^T \text{tril}\,(W, -l)^T = C^T + B^T \text{triu}\left(W^T, l\right).$$

Hence Lemma 1.93 implies the requested result. ■

THEOREM 1.95 *Assume that A and $A + \delta_A$ are nonsingular and have LU factorizations $A = L_1 U$ and $A + \delta_A = (L_1 + \delta_{L_1})(U + \delta_U)$, respectively. Let $B = L_1^{-1} \delta_A U^{-1}$. The exact perturbation terms δ_{L_1} and δ_U are then given by $\delta_{L_1} = L_1\, tril^*(F)$ and $\delta_U = triu\,(G)\, U$, where F and G are the unique solutions of the equations $F = B - B\,triu\,(F)$ and $G = B - tril^*(G)\, B$, respectively.*

Proof. Let $X_1 = L_1^{-1} \delta_{L_1}$, $Y = \delta_U U^{-1}$ and write

$$L_1^{-1}(A + \delta_A)\, U^{-1} = L_1^{-1}(L_1 + \delta_{L_1})(U + \delta_U)\, U^{-1} \tag{1.113}$$

in the form $I + B = (I + X_1)(I + Y) = I + X_1 + Y + X_1 Y$, where $B = L_1^{-1} \delta_A U^{-1}$. Observe that X_1 is strict lower triangular and Y is upper triangular. Hence $I + X_1$ is unit lower triangular and $I + Y$ is upper triangular and both provide the unique L_1U factorization of $I+B$. Note that $I+B$ and $I+Y$ are nonsingular by the initial assumptions. From the identity

$$B = X_1 + Y + X_1Y \tag{1.114}$$

it follows that $F := B\,(I+Y)^{-1} = Y\,(I+Y)^{-1} + X_1$, where $Y\,(I+Y)^{-1}$ is upper triangular. Hence $\text{tril}^*(F) = X_1$ and $\text{triu}(F) = Y\,(I+Y)^{-1}$. The relation $(I+Y)^{-1} = I - Y\,(I+Y)^{-1}$ implies $B\,(I+Y)^{-1} = B - BY\,(I+Y)^{-1}$, which can be written as

$$F = B - B\text{triu}(F). \tag{1.115}$$

Hence the exact perturbation term is given by $\delta_{L_1} = L_1X_1 = L_1\text{tril}^*(F)$. Next we derive δ_U. In fact, relation (1.114) also yields

$$G := (I+X_1)^{-1}\,B = (I+X_1)^{-1}\,X_1 + Y,$$

where $(I+X_1)^{-1}\,X_1$ is strict lower triangular. Thus we have the relations $\text{tril}^*(G) = (I+X_1)^{-1}\,X_1$ and $\text{triu}(G) = Y$. As $(I+X_1)^{-1} = I - (I+X_1)^{-1}\,X_1$ we have $(I+X_1)^{-1}\,B = B - (I+X_1)^{-1}\,X_1B$, which can be written as

$$G = B - \text{tril}^*(G)\,B. \tag{1.116}$$

The exact perturbation term δ_U is then given by $\delta_U = YU = \text{triu}(G)\,U$. Thus we proved the theorem. ■

REMARK 1.96 *Lemma 1.93 yields* $Fe_k = \left(I + BI^{(k)}\right)^{-1} Be_k$, *while Corollary 1.94 implies* $e_k^TG = e_k^TB\left(I + I^{(k-1)}B\right)^{-1}$.

Note that the only assumptions are that $I+B$ be nonsingular and have LU factorization. By setting $\delta_A = L_1\Delta_AU$ we obtain $B = \Delta_A$. Hence F and G become independent of A.

THEOREM 1.97 *Assume that A and $A+\delta_A$ are nonsingular and have LDU factorizations $A = L_1DU_1$ and $A+\delta_A = (L_1+\delta_{L_1})\,(D+\delta_D)\,(U_1+\delta_{U_1})$, respectively. Let $B = L_1^{-1}\delta_AU^{-1}$ and $\widetilde{B} = L^{-1}\delta_AU_1^{-1}$. Then the exact perturbation terms δ_{L_1}, δ_D and δ_{U_1} are given by $\delta_{L_1} = L_1\,\text{tril}^*(F)$, $\delta_D = \text{diag}(G)\,D$ and $\delta_{U_1} = \text{triu}^*\left(\widetilde{F}\right)U_1$, where F, G and $\widetilde{F}$ are the unique solutions of the equations $F = B - B\,\text{triu}(F)$, $G = B - \text{tril}^*(G)\,B$ and $F = \widetilde{B} - \text{tril}(F)\,\widetilde{B}$, respectively.*

Proof. We have to derive only δ_{U_1} and δ_D. From the relation

$$A+\delta_A = (L_1+\delta_{L_1})\,(U+\delta_U) = (L_1+\delta_{L_1})\,(D+\delta_D)\,(U_1+\delta_{U_1})$$

we obtain

$$\delta_U = (D + \delta_D)(U_1 + \delta_{U_1}) - DU_1 = \delta_D U_1 + (D + \delta_D)\,\delta_{U_1},$$

where $(D + \delta_D)\,\delta_{U_1}$ is strict upper triangular and U_1 is unit upper triangular. Hence $\delta_D = \text{diag}(\delta_D U_1) = \text{diag}(\delta_U)$. The relation $\delta_U = \text{triu}(G)\,U$, where U is upper triangular, implies

$$\text{diag}(\delta_U) = \text{diag}(G)\,\text{diag}(U) = \text{diag}(G)\,D.$$

If $A + \delta_A$ is decomposed in the LU_1 form $A + \delta_A = (L + \delta_L)(U_1 + \delta_{U_1})$, then $A^T + \delta_A^T = \left(U_1^T + \delta_{U_1}^T\right)\left(L^T + \delta_L^T\right)$ is an L_1U factorization. Let $\widetilde{B} = L^{-1}\delta_A U_1^{-1}$. Theorem 1.95 implies $\delta_{U_1}^T = U_1^T Y_1^T$, where $Y_1^T = \text{tril}^*\left(\widetilde{F}^T\right)$ and $\widetilde{F}^T = \widetilde{B}^T - \widetilde{B}^T \text{triu}\left(\widetilde{F}^T\right)$. Transposing the relations again we easily obtain that $\delta_{U_1} = \text{triu}^*\left(\widetilde{F}\right) U_1$, where $\widetilde{F}$ is the unique solution of the equation

$$F = \widetilde{B} - \text{tril}(F)\,\widetilde{B}. \tag{1.117}$$

This proves the theorem. ■

We point out that the term δ_{U_1} can be dropped in the case of the LDL^T factorization.

The componentwise perturbation bounds are obtained by estimating the solutions of equations (1.115), (1.116) and (1.117). We use the following simple observations. If $\rho(|B|) < 1$, then $I - |B|$ is an M-matrix and

$$\left|(I + B)^{-1}\right| \leq (I - |B|)^{-1}. \tag{1.118}$$

As $\rho(B) \leq \rho(|B|)$ and $\rho\left(|B|\,I^{(k)}\right) \leq \rho\left(|B|\,I^{(k+1)}\right) \leq \rho(|B|)$ similar statements hold for the matrices $BI^{(k)}$. For $\rho(|B|) < 1$, $I - |B|\,I^{(k)}$ is also an M-matrix ($k = 0, 1, \ldots, n$) and

$$I \leq \left(I - |B|\,I^{(k)}\right)^{-1} \leq \left(I - |B|\,I^{(k+1)}\right)^{-1} \leq (I - |B|)^{-1}. \tag{1.119}$$

THEOREM 1.98 *Assume that A and $A + \delta_A$ are nonsingular and have LU factorizations $A = L_1U$ and $A + \delta_A = (L_1 + \delta_{L_1})(U + \delta_U)$, respectively. Let $B = L_1^{-1}\delta_A U^{-1}$. Then*

$$|\delta_{L_1}| \leq |L_1|\, \textit{tril}^*(|F|), \quad |\delta_U| \leq \textit{triu}(|G|)\,|U|, \tag{1.120}$$

where F and G are given by equations (1.115) and (1.116), respectively. If $\rho(|B|) < 1$, then

$$|\delta_{L_1}| \leq |L_1|\, \textit{tril}^*\left(F^{b,1}\right), \quad |\delta_U| \leq \textit{triu}\left(G^{b,1}\right)|U|, \tag{1.121}$$

where

$$F^{b,1}e_k = \left(I - |B|\, I^{(k)}\right)^{-1} |B|\, e_k, \quad e_k^T G^{b,1} = e_k^T |B| \left(I - I^{(k-1)} |B|\right)^{-1} \tag{1.122}$$

for $k = 1, \ldots, n$.

Proof. Theorem 1.95 clearly implies (1.120). Matrices F and G have the forms $Fe_k = \left(I + BI^{(k)}\right)^{-1} Be_k$ and $e_k^T G = e_k^T B \left(I + I^{(k-1)}B\right)^{-1}$ $(k = 1, \ldots, n)$. Hence condition $\rho(|B|) < 1$ implies that

$$|F|\, e_k \leq \left|\left(I + BI^{(k)}\right)^{-1}\right| |B|\, e_k \leq \left(I - |B|\, I^{(k)}\right)^{-1} |B|\, e_k = F^{b,1}e_k \tag{1.123}$$

and

$$e_k^T |G| \leq e_k^T |B| \left|\left(I + I^{(k-1)}B\right)^{-1}\right| \leq e_k^T |B| \left(I - I^{(k-1)} |B|\right)^{-1} = e_k^T G^{b,1}. \tag{1.124}$$

■

REMARK 1.99 *It follows from Lemma 1.93 that matrices $F^{b,1}$ and $G^{b,1}$ are the unique solutions of the equations*

$$F = |B| + |B|\, triu(F), \quad G = |B| + tril^*(G)\, |B| \tag{1.125}$$

respectively.

Inequality (1.119) implies that

$$F^{b,1} \leq (I - |B|)^{-1} |B| = F^{b,2}, \quad G^{b,1} \leq |B| (I - |B|)^{-1} = G^{b,2}. \tag{1.126}$$

The resulting weaker estimates

$$|\delta_{L_1}| \leq |L_1|\, \mathrm{tril}^*\left(F^{b,2}\right), \quad |\delta_U| \leq \left|\mathrm{triu}\left(G^{b,2}\right)\right| |U| \tag{1.127}$$

are due to Sun [333]. Note that $F^{b,2} = G^{b,2}$.

THEOREM 1.100 *Assume that A and $A + \delta_A$ are nonsingular and have LDU factorizations $A = L_1 D U_1$ and $A + \delta_A = (L_1 + \delta_{L_1})(D + \delta_D)(U_1 + \delta_{U_1})$, respectively. Let $B = L_1^{-1}\delta_A U^{-1}$ and $\widetilde{B} = L^{-1}\delta_A U_1^{-1}$. Then*

$$|\delta_{L_1}| \leq |L_1|\, tril^*(|F|)\,, \; |\delta_D| \leq diag(|G|)\, |D|\,, \; |\delta_{U_1}| \leq triu^*\left(\left|\widetilde{F}\right|\right) |U_1|\,, \tag{1.128}$$

where F, G and $\widetilde{F}$ are given by equations (1.115), (1.116) and (1.117), respectively. If $\rho(|B|) < 1$ and $\rho\left(\left|\widetilde{B}\right|\right) < 1$, then

$$|\delta_{L_1}| \leq |L_1|\, tril^*\left(F^{b,1}\right),\; |\delta_D| \leq diag\left(G^{b,1}\right) |D|\,,\; |\delta_{U_1}| \leq triu^*\left(\widetilde{F}^{b,1}\right) |U_1|\,, \tag{1.129}$$

where $F^{b,1}$ and $G^{b,1}$ are given by (1.122) and

$$e_k^T \widetilde{F}^{b,1} = e_k^T \left|\widetilde{B}\right| \left(I - I^{(k)} \left|\widetilde{B}\right|\right)^{-1} \tag{1.130}$$

for $k = 1, \ldots, n$.

Proof. Theorem 1.97 implies (1.128). The first two bounds of (1.129) follow from Theorem 1.98. We have to prove only the last bound. Matrix $\widetilde{F}$ has the form $e_k^T \widetilde{F} = e_k^T \widetilde{B} \left(I + I^{(k)} \widetilde{B}\right)^{-1}$ $(k = 1, \ldots, n)$. Condition $\rho\left(\left|\widetilde{B}\right|\right) < 1$ implies

$$e_k^T \left|\widetilde{F}\right| \leq e_k^T \left|\widetilde{B}\right| \left|\left(I + I^{(k)} \widetilde{B}\right)^{-1}\right| \leq e_k^T \left|\widetilde{B}\right| \left(I - I^{(k)} \left|\widetilde{B}\right|\right)^{-1} = e_k^T \widetilde{F}^{b,1}. \tag{1.131}$$

■

REMARK 1.101 *Matrix $\widetilde{F}^{b,1}$ is the unique solution of equation*

$$F = \left|\widetilde{B}\right| + tril(F) \left|\widetilde{B}\right|. \tag{1.132}$$

Inequality (1.119) implies $\widetilde{F}^{b,1} \leq \left|\widetilde{B}\right| \left(I - \left|\widetilde{B}\right|\right)^{-1}$ for $\rho\left(\left|\widetilde{B}\right|\right) < 1$.

REMARK 1.102 *The estimates $F^{b,1}$, $G^{b,1}$ and $\widetilde{F}^{b,1}$ are sharp. If $B \leq 0$ and $\widetilde{B} \leq 0$, then $|F| = F^{b,1}$, $|G| = G^{b,1}$ and $\widetilde{F}^{b,1} = \left|\widetilde{F}\right|$. Such situation ($B \leq 0$, $\widetilde{B} \leq 0$) occurs, if A is an M-matrix and $\delta_A \leq 0$.*

Further results and norm bounds can be found in [146] and [148]. The norm bounds are based on various estimates of the projection $I_{(k)} \left(I + BI^{(k)}\right)^{-1}$, which plays a key role in the exact perturbation terms [148].

9. The Gram matrix

Assume that $\langle \cdot, \cdot \rangle : \mathbb{F}^n \times \mathbb{F}^n \to \mathbb{F}$ is an inner product and let $x_1, x_2, \ldots, x_n \in \mathbb{F}^n$. The $n \times n$ matrix

$$G = \left[\langle x_i, x_j \rangle\right]_{i,j=1}^n \tag{1.133}$$

is known as the Gram matrix of $x_1, x_2, \ldots, x_n$. Its determinant

$$g = g\left(x_1, \ldots, x_n\right) = \det\left(G\right) \tag{1.134}$$

is known as the Gram determinant of the elements.

THEOREM 1.103 *Let $x_1, x_2, \ldots, x_n \neq 0$. Then*

$$0 \leq g\left(x_1, \ldots, x_n\right) \leq \left\|x_1\right\|^2 \left\|x_2\right\|^2 \cdots \left\|x_n\right\|^2. \tag{1.135}$$

The lower extreme occurs if and only if the vectors x_i are dependent. The upper extreme occurs if and only if the vectors are orthogonal. If the vectors are normalized ($\|x_i\| = 1$), then we have

$$0 \leq g \leq 1. \tag{1.136}$$

THEOREM 1.104 *The Gram determinant $g(x_1, \dots, x_n)$ satisfies the following properties:*
(i) g is a symmetric function of its arguments;
(ii) $g(\tau_1 x_1, \tau_2 x_2, \dots, \tau_n x_n) = |\tau_1|^2 |\tau_2|^2 \cdots |\tau_n|^2 g(x_1, \dots, x_n)$;
(iii) $g(x_1, \dots, x_{j-1}, x_j + \sigma x_k, x_{j+1}, \dots, x_n) = g(x_1, \dots, x_n)$, $j \neq k$;
(iv) $g^{\frac{1}{2}}(x_1 + \widehat{x}_1, x_2, \dots, x_n) \leq g^{\frac{1}{2}}(x_1, x_2, \dots, x_n) + g^{\frac{1}{2}}(\widehat{x}_1, x_2, \dots, x_n)$;
(v) $g(x_1, x_2, \dots, x_n) \leq g(x_1, \dots, x_p)\, g(x_{p+1}, \dots, x_n)$, $1 \leq p < n$.
Equality holds in (v) if and only if

$$\langle x_i, x_j \rangle = 0, \quad 1 \leq i \leq p, \quad p+1 \leq j \leq n. \tag{1.137}$$

For geometric interpretation and proofs we refer to [150], [95] and [57]. We only note that if $X \in \mathbb{C}^{n \times k}$ $(k \leq n)$ and $X = U\Sigma V^H$ is a singular value decomposition with singular values $\sigma_1, \dots, \sigma_k$, then $g(x_1, \dots, x_k) = \det\left(X^H X\right) = \sigma_1^2 \sigma_2^2 \dots \sigma_k^2$. We also obtain the simple estimate

$$\sigma_1 \sigma_2 \dots \sigma_k \leq \|x_1\| \, \|x_2\| \cdots \|x_k\|. \tag{1.138}$$

Chapter 2

PROJECTIONS

This chapter is a collection of various results on projections, where the well-known facts are supplemented by less-known ones. The material is selected so that it supports the subsequent parts of the book. Several researchers, mainly in the last decade, investigated the properties of projections and achieved a large number of interesting and important new results. Since we could include only a fraction of these, we encourage the interested reader to consult with the cited works as well.

1. Basic definitions and properties

We give equivalent definitions of the oblique and orthogonal projections. Then we investigate some basic properties and operations.

DEFINITION 2.1 *A matrix $A \in \mathbb{R}^{n\times n}$ is called projection (idempotent), if and only if $A^2 = A$.*

DEFINITION 2.2 *Let $M, N \subset \mathbb{R}^n$ be complementary subspaces. For each $z \in \mathbb{R}^n$ there exists a unique decomposition in the form*

$$z = x + y \quad (x \in M,\ y \in N).$$

Transformation $P_{M,N}$ defined by $P_{M,N}z := x$ is called projection onto M along N. Transformation $P_{N,M}$ defined by $P_{N,M}z := y$ is called projection onto N along M. Vector x is the projection of z onto M along N and vector y is the projection of z onto N along M.

DEFINITION 2.3 *The projection P onto M along $M^\perp$ is called orthogonal.*

DEFINITION 2.4 *Let $M \subset \mathbb{R}^n$ be a subspace. $P \in \mathbb{R}^{n\times n}$ is the orthogonal projection onto M if (a) $\mathcal{R}(P) = M$, (b) $P^2 = P$, (c) $P^T = P$.*

The subspace M uniquely determines $M^{\perp}$. Hence we speak of orthogonal projection onto M and denote it by P_M. General projections (idempotents) are called oblique projections. Orthogonal projections are often called symmetric projections .

The following theorem establishes the equivalence of Definitions 2.1 and 2.2.

THEOREM 2.5 *Transformation P is a projection onto some subspace, if and only if $P^2 = P$.*

Proof. If P is a projection onto M along N, $z = x + y$, $x \in M$, $y \in N$, then $P^2 z = PPz = Px = x = Pz$. Conversely, assume that $P^2 = P$. Let $N = \{z \in \mathbb{R}^n | Pz = 0\}$ and $M = \{z \in \mathbb{R}^n | Pz = z\}$. It is clear that M and N are subspaces. If $z \in M \cap N$, then $Pz = z$ and $Pz = 0$ imply that $z = 0$ and $M \cap N = \{0\}$. For arbitrary $z \in \mathbb{R}^n$,

$$z = Pz + (I - P)\, z.$$

Let $Pz = x$ and $(I - P)\, z = y$. As $Px = P^2 z = Pz = x$ and $Py = P(I - P)\, z = Pz - P^2 z = 0$, one obtains that $x \in M$ and $y \in N$. This proves that $\mathbb{R}^n = M \oplus N$ and P is the projection onto M along N. ∎

A slightly different formulation of this result is given in

THEOREM 2.6 *If $H \in \mathbb{R}^{n \times n}$ is an arbitrary idempotent matrix, then $\mathcal{R}(H)$ and $\mathcal{N}(H)$ are complementary subspaces and*

$$H = P_{\mathcal{R}(H),\mathcal{N}(H)}.$$

Conversely, if L and M are complementary subspaces, then there exists a unique idempotent matrix $P_{L,M}$ such that $\mathcal{R}(P_{L,M}) = L$ and $\mathcal{N}(P_{L,M}) = M$.

THEOREM 2.7 *If P is a projection onto M along N, then M and N are the solution sets of the equations $Pz = z$ and $Pz = 0$, respectively.*

THEOREM 2.8 *P is projection, if and only if $I - P$ is also projection; If P is projection onto M along N, then $I - P$ is projection onto N along M.*

THEOREM 2.9 *If P is a projection with $\mathcal{R}(P)$ and $\mathcal{N}(P)$, then P^T is a projection with $\mathcal{R}(P^T) = \mathcal{N}^{\perp}(P)$ and $\mathcal{N}(P^T) = \mathcal{R}^{\perp}(P)$.*

Proof. As $P^2 = P$ implies $(PP)^T = P^T P^T = (P^T)^2 = P^T$, the transpose of P is also projection. From Lemma 1.3 it follows that $\mathcal{R}(P^T) = \mathcal{N}^{\perp}(P)$ and $\mathcal{N}(P^T) = \mathcal{R}^{\perp}(P)$. ∎

THEOREM 2.10 *Any projection P may have only 0 and 1 eigenvalues.*

Proof. If λ is an eigenvalue of P and $x \neq 0$ is the corresponding eigenvector, then $\lambda x = Px = P^2 x = \lambda^2 x$ implies that $\lambda = \lambda^2$. Hence λ may be 0 or 1. ■

THEOREM 2.11 *Any projection $P \in \mathbb{R}^{n\times n}$ is diagonalizable by similarity transformation.*

Proof. If P is idempotent, then its minimal polynomial $x^2 - x = 0$ has only the single roots 0 and 1. Therefore, P is diagonalizable in the form

$$P = T\text{diag}(\overbrace{1,\ldots,1}^{r},\overbrace{0,\ldots,0}^{n-r})T^{-1}, \tag{2.1}$$

where $r = \text{rank}(P)$. Other proof is obtained from the Jordan form $P = T\text{diag}(J_1,\ldots,J_s)T^{-1}$. As $P^2 = T\text{diag}\left(J_1^2,\ldots,J_s^2\right)T^{-1}$, the equation $P^2 = P$ holds if and only if $J_i^2 = J_i$ for all i. This is true if J_i is 1 or 0. ■

Equation (2.1) and $\text{tr}(SAS^{-1}) = \text{tr}(A)$ imply

COROLLARY 2.12 *For any projection $P \in \mathbb{R}^{n\times n}$, tr$(P)$ =rank(P).*

The following theorem summarizes the basic operations with projections.

THEOREM 2.13 *Suppose that P_1 is projection onto M_1 along N_1 and P_2 is projection onto M_2 along N_2. Then*
(i) $P = P_1 + P_2$ is projection if and only if $P_1P_2 = P_2P_1 = 0$; if $P_1P_2 = P_2P_1 = 0$, then P is the projection onto M along N, where $M = M_1 \oplus M_2$ and $N = N_1 \cap N_2$.
(ii) $P = P_1 - P_2$ is projection, if and only if $P_1P_2 = P_2P_1 = P_2$; if $P_1P_2 = P_2P_1 = P_2$, then P is the projection onto $M = M_1 \cap N_2$ along $N = N_1 \oplus M_2$.
(iii) If $P_1P_2 = P_2P_1 = P$, then P is the projection onto $M = M_1 \cap M_2$ along $N = N_1 + N_2$.

Proof. (i) $(P_1 + P_2)^2 = P_1 + P_2 \Leftrightarrow P_1P_2 + P_2P_1 = 0$. Multiply the last equation by P_1 on the left and also on the right: $P_1P_2 + P_1P_2P_1 = 0$, $P_1P_2P_1 + P_2P_1 = 0$. By subtraction one obtains $P_1P_2 - P_2P_1 = 0$, implying that P_1 and P_2 are commutative. Hence $P_1P_2 + P_2P_1 = 0$ implies that $P_1P_2 = P_2P_1 = 0$. Conversely, if the latter conditions hold, then P is obviously a projection. Assume that P is projection. If $z \in M_1 \cap M_2$, i.e., $P_1z = P_2z = z$, then $z = P_1z = P_1(P_2z) = (P_1P_2)z = 0$, which implies that $M_1 \cap M_2 = \{0\}$. Let $M = \{z|Pz = z\}$ and $N = \{z|Pz = 0\}$. Let $z = x_1 + y_1 = x_2 + y_2$, where $x_1 = P_1z \in M_1$, $y_1 = (I - P_1)z \in N_1$, $x_2 = P_2z \in M_2$ and $y_2 = (I - P_2)z \in N_2$. If $z \in M$, i.e., $P_1z + P_2z = z$, then

$$z = P_1(x_2 + y_2) + P_2(x_1 + y_1) = P_1y_2 + P_2y_1.$$

Since $P_1y_2 \in M_1$ and $P_2y_1 \in M_2$ we obtain that $M \subset M_1 + M_2$. Conversely, if $z = m_1 + m_2$ ($m_1 \in M_1$, $m_2 \in M_2$), then $(P_1 + P_2)z = P_1m_1 + P_2m_2 =$

$m_1+m_2 = z \in M$. Thus $M_1+M_2 \supseteq M$, implying that $M_1+M_2 = M$. Hence we proved that $M = M_1 \oplus M_2$. If $z \in N_1 \cap N_2$ then $Pz = (P_1 + P_2) z = 0$ and $z \in N$. Conversely, $Pz = P_1 z + P_2 z = 0$ implies $P_1 z + P_1 P_2 z = 0$ and $P_2 P_1 z + P_2 z = 0$. As $P_1 P_2 z = P_2 P_1 z = 0$ for all z we obtain that $P_1 z = P_2 z = 0$ and $x \in N_1 \cap N_2$. Thus we proved that $N = N_1 \cap N_2$.
(ii) $P = P_1 - P_2$ is a projection if and only if $I - (P_1 - P_2) = (I - P_1) + P_2$ is projection. Using (i) and the fact that $I - P_1$ is projection onto N_1 along M_1 we obtain the necessary and sufficient condition

$$(I - P_1) P_2 = P_2 (I - P_1) = 0,$$

under which $(I - P_1) + P_2$ is the projection onto $N_1 \oplus M_2$ along $M_1 \cap N_2$. Since the relation above is equivalent with $P_1 P_2 = P_2 P_1 = P_2$, statement (ii) is proved.
(iii) If $P = P_1 P_2 = P_2 P_1$, then

$$P^2 = (P_1 P_2)^2 = P_1 P_2 P_1 P_2 = P_1 P_1 P_2 P_2 = P_1 P_2 = P.$$

Hence P is projection. If $Pz = z$ $(z \in M)$, then

$$P_1 z = P_1 P z = P_1 P_1 P_2 z = P_1 P_2 z = z$$

and

$$P_2 z = P_2 P z = P_2 P_1 P_2 z = P_2 P_1 z = z.$$

Thus $z \in M_1 \cap M_2$. If $z \in M_1 \cap M_2$, i.e., $P_1 z = z = P_2 z$, then $Pz = P_1 P_2 z = P_1 z = z \in M$. Hence $M = M_1 \cap M_2$ is proved. Assume now that $z \in N$, i.e., $Pz = P_1 P_2 z = P_2 P_1 z = 0$. This implies that $P_2 z \in N_1$ and $P_1 z \in N_2$. Since $z = P_2 z + (I - P_2) z$ and $(I - P_2) z \in N_2$ we obtain $z \in N_1 + N_2$. If $z = n_1 + n_2$ $(n_1 \in N_1,\ n_2 \in N_2)$, then $P_1 P_2 z = P_1 P_2 n_1 = P_2 P_1 n_1 = P_2 0 = 0$ implies that $z \in N$. Thus we proved that $N = N_1 + N_2$. ∎

Later we extend statement (i) for more than two projections. The condition of statement (iii) is only sufficient.

1.1 Properties of orthogonal projections

An elementary property of orthogonal projections is the following. If P is the orthogonal projection onto a subspace $S \subset \mathbb{R}^n$, then, by definition, $Px \in S$ and $(I - P) x \in S^{\perp}$ hold for all $x \in \mathbb{R}^n$. The converse is also true and characterizes the orthogonal projections [362].

THEOREM 2.14 *Assume that $P \in \mathbb{R}^{n \times n}$ and $S \subset \mathbb{R}^n$ is a nonempty subset such that $Px \in S$ and $x - Px \in S^{\perp}$ hold for all $x \in \mathbb{R}^n$. Then P is the orthogonal projection on S.*

Proof. Let $x \in S$. Then $Px \in S$ and $x - Px \in S^{\perp}$. Since $x \perp (x - Px)$ and $Px \perp (x - Px)$, we have $(x - Px) \perp (x - Px)$ and $Px = x$. Since $Py \in$

S for all $y \in \mathbb{R}^n$, we obtain that $P^2 y = P(Py) = Py$. Thus P is a projection onto S. P is symmetric because

$$\begin{aligned} x^T P y &= ((x - Px) + Px)^T Py = (Px)^T Py \\ &= (Px)^T ((Py - y) + y) \\ &= x^T P^T y \end{aligned}$$

holds for all $x, y \in \mathbb{R}^n$. ∎

The orthogonal projections provide best approximations in the following sense. Let $u \in \mathbb{R}^n$ and $\emptyset \neq M \subset \mathbb{R}^n$ be arbitrary. The distance of vector u from the set M is defined by

$$d(u, M) = \inf_{v \in M} \|u - v\|_2 . \tag{2.2}$$

THEOREM 2.15 *If* $0 \neq S \subset \mathbb{R}^n$ *is a subspace, then*

$$d(u, S) = \|(I - P_S) u\|_2 \quad (u \in \mathbb{R}^n). \tag{2.3}$$

Proof. For any $x \in S$, $x^T (I - P_S) v = x^T P_S (I - P_S) v = 0$. For arbitrary $v \in S$, $u - v = (I - P_S) u + (P_S u - v)$, $(I - P_S) u \perp P_S u - v \in S$ and

$$\|u - v\|_2^2 = \|(I - P_S) u\|_2^2 + \|P_S u - v\|_2^2 \geq \|(I - P_S) u\|_2^2 .$$

∎

Hence $P_S v$ gives the nearest point (best approximation) to v from S. We can also observe that $(I - P_S) v \perp S$. It is the best approximation property that makes the orthogonal projections important in many applications (see, e.g., [95]).

The following observation gives another extremal property of the orthogonal projections (see, e.g., [194]). Let P be any projection with $\mathcal{R}(P) = S$. Then $(x, Px)_{\measuredangle} \geq (x, P_S x)_{\measuredangle}$ holds provided that $x, Px \neq 0$. This minimal angle property of the orthogonal projection follows from the inequality

$$\begin{aligned} \cos(x, Px)_{\measuredangle} &= \frac{x^T P x}{\|x\|_2 \|Px\|_2} = \frac{x^T P_S P x}{\|x\|_2 \|Px\|_2} \\ &\leq \frac{\|P_S x\|_2}{\|x\|_2} = \frac{x^T P_S x}{\|x\|_2 \|P_S x\|_2} = \cos(x, P_S x)_{\measuredangle} . \end{aligned}$$

The equivalence of Definitions 2.3 and 2.4 is given in

THEOREM 2.16 *The linear transformation* P *is an orthogonal projection if and only if* $P = P^2 = P^T$. *Moreover the orthogonal projections are positive semidefinite and satisfy* $\|Px\|_2 \leq \|x\|_2$ *for all* x.

Proof. As $P = P^2$ is the if and only if condition for P to be a projection we need only to prove the rest. If $P = P^T$, then by the relations $P = P_{R,N}$

and $P^T = P_{N^\perp,R^\perp}$ we obtain that $N = R^\perp$ proving that P is orthogonal. If P is orthogonal, then $P = P_{R,R^\perp}$. As $P^T = P_{R^{\perp\perp},R^\perp} = P_{R,R^\perp}$ we proved that $P = P^T$. P is positive semidefinite: for all $x \in \mathbb{R}^n$,

$$x^T P x = x^T P^2 x = x^T P^T P x = \|Px\|_2^2 \geq 0.$$

P is orthogonal if and only if $I - P$ is also orthogonal. So $I - P$ is also positive semidefinite. P is contractive: for all $x \in \mathbb{R}^n$,

$$\|x\|_2^2 - \|Px\|_2^2 = x^T x - x^T P x = x^T (I - P) x \geq 0.$$

■

THEOREM 2.17 *If $P = P^2$ and $\|Px\|_2 \leq \|x\|_2$ for all $x \in \mathbb{R}^n$, then $P = P^T$.*

Proof. Let $P = P_{R,N}$. We prove that $R \perp N$. Let $x \in N^\perp$. Then $y = Px - x \in N$ follows from $Py = P^2x - Px = 0$. Consequently $Px = x + y$ and $x \perp y$. Thus we have

$$\|x\|_2^2 \geq \|Px\|_2^2 = \|x\|_2^2 + \|y\|_2^2 \geq \|x\|_2^2,$$

which implies $y = 0$. Thus $Px = x \in R$ and $N^\perp \subset R$. Conversely, let $z \in R$, i.e., $Pz = z$. Let $x \in N^\perp$, $y \in N$ be such that $z = x + y$. Then $z = Pz = Px + Py = Px$. As $N^\perp \subset R$ we have $Px = x$ and $z = x \in N^\perp$. Thus $R \subset N^\perp$ and $R = N^\perp$. ■

COROLLARY 2.18 *If $P \neq 0$ is an orthogonal projection, then $\|P\|_2 = 1$.*

Proof. From Theorem 2.16 it follows that $\|P\|_2 \leq 1$. On the other hand, $\|P\| \geq \max_{1 \leq i \leq n} |\lambda_i| = 1$. ■

DEFINITION 2.19 *Let P and Q be orthogonal projections onto M and N, respectively. P is called orthogonal to Q if $PQ = 0$.*

THEOREM 2.20 *The projections $P = P_M$ and $Q = P_N$ are orthogonal to each other if and only if $M \perp N$.*

Proof. If $PQ = 0$, then for all $x \in M$ and $y \in N$ we have $x^T y = x^T P^T Q y = x^T P Q y = 0$ proving that $M \perp N$. Conversely, if $N \subset M^\perp$ (i.e., $M \perp N$), then for all $x \in \mathbb{R}^n$, $Qx \in N \subset M^\perp$, $PQx = 0$. ■

THEOREM 2.21 *Let $P_1, \ldots, P_n$ be orthogonal projections ($P_i = P_{M_i}$). Then $P = P_1 + \cdots + P_n$ is an orthogonal projection, if and only if $P_i P_j = 0$, for all $i \neq j$.*

Proof. The sufficiency is trivial. Assume that P is orthogonal. Let $x \in M_i$. Then

$$\|x\|_2^2 \geq \|Px\|_2^2 = x^T P x = x^T \sum_{j=1}^n P_j x = \sum_{j=1}^n \|P_j x\|_2^2 \geq \|P_i x\|_2^2 = \|x\|_2^2.$$

So

$$\sum_{j=1}^{n} \|P_j x\|_2^2 = \|P_i x\|_2^2 ,$$

which implies that $P_j x = 0$ for all $j \neq i$. Consider now an arbitrary $z \in \mathbb{R}^n$ and let $z = x + y$ $(x \in M_i,\ y \in M_i^{\perp})$. Then

$$P_i z = P_i x + P_i y = P_i x = x \Rightarrow P_j P_i z = P_j P_i x = P_j x = 0,$$

which was to be proved. ■

The relation $P_j x = 0$ implies the inclusion $M_i \subset M_j^{\perp}$ for all $j \neq i$. Theorem 2.21 will be generalized in Theorem 2.50.

DEFINITION 2.22 *$A \geq B$ if and only if $A - B$ positive semidefinite.*

THEOREM 2.23 *If $E = P_M$ and $F = P_N$ are orthogonal projections, then the following statements are equivalents:*
(i) $E \leq F$;
(ii) $\|Ex\| \leq \|Fx\|$ for all $x \in \mathbb{R}^n$;
(iii) $M \subset N$;
(iva) $FE = E$;
(ivb) $EF = E$.

Proof. (i)⇒(ii). If $E \leq F$, then for all $x \in \mathbb{R}^n$,

$$0 \leq x^T (F - E)\, x = x^T F x - x^T E x = \|Fx\|_2^2 - \|Ex\|_2^2 .$$

(ii)⇒(iii). For all x, $\|Ex\|_2 \leq \|Fx\|_2$. Let $x \in M$. Then

$$\|x\|_2 \geq \|Fx\|_2 \geq \|Ex\|_2 = \|x\|_2$$

implies

$$\|Fx\|_2 = \|x\|_2 \Leftrightarrow x^T x - x^T F x = x^T (I - F)\, x = \|(I - F)\, x\|_2^2 = 0.$$

Hence $x = Fx$, $x \in N$ and $M \subset N$.
(iii)⇒(iva). If $M \subset N$, then $Ex \in M \subset N$ for all x. Thus $FEx = Ex$.
(iva)⇒(ivb). $FE = E \Rightarrow (FE)^T = E^T F^T = EF = E^T = E$.
(iv)⇒(i). If $EF = FE = E$, then for all x

$$x^T F x - x^T E x = x^T F x - x^T E F x = x^T (I - E)\, F x.$$

The projections $I - E$ and F are commuting:

$$(I - E)\, F = F - EF = F\,(I - E) = F - FE.$$

So $G = F\,(I - E)$ is also a projection (see Theorem 2.13). It is orthogonal:

$$(F\,(I - E))^T = (I - E)\, F = F\,(I - E)\, .$$

Thus

$$x^T F x - x^T E x = x^T G x = \|Gx\|_2^2 \geq 0$$

proving statement (i). ■

Theorem 2.13 can be refined for orthogonal projections (see, e.g., [178]). We can also prove the following result.

THEOREM 2.24 *Assume that E and F are orthogonal projections and $EF = FE$. Then $E + F - EF$ is also an orthogonal projection.*

Proof. It is straightforward to show that $(E + F - EF)^T = E + F - EF$ and $(E + F - EF)^2 = E + F - EF$. ■

2. Representations of projections

We investigate different representations of projections, which are important either in characterizing the properties of projections or in applications (for other representations see, e.g., [301], [243] or [350]).

THEOREM 2.25 *(Egerváry). Let $P \in \mathbb{R}^{n\times n}$ be a projection matrix of rank r and let*

$$P = VW^T \quad \left(V, W \in \mathbb{R}^{n\times r}\right), \tag{2.4}$$

be any full rank factorization. Then the column vectors of V and W are biorthogonal, that is $W^T V = I_r$.

Proof. Using (2.4) $P^2 - P = 0$ can be written as

$$V\left(W^T V - I_r\right) W^T = 0.$$

Since V and W^T are of rank r they have left and right inverses, respectively. Multiplying the equation by the left and right inverses we obtain, that $W^T V - I_r = 0$, which was to be proved. ■

It is clear that $\mathcal{R}(P) = \mathcal{R}(V)$ and $\mathcal{N}(P) = \mathcal{N}\left(W^T\right) = \mathcal{R}^\perp(W)$.

THEOREM 2.26 *Assume that subspaces $M, N \subset \mathbb{R}^n$ are complementary and let the columns of $V \in \mathbb{R}^{n\times r}$ and $W \in \mathbb{R}^{n\times r}$ form a basis for M and $N^\perp$, respectively. Then*

$$P_{M,N} = V\left(W^T V\right)^{-1} W^T. \tag{2.5}$$

Proof. It is obvious that $P^2 = P$, $\mathcal{R}(P) = \mathcal{R}(V)$ and $\mathcal{R}\left(P^T\right) = \mathcal{R}(W)$. So we have to prove only that $W^T V \in \mathbb{R}^{r\times r}$ is nonsingular. Assume the contrary. If $\det\left(W^T V\right) = 0$, then there exists a vector $z \neq 0$ such that $W^T V z = 0$. As $0 \neq Vz \in M$ and $Vz \in \mathcal{R}^\perp(W) = N$ we obtained that $M \cap N \neq \{0\}$, which is a contradiction. ■

Note that either V and $\left(W^T V\right)^{-1} W^T$ or $V\left(W^T V\right)^{-1}$ and W are biorthogonal pairs.

COROLLARY 2.27 *If the columns of $V = [v_1, \ldots, v_k]$ form an orthonormal basis for a subspace $M \subset \mathbb{R}^n$, then*

$$P_M = VV^T \tag{2.6}$$

is the unique orthogonal projection onto S.

The previous statements give $\mathcal{R}(P)$, $\mathcal{N}^{\perp}(P)$ representations of P because $\mathcal{R}(P) = \mathcal{R}(V)$ and $\mathcal{N}(P) = \mathcal{R}^{\perp}(W)$. Next we derive an $\mathcal{R}(P)$, $\mathcal{N}(P)$ representation.

LEMMA 2.28 *Let $P \in \mathbb{R}^{n\times n}$ be a projection of rank $r \geq 1$, $P = X_1Y_1^T$ and $I - P = X_2Y_2^T$ full rank factorizations. Then $X = [X_1, X_2]$ and $Y = [Y_1, Y_2]$ are biorthogonal, that is $Y^TX = I$.*

Proof. Theorem 2.25 implies that $Y_1^TX_1 = I_r$ and $Y_2^TX_2 = I_{n-r}$. By definition

$$P(I-P) = X_1Y_1^TX_2Y_2^T = 0.$$

The matrices X_1 and Y_2^T have left and right inverses, respectively. Multiplying the equation by these left and right inverses we obtain, that $Y_1^TX_2 = 0$. Similarly, we get

$$(I-P)P = X_2Y_2^TX_1Y_1^T = 0$$

and $Y_2^TX_1 = 0$. Thus we have

$$\begin{bmatrix} Y_1^T \\ Y_2^T \end{bmatrix} [X_1, X_2] = \begin{bmatrix} I_r & 0 \\ 0 & I_{n-r} \end{bmatrix},$$

which is the stated biorthogonality relation. ■

Observe that $X = [X_1, X_2] \in \mathbb{R}^{n\times n}$ is nonsingular, $\mathcal{R}(P) = \mathcal{R}(X_1)$ and $\mathcal{N}(P) = \mathcal{R}(I-P) = \mathcal{R}(X_2)$. In Householder notation we just obtained that

$$P = P_{\mathcal{R}(X^{|r}),\mathcal{R}(X^{n-r|})}, \quad I - P = P_{\mathcal{R}(X^{n-r|}),\mathcal{R}(X^{|r})}. \tag{2.7}$$

LEMMA 2.29 *If $P = P_{\mathcal{R}(X^{|r}),\mathcal{R}(X^{n-r|})}$ for some $X \in \mathbb{R}^{n\times n}$, then P and $I - P$ can be written in the form*

$$P = X^{|r}\left(X^{-1}\right)^{\overline{r}}, \quad I - P = X^{n-r|}\left(X^{-1}\right)^{\underline{n-r}} \tag{2.8}$$

Proof. Since

$$\begin{bmatrix} Y_1^T \\ Y_2^T \end{bmatrix} = [X_1, X_2]^{-1} = X^{-1},$$

then $Y_1^T = \left(X^{-1}\right)^{\overline{r}}$ and $Y_2^T = \left(X^{-1}\right)^{\underline{n-r}}$ follow. ■

Representation (2.8) is taken from [325]. The following representation is due to Ipsen and Meyer [212].

LEMMA 2.30 *Assume that subspaces $M, N \subset \mathbb{R}^n$ are complementary, $U = [U_1, U_2]$ and $V = [V_1, V_2]$ are orthogonal matrices such that U_1 and U_2 are bases for M and $M^\perp$, respectively, and V_1 and V_2 are bases for N and $N^\perp$, respectively. Then*

$$P_{M,N} = U \begin{bmatrix} 0 & \left(V_2^T U_1\right)^{-1} \\ 0 & 0 \end{bmatrix} V^T. \tag{2.9}$$

Proof. Since $P = U_1 \left(V_2^T U_1\right)^{-1} V_2^T$ and

$$U^T P V = \begin{bmatrix} U_1^T P V_1 & U_1^T P V_2 \\ U_2^T P V_1 & U_2^T P V_2 \end{bmatrix} = \begin{bmatrix} 0 & \left(V_2^T U_1\right)^{-1} \\ 0 & 0 \end{bmatrix},$$

we easily obtain the required representation. ∎

The orthogonal matrices U and V can be chosen such that we obtain the singular value decomposition of P (see Section 2.5).

THEOREM 2.31 *If $P \in \mathbb{R}^{n\times n}$ is a projection of rank k, then there is an orthogonal matrix V such that*

$$P = V \begin{bmatrix} I_k & B \\ 0 & 0_{n-k} \end{bmatrix} V^T. \tag{2.10}$$

Proof. By the Schur theorem there is an orthogonal matrix V such that $V^T P V = Z$ is upper triangular. We can assume that Z is partitioned such that

$$Z = \begin{bmatrix} A & B \\ 0 & C \end{bmatrix}, \quad A \in \mathbb{R}^{k\times k}, \quad C \in \mathbb{R}^{(n-k)\times(n-k)},$$

and A contains only 1 on the main diagonal. Since Z is idempotent, we get from $Z^2 = Z$, the relations

$$A^2 = A, \quad C^2 = C,$$

showing that A and C are also idempotent. Since A has only the eigenvalue 1, and C only the eigenvalue 0, we conclude that

$$A = I_k, \quad C = 0.$$

Thus

$$Z = \begin{bmatrix} I_k & B \\ 0 & 0 \end{bmatrix},$$

where B is arbitrary. ∎

The above representation is used in many fields (for use in approximation theory, see Motte [260]).

Observe that P is an orthogonal projection if and only if $B = 0$. Let us write

$$X = \begin{bmatrix} I_k & -B \\ 0 & I_{n-k} \end{bmatrix}, \tag{2.11}$$

so that we have

$$ZX = X\Lambda, \quad \Lambda = \begin{bmatrix} I_k & 0 \\ 0 & 0 \end{bmatrix}.$$

The eigenvectors of Z are the columns of X. Since X is nonsingular, the eigenvectors of Z are linearly independent. In general, $B \neq 0$, X is not orthogonal and the eigenvectors are neither orthogonal. It is also obvious that

$$\mathcal{R}(Z) = \mathcal{R}(e_1, \ldots, e_k), \quad \mathcal{N}(Z) = \mathcal{R}\left(\begin{bmatrix} -B \\ I_{n-k} \end{bmatrix}\right) \tag{2.12}$$

and

$$\mathcal{R}(P) = \mathcal{R}\left(V^{|k}\right), \quad \mathcal{N}(P) = \mathcal{R}\left(V\begin{bmatrix} -B \\ I_{n-k} \end{bmatrix}\right). \tag{2.13}$$

THEOREM 2.32 *If $P \in \mathbb{R}^{n\times n}$ has the representation of Theorem 2.31, then its Jordan normal form is given by*

$$P = VX\begin{bmatrix} I_k & 0 \\ 0 & 0 \end{bmatrix}X^{-1}V^T, \tag{2.14}$$

where

$$X^{-1} = \begin{bmatrix} I_k & B \\ 0 & I_{n-k} \end{bmatrix}. \tag{2.15}$$

Proof. The result follows from the relation $Z = X\Lambda X^{-1}$. ∎

DEFINITION 2.33 *A matrix $H \in \mathbb{R}^{n\times n}$ of rank r is said to be in Hermite canonical (or echelon) form if it satisfies the following four conditions: (i) $h_{ij} = 0$ if $i > j$; (ii) h_{ii} is either 0 or 1; (iii) If $h_{ii} = 0$, then $h_{ik} = 0$ for every k, $1 \le k \le n$; (iv) If $h_{ii} = 1$, then $h_{ki} = 0$ for every $k \neq i$. Alternatively, H is in the Hermite canonical form if there exists a permutation matrix Π such that*

$$\Pi H \Pi^T = \begin{bmatrix} I_r & B \\ 0 & 0 \end{bmatrix},$$

where B can be arbitrary.

THEOREM 2.34 *The matrix P is a projection if and only if it is similar to a Hermite canonical form.*

Proof. A matrix H in Hermite canonical form is necessarily idempotent. Theorem 2.31 shows that all projections are similar to a Hermite canonical form. ∎

Dokovic [108] proved the following interesting canonical form.

THEOREM 2.35 *(Dokovic). Let $P \in \mathbb{C}^{n\times n}$ be a projector. Then there is a unitary similarity that reduces P to the block diagonal form*

$$\begin{bmatrix} \begin{bmatrix} 1 & \sigma_1 \\ 0 & 0 \end{bmatrix} & & & & \\ & \ddots & & & \\ & & \begin{bmatrix} 1 & \sigma_k \\ 0 & 0 \end{bmatrix} & & \\ & & & I_m & \\ & & & & 0_s \end{bmatrix}, \tag{2.16}$$

where $\sigma_1 \geq \cdots \geq \sigma_k > 0$; I_m and 0_s are the identity and zero matrices of the corresponding order; and the numbers $\sigma_1, \ldots, \sigma_k$, k, m, and s are uniquely defined by the projector P.

Ikramov [210] showed that the numbers $\sigma_1, \ldots, \sigma_k$ are the tangents of the principal angles between the range spaces of P and the adjoint operator P^H. Ikramov [211] also gives an application of this result to projector norms.

The following representation of orthogonal projections is based on the singular value decomposition [165].

THEOREM 2.36 *Assume that $A \in \mathbb{R}^{m\times n}$ has the following singular value decomposition*

$$A = \Big[\underbrace{U_r}_{r}, \underbrace{\widetilde{U}_r}_{m-r}\Big]\Sigma\Big[\underbrace{V_r}_{r}, \underbrace{\widetilde{V}_r}_{n-r}\Big]^T, \tag{2.17}$$

where $r = rank(A)$. Then
(i) $V_rV_r^T$ is the orthogonal projection onto $\mathcal{N}(A)^\perp = \mathcal{R}(A^T)$;
(ii) $\widetilde{V}_r\widetilde{V}_r^T$ is the orthogonal projection onto $\mathcal{N}(A)$;
(iii) $U_rU_r^T$ is the orthogonal projection onto $\mathcal{R}(A)$;
(iv) $\widetilde{U}_r\widetilde{U}_r^T$ is the orthogonal projection onto $\mathcal{R}^\perp(A) = \mathcal{N}(A^T)$.

Finally we investigate representations of nonnegative projections. It is clear that either $P = UV^T$ or $P = \Pi \begin{bmatrix} I_k & B \\ 0 & 0 \end{bmatrix} \Pi^T$ are nonnegative projections if $U, V \geq 0$, $V^TU = I$ and $B \geq 0$. If $P \geq 0$ and P has rank one, then it is easy to see, that P can be represented in the form $P = uv^T$, where $u, v \in \mathbb{R}^{n\times n}$, $u, v \geq 0$ and $v^Tu = 1$. The question is how we can characterize all nonnegative projections. Flor [125] proved the following result.

THEOREM 2.37 *Let P be a nonnegative idempotent matrix of rank k. There exists a permutation matrix Π such that*

$$\Pi P \Pi^T = \begin{bmatrix} J & JT & 0 & 0 \\ 0 & 0 & 0 & 0 \\ SJ & SJT & 0 & 0 \\ 0 & 0 & 0 & 0 \end{bmatrix}, \tag{2.18}$$

$$J = \begin{bmatrix} J_1 & & & 0 \\ & J_2 & & \\ & & \ddots & \\ 0 & 0 & & J_k \end{bmatrix}, \tag{2.19}$$

where the J_j are nonnegative idempotent matrices of rank 1, that is matrices $x_j y_j^T$, $x_j \geq 0$, $y_j \geq 0$, $y_j^T x_j = 1$. Conversely, while S and T are of appropriate sizes, every matrix of the above-mentioned form is idempotent and of rank k.

Bapat, Jain and Snyder [30] use Flor's theorem to describe the structure of nonnegative matrices which are dominated by a given nonnegative projection matrix under the minus or rank-subtractivity partial order . If $A, B \in \mathbb{R}^{m\times n}$, then A is dominated by B in the minus (rank-subtractivity) partial order, denoted by $A \leq^- B$ ($A \overset{rs}{\leq} B$), if rank$(B - A)$ =rank(B) −rank(A). Bapat, Jain and Snyder show that if $A, E \in \mathbb{R}^{n\times n}$ are nonnegative, $E^2 = E$ and $A \leq^- E$, then there exists a permutation matrix Π such that $\Pi E \Pi^T$ and $\Pi A \Pi^T$ have the same block structure (2.18).

The following theorem was proved by Belitskii and Lyubich [36].

THEOREM 2.38 *The general form of a nonnegative projection is*

$$P = (A + B)\, C^T, \tag{2.20}$$

where r = rank(P), $A, B, C \in \mathbb{R}^{n\times r}$, $A, B, C \geq 0$, $A^T A = I$, $C^T A = I$, $B^T A = 0$ and $B^T C = 0$.

Further results on nonnegative idempotent matrices can be found in DeMarr [100]. Graph-theoretic characterizations of nonnegative idempotent matrices are given by Lewin [233].

If the nonnegative projections are restricted to the class of stochastic matrices, then they have a particularly simple structure. A matrix $A \in \mathbb{R}^{n\times n}$ is called stochastic , if $A \geq 0$ and $\sum_{j=1}^n a_{ij} = 1$ for all $i = 1, \ldots, n$. The matrix A is doubly-stochastic if $A \geq 0$ and $\sum_{j=1}^n a_{ij} = \sum_{i=1}^n a_{ij} = 1$ for all $i, j = 1, \ldots, n$. A matrix A is reducible if there is a permutation matrix Π such that

$$\Pi^T A \Pi = \begin{bmatrix} X & 0 \\ Y & Z \end{bmatrix}$$

with X and Z both square blocks. Otherwise, the matrix is called irreducible . For doubly-stochastic projection matrices Farahat [119] proved the following result.

THEOREM 2.39 *(Farahat). If $P \in \mathbb{R}^{n\times n}$ is an irreducible doubly-stochastic projection, then every element of P is equal to $1/n$.*

Proof. The eigenvalues of P are 1 or 0, and the number of eigenvalues equal to 1 is the rank of P. If P is irreducible, then it follows from the Perron-Frobenius theorem on nonnegative matrices (cf. [203] or [43]) that 1 is the simple eigenvalue of P and hence that P has rank 1. The result now follows easily. ■

If P is only a doubly-stochastic projection, then it is permutationally similar to a block diagonal matrix with irreducible doubly-stochastic projections in the diagonal [316].

THEOREM 2.40 *(Schwarz). Any doubly-stochastic projection $P \in \mathbb{R}^{n\times n}$ is of the form $P = \Pi^T U \Pi$, where Π is a permutation matrix and U is a matrix of the form*

$$U = \begin{bmatrix} Q_1 & 0 & \dots & 0 \\ 0 & Q_2 & \dots & 0 \\ \vdots & & & \\ 0 & 0 & \dots & Q_s \end{bmatrix}. \tag{2.21}$$

Here $Q_i \in \mathbb{R}^{r_i\times r_i}$ with all elements equal to $1/r_i$ and $\sum_{i=1}^{s} r_i = n$. Conversely, every matrix of this form is a doubly-stochastic projection of rank s.

Proof. The statement follows from the Farahat result and the fact that every doubly-stochastic matrix is permutationally similar to a block diagonal matrix having irreducible doubly-stochastic matrices in the diagonal. ■

It follows that the number of doubly-stochastic projections of order n is finite. Further characterizations of stochastic projections can be found in Schwarz [315].

3. Operations with projections

In this section we refine and complement the results of Section 2.1. We first study the operations with two projections. It is remarkable that most of the cited results were published in the last decade. Then we investigate the sum of k projections for $k \geq 2$.

3.1 Operations with two projections

There are several new results that complement the well-known statements of Theorem 2.13. We start with the result of Baksalary and Baksalary [25].

THEOREM 2.41 *Given two different nonzero projections P_1 and P_2, let P be their linear combination of the form*

$$P = c_1 P_1 + c_2 P_2 \quad (c_1, c_2 \in \mathbb{R},\ c_1, c_2 \neq 0)\,. \tag{2.22}$$

Then there are exactly four situations, where P is a projection:
(a) $P_1P_2 = P_2P_1$ holds along with either one of the following sets of conditions:
(i) $c_1 = 1,\ c_2 = 1,\ P_1P_2 = 0$;
(ii) $c_1 = 1,\ c_2 = -1,\ P_1P_2 = P_2$;
(iii) $c_1 = -1,\ c_2 = 1,\ P_1P_2 = P_1$;
(b) $P_1P_2 \neq P_2P_1$ holds along with the conditions $c_2 = 1 - c_1$, $(P_1 - P_2)^2 = 0$.

Proof. P is idempotent if and only if

$$c_1(1-c_1)P_1 - c_1c_2P_1P_2 - c_1c_2P_2P_1 + c_2(1-c_2)P_2 = 0. \tag{2.23}$$

Hence the sufficiency follows easily. We first prove the necessity in case (a). Assume that $P_1P_2 = P_2P_1$. Multiply equation (2.23) on the left by P_1 and also by P_2. This gives the equations

$$c_1(1-c_1)P_1 + c_2(1-2c_1-c_2)P_1P_2 = 0 \tag{2:24}$$

and

$$c_2(1-c_2)P_2 + c_1(1-c_1-2c_2)P_1P_2 = 0. \tag{2.25}$$

Multiplying (2.24) on the right by P_2 or (2.25) on the left by P_1 yields

$$\left[c_1 + c_2 - (c_1+c_2)^2\right]P_1P_2 = 0. \tag{2.26}$$

If $P_1P_2 = 0$, then equations (2.24) and (2.25) imply that if both c_1 and c_2 must be equal to 1, which is the situation (i). On the other hand, if $P_1P_2 \neq 0$, then (2.26) implies that either

$$c_1 + c_2 = 1 \tag{2.27}$$

or

$$c_1 + c_2 = 0. \tag{2.28}$$

Under the assumption (2.27) equalities (2.24) and (2.25) can be reduced to

$$c_1c_2(P_1 - P_1P_2) = 0, \quad c_1c_2(P_2 - P_1P_2) = 0. \tag{2.29}$$

Assumption (2.28) similarly implies the equations

$$c_1(1-c_1)(P_1 - P_1P_2) = 0, \quad c_1(1+c_1)(P_2 - P_1P_2) = 0. \tag{2.30}$$

Since $P_1 \neq P_2$, the equalities $P_1P_2 = P_1$ and $P_1P_2 = P_2$ cannot hold simultaneously. Consequently, since $c_1 \neq 0$ and $c_2 \neq 0$, it follows that the pair of conditions (2.29) can never be fulfilled, whereas the pair of conditions (2.30) is fulfilled if and only if either $c_1 = 1$ (implying $c_2 = -1$) and $P_1P_2 = P_2$, which is the situation (ii), or $c_1 = -1$ (implying $c_2 = 1$) and $P_1P_2 = P_1$, which is the situation (iii). We prove the necessity in case (b) when $P_1P_2 \neq P_2P_1$ is assumed. Multiplying equation (2.23) on the left and on the right by P_1, respectively, we have the equations

$$c_1(1-c_1)P_1 + c_2(1-c_1-c_2)P_1P_2 - c_1c_2P_1P_2P_1 = 0 \tag{2.31}$$

and

$$c_1(1-c_1)P_1 + c_2(1-c_1-c_2)P_2P_1 - c_1c_2P_1P_2P_1 = 0. \tag{2.32}$$

Hence

$$c_2(1-c_1-c_2)(P_1P_2 - P_2P_1) = 0. \tag{2.33}$$

The assumptions $c_2 \neq 0$ and $P_1P_2 \neq P_2P_1$ imply that $c_1 + c_2 = 1$. Substituting this into (2.23) simplifies the latter to the equality

$$c_1c_2(P_1 - P_1P_2 - P_2P_1 + P_2) = c_1c_2(P_1 - P_2)^2 = 0.$$

Since $c_1c_2 \neq 0$, the equality hold if and only if $(P_1 - P_2)^2 = 0$. ■

It is interesting to observe the following relationship.

COROLLARY 2.42 *Under the assumption of the theorem, a necessary condition for $P = c_1P_1 + c_2P_2$ to be a projection matrix is that each of the products P_1P_2 and P_2P_1 be a projection.*

Proof. In case (a) the corollary is obvious. In case (b) we have

$$P_1 - P_1P_2 - P_2P_1 + P_2 = 0.$$

Multiplying on the left by P_1 leads to the equality $P_1 = P_1P_2P_1$, which implies the idempotency of both P_1P_2 and P_2P_1. ■

Statement (iii) of Theorem 2.13 says that the product P_1P_2 is a projector if P_1 and P_2 commute, i.e., $P_1P_2 = P_2P_1$. This condition is only sufficient. Groß and Trenkler [170] derived several necessary and sufficient conditions under which the product P_1P_2 is a projector.

THEOREM 2.43 *Let P_1 and P_2 be projectors. The following statements are equivalent.*

(i) P_1P_2 *is a projector;*

(ii) $\mathcal{R}(P_1P_2) = \mathcal{R}(P_1) \cap (\mathcal{R}(P_2) \oplus [\mathcal{N}(P_1) \cap \mathcal{N}(P_2)])$;

(iii) $\mathcal{R}(I - P_1P_2) = \mathcal{N}(P_2) \oplus [\mathcal{N}(P_1) \cap \mathcal{R}(P_2)]$.

Groß and Trenkler also gave necessary and sufficient conditions for the commutativity of two projectors.

THEOREM 2.44 *Let P_1 and P_2 be projectors. The following statements are equivalent.*
(i) $P_1P_2 = P_2P_1$;
(ii) P_1P_2 *is the projector onto* $\mathcal{R}(P_1) \cap \mathcal{R}(P_2)$ *along* $\mathcal{N}(P_1) + \mathcal{N}(P_2)$ *and* $\text{rank}(P_1P_2) = \text{rank}(P_2P_1)$;
(iii) $P_1+P_2-P_1P_2$ *is the projector onto* $\mathcal{R}(P_1)+\mathcal{R}(P_2)$ *along* $\mathcal{N}(P_1)\cap\mathcal{N}(P_2)$ *and* $\text{rank}(P_1P_2) = \text{rank}(P_2P_1)$.

An interesting consequence of the above results is the following characterization of $P_2 - P_1$ and $P_1 + P_2$.

THEOREM 2.45 *Let P_1 and P_2 be projectors. The following statements hold:*
(a) P_2-P_1 *is a projector if and only if* P_1 *is the projector onto* $\mathcal{R}(P_1)\cap\mathcal{R}(P_2)$ *along* $\mathcal{N}(P_1) + \mathcal{N}(P_2)$;
(b) P_1+P_2 *is a projector if and only if* P_1 *is the projector onto* $\mathcal{R}(P_1)\cap\mathcal{N}(P_2)$ *along* $\mathcal{N}(P_1) + \mathcal{R}(P_2)$.

Further we mention that quite numerous results are given by Takane and Yanai [334], and Baksalary and Baksalary [26].

For the product of orthogonal projections Groß [168] derived the following canonical form.

THEOREM 2.46 *Let $P_1, P_2 \in \mathbb{C}^{n\times n}$ be orthogonal projectors. Then there exists a unitary matrix U such that*

$$P_1P_2 = U \begin{bmatrix} T & X & 0 \\ 0 & 0 & 0 \\ 0 & 0 & I \end{bmatrix} U^H, \tag{2.34}$$

where T is diagonal with all diagonal elements in the open interval $(0,1)$, and $XX^H = T - T^2$.

Various rank equalities are also used to characterize the projection operations. Groß and Trenkler [171] gave formulas for the rank of $P_1 - P_2$ and $P_1 + P_2$. Tian and Styan [340] also established rank equalities for the difference, the sum and the product of idempotent matrices.

The eigenvalue distribution of $P_1 - P_2$ and P_1P_2 is also of interest. The following result is easy to prove [16].

LEMMA 2.47 *Let $P_1, P_2 \in \mathbb{C}^{n\times n}$ be orthogonal projections.*
(i) If λ is an eigenvalue of $P_1 - P_2$, then $-1 \le \lambda \le 1$;
(ii) If σ is an eigenvalue of P_1P_2, then σ is real and $0 \le \sigma \le 1$.

Proof. (i) Let x be a unit eigenvector of λ. Then $\lambda = x^H P_1 x - x^H P_2 x$. Since both quantities are in $[0,1]$, their difference lies in $[-1,1]$.
(ii) Since $\|P_1\| = \|P_2\| = 1$, it follows that $|\sigma| \leq 1$. The equality $P_1 P_2 x = \sigma x$ implies $P_1 P_2 x = \sigma P_1 x$. Hence $x = P_1 x$ and $x^T P_1 P_2 x = x^H P_2 x = \sigma x^H x$, which implies $\sigma = x^H P_2 x / x^H x \geq 0$. ■

Further results are given in Anderson, Harner and Trapp [16], Nelson and Neumann [264] and Groß [168]. Here we quote the following result of Nelson and Neumann [264].

THEOREM 2.48 *Suppose that* $\lambda_1, \ldots, \lambda_n \in [0,1]$. *Let*

$$S = \{i \mid \lambda_i = 0,\ 1 \leq i \leq n\}$$

and

$$T = \{i \mid 0 < \lambda_i < 1,\ 1 \leq i \leq n\}.$$

Then there exist two $n \times n$ *orthogonal projections* P_1 *and* P_2 *such that*

$$\sigma(P_1 P_2) = \{\lambda_i\}_{i=1}^n$$

if and only if the cardinalities of S *and* T *satisfy* $|T| \leq |S|$.

3.2 Sums of projections

The sum of k projections ($k \geq 2$) was investigated by several authors [167], [85], [29], [239], [97], [223], [107], [336], [242]. The research on the sums of projections was initiated by a theorem of Cochran in statistics (see, e.g., [167]). The following, which became a classic result, generalizes Cochran's theorem by allowing nonsymmetric matrices P_i. For symmetric matrices it was essentially proved by Craig and Hotelling (see Graybill and Marsaglia [167]).

THEOREM 2.49 *Let* $P_i \in \mathbb{R}^{n \times n}$ *and* $rank(P_i) = r_i$ $(i = 1, \ldots, m)$. *If the relation* $\sum_{i=1}^m P_i = I$ *holds, then the following statements are equivalent:*
(i) $P_i^2 = P_i$, $i = 1, \ldots, m$;
(ii) $\sum_{i=1}^m r_i = n$;
(iii) $P_i P_j = 0$ *for all* $i \neq j$.

Proof. We prove first that (a) implies (b). As $\mathrm{tr}(P_i) = \mathrm{rank}(P_i) = r_i$ and $\mathrm{tr}(I) = \sum_{i=1}^m \mathrm{tr}(P_i)$ we have $n = \sum_{i=1}^m r_i$. We show now that (b) implies (c). Let V_i be the range of P_i. Since $\sum_{i=1}^m P_i = I$ we must have $\sum_{i=1}^m V_i = \mathbb{R}^n$. By (b) we have $\sum_{i=1}^m \dim(V_i) = \dim(\mathbb{R}^n)$ implying that $\mathbb{R}^n = V_1 \oplus \cdots \oplus V_m$. Using $I = \sum_{i=1}^m P_i$ we get

$$\left(\sum_{i=1}^m P_i\right) P_j x = P_j x$$

for all $x \in \mathbb{R}^n$ and for all $j = 1, \ldots, m$. Since $P_i P_j x \in V_i$, $P_j x \in V_j$ and the sum of V_i's is direct we must have

$$P_i P_j x = 0$$

for all $x \in \mathbb{R}^n$ and $i \neq j$. This means that $P_i P_j = 0$ $(i \neq j)$. Finally we show that (c) implies (a). Multiplying $\sum_{i=1}^m P_i = I$ by P_j we get $P_j^2 = P_j$ and the proof is complete. ∎

For another interesting proof see, e.g., [243]. The theorem no longer holds if $\sum_{i=1}^k P_i = I$ is not assumed. Instead we have the result of Khatri ([300], pp. 111, 112) which is presented with a new proof that exploits Theorem 2.25 of Egerváry. For symmetric idempotents a similar result was proved by Graybill and Marsaglia [167], Banerjee [29] and Loynes [239].

THEOREM 2.50 *(Khatri). Let P_i, $i = 1, \ldots, k$, be square matrices of order n and $P = \sum_{i=1}^k P_i$. Consider the following four statements:*
(i) $P_i^2 = P_i$, $i = 1, \ldots, k$;
(ii) $P_i P_j = 0$ for all $i \neq j$, rank $(P_i^2) =$ rank (P_i) for all i;
(iii) $P^2 = P$;
(iv) rank $(P) = \sum_{i=1}^k$ rank (P_i).
Then any two of the three conditions (i), (ii), (iii) imply all four. (iii) and (iv) imply (i) and (ii).

Proof. (a) (iii) and (iv) imply (i) and (ii). Let rank $(P) = r$ and rank $(P_i) = r_i$. Furthermore let P_i have the full rank factorization $P_i = B_i C_i$ with $B_i \in \mathbb{R}^{n \times r_i}$, $C_i \in \mathbb{R}^{r_i \times n}$ and rank $(B_i) =$ rank $(C_i) = r_i$. Then

$$P = \sum_{i=1}^k P_i = \sum_{i=1}^k B_i C_i = [B_1, \ldots, B_k] \begin{bmatrix} C_1 \\ \vdots \\ C_k \end{bmatrix}$$

holds with

$$\text{rank}\,([B_1, \ldots, B_k]) = \text{rank}\left(\begin{bmatrix} C_1 \\ \vdots \\ C_k \end{bmatrix} \right) = \sum_{i=1}^k r_i = r.$$

The matrix P is a projection if and only if $P^2 - P = 0$, that is, if

$$P^2 - P = [B_1, \ldots, B_k] \left(\begin{bmatrix} C_1 \\ \vdots \\ C_k \end{bmatrix} [B_1, \ldots, B_k] - I_r \right) \begin{bmatrix} C_1 \\ \vdots \\ C_k \end{bmatrix} = 0.$$

The rank condition implies that the matrices

$$[B_1, \ldots, B_k], \quad \begin{bmatrix} C_1 \\ \vdots \\ C_k \end{bmatrix}$$

have left and right inverse, respectively. Multiplying the above equation by the corresponding left and right inverses we obtain that P is a projection if and only if

$$\begin{bmatrix} C_1 \\ \vdots \\ C_k \end{bmatrix} [B_1, \ldots, B_k] - I_r = 0.$$

This implies that $C_i B_i = I_{r_i}$ and $C_i B_j = 0$ for all $i \neq j$. Consequently $P_i^2 = B_i C_i B_i C_i = P_i$ and $P_i P_j = B_i C_i B_j C_j = 0$ for $i \neq j$.
(b) (i) and (ii) imply (iii) as

$$P^2 = \sum_{i=1}^{k} \sum_{j=1}^{k} P_i P_j = \sum_{i=1}^{n} P_i^2 = \sum_{i=1}^{n} P_i.$$

(c) (i) and (iii) imply (iv). As $\operatorname{tr}(P) = \operatorname{rank}(P)$ holds for any projection P, the relation

$$\operatorname{tr}(P) = \operatorname{tr}\left(\sum_{i=1}^{k} P_i\right) = \sum_{i=1}^{k} \operatorname{tr}(P_i)$$

implies (iv).
(d) (ii) and (iii) imply (i). It is obvious that (iii) and $P_i P_j = 0$ $(i \neq j)$ imply

$$\sum_{i=1}^{k} P_i^2 = \sum_{i=1}^{k} P_i.$$

Multiplying this by P_j we obtain $P_j^3 = P_j^2$. Taking the full rank factorization $P_j = B_j C_j$ we obtain the equation

$$B_j C_j (I - B_j C_j) B_j C_j = 0.$$

The matrices B_j and C_j have left and right inverse, respectively. Hence this equation is equivalent with

$$C_j (I - B_j C_j) B_j = C_j B_j (I - C_j B_j) = 0.$$

As $r_j = \operatorname{rank}\left(P_j^2\right) = \operatorname{rank}(B_j C_j B_j C_j) \leq \operatorname{rank}(C_j B_j) \leq r_j$, $\operatorname{rank}(C_j B_j) = r_j$ implying that $C_j B_j \in \mathbb{R}^{r_j \times r_j}$ is invertible and $I_{r_j} = C_j B_j$. Therefore P_j is idempotent. ■

In general, condition (ii) can not be weakened (see, e.g., [300], p. 112). Let

$$P_1 = \begin{bmatrix} 1 & 0 & 0 & 0 \\ 0 & 0 & 0 & 0 \\ 0 & 0 & 0 & 0 \\ 0 & 0 & 1 & 0 \end{bmatrix}, \quad P_2 = \begin{bmatrix} 0 & 0 & 0 & 0 \\ 0 & 1 & 0 & 0 \\ 0 & 0 & 0 & 0 \\ 0 & 0 & -1 & 0 \end{bmatrix}.$$

Then P_1+P_2 is idempotent, $P_1P_2 = P_2P_1 = 0$, while P_1 and P_2 are not idempotent. If all matrices P_i are symmetric, then condition (ii) can be weakened to (ii*) $P_iP_j = 0, \quad i \neq j$ (see Graybill and Marsaglia [167]). Various parts of Theorem 2.50 were also proved by Djokovic [107], Chipman and Rao [85] and others (see Kestelman [223]).

The rank additivity property (iv) was investigated by Marsaglia and Styan [244] and Carlson [76].

An interesting variation of the sum theorems is given by Mäkeläinen and Styan [242].

THEOREM 2.51 *(i) Let $P_1, \ldots, P_k \in \mathbb{R}^{n\times n}$ be idempotent matrices, not necessarily symmetric, and let F be a diagonalizable matrix with nonnegative eigenvalues such that*

$$F + \sum_{i=1}^{k} P_i = I_n.$$

Then F is idempotent.
(ii) Let $P_1, \ldots, P_k, E \in \mathbb{R}^{n\times n}$ be idempotent matrices, not necessarily symmetric, and let F have nonnegative eigenvalues and rank(F) =rank(F^2). If

$$F + \sum_{i=1}^{k} P_i = E,$$

then F is idempotent.

Let $\mathbb{F}$ be a field of characteristic zero. Hartwig and Putcha [189] proved that a matrix $A \in \mathbb{F}^{n\times n}$ is a sum of idempotent matrices, exactly when the trace of A is an integer at least as large as rank(A). Laurie, Mathes and Radjavi [231] investigated the question of how many idempotent matrices are needed to sum to a given A.

4. Generalized inverses and projections

Projections and generalized inverses are closely interrelated. We show three connections between projections and generalized inverses. We first recall Moore's original definition of the Moore-Penrose pseudo inverse.

DEFINITION 2.52 *(Moore). If $A \in \mathbb{R}^{m\times n}$, then the generalized inverse of A is defined to be the unique matrix A^+ such that*

(a) $AA^+ = P_{\mathcal{R}(A)}$;
(b) $A^+A = P_{\mathcal{R}(A^+)} = P_{\mathcal{R}(A^T)}$.

The equivalence of the Moore and Penrose definitions is shown in many books (see, e.g., [75]). The Urquhart theorem implies that we can represent both projections with (i,j)-inverses as

$$P_{\mathcal{R}(A^T)} = A^+A = A^{(1,4)}A \tag{2.35}$$

and

$$P_{\mathcal{R}(A)} = AA^+ = AA^{(1,3)}. \tag{2.36}$$

For any matrix $A \in \mathbb{C}^{m\times n}$ of rank r and for any g-inverse $G \in A\{1\}$ both AG and GA are projections with rank r and therefore

$$AG = P_{\mathcal{R}(A),S}, \quad GA = P_{T,\mathcal{N}(A)}, \tag{2.37}$$

where $\mathcal{N}(G) \subset S \subset \mathbb{C}^m$ and $T \subset \mathcal{R}(G) \subset \mathbb{C}^n$ are some subspaces complementary to $\mathcal{R}(A)$ and $\mathcal{N}(A)$, respectively. Conversely, for arbitrary subspaces S and T complementary to $\mathcal{R}(A)$ and $\mathcal{N}(A)$, respectively, there exists a g-inverse $G \in A\{1\}$ satisfying (2.37). The earliest result of this type is due to Egerváry [115].

THEOREM 2.53 *(Egerváry). Let $A \in \mathbb{C}^{m\times n}$ $(\operatorname{rank}(A) = r)$ have the full rank factorization $A = XY$. If the matrix $G \in \mathbb{C}^{n\times m}$ has rank r and satisfies the equations*

$$(AG)^2 = AG, \quad (GA)^2 = GA, \tag{2.38}$$

then it can be given in the form

$$G = M(YM)^{-1}(QX)^{-1}Q = M(QAM)^{-1}Q, \tag{2.39}$$

where M and Q are arbitrary satisfying $\det(YM) \neq 0$ *and* $\det(QX) \neq 0$.

The following theorem is proved in Ben-Israel and Greville [40], where other related results can also be found.

THEOREM 2.54 *Let $A \in \mathbb{C}^{m\times n}$ have rank r, and let S and T be arbitrary subspaces complementary to $\mathcal{R}(A)$ and $\mathcal{N}(A)$, respectively. Then:*
(i) The general solution of (2.37) is

$$G = P_{T,\mathcal{N}(A)}A^{(1)}P_{\mathcal{R}(A),S} + \left(I_n - A^{(1)}A\right)Y\left(I_m - AA^{(1)}\right), \tag{2.40}$$

where $A^{(1)} \in A\{1\}$ is fixed but arbitrary, and $Y \in \mathbb{C}^{n\times m}$ is arbitrary.
(ii) The matrix

$$A_{T,S}^{(1,2)} = P_{T,\mathcal{N}(A)}A^{(1)}P_{\mathcal{R}(A),S} \tag{2.41}$$

is the unique $(1,2)$*-inverse of* A *having range* T *and null space* S.

Another interesting representation of oblique projections can be given using the group inverse. Let $A \in \mathbb{C}^{n\times n}$ have a group inverse $A^{\#} \in \mathbb{C}^{n\times n}$. Then

$$A^{\#}A = AA^{\#} = P_{\mathcal{R}(A),\mathcal{N}(A)}. \tag{2.42}$$

Let $A = FG$ be a full rank factorization. Then $A^{\#} = F\left(GF\right)^{-2}G$ and $AA^{\#} = F\left(GF\right)^{-1}G$ is indeed a projection and also a full rank factorization of $AA^{\#}$. Hence $\mathcal{R}\left(AA^{\#}\right) = \mathcal{R}\left(F\right) = \mathcal{R}\left(A\right)$ and $\mathcal{N}\left(AA^{\#}\right) = \mathcal{N}\left(G\right) = \mathcal{N}\left(A\right)$.

5. The gap between two subspaces and norms of projections

Using the singular value decomposition we determine the unitarily invariant norm of oblique projections. The same technique gives the gap between subspaces and other related results. Throughout the section we use unitarily invariant matrix norms with symmetric gauge function ϕ and the principal angles between subspaces. The taken approach is based on [143].

Let $M, N \subset \mathbb{R}^n$ be subspaces with $p = \dim(M) \geq \dim(N) = q$. Denote the principal angles between M and N by θ_i $(i = 1,\ldots,q)$. Let $U = [U_1, U_2] \in \mathbb{R}^{n\times n}$ be orthogonal such that $U_1 \in \mathbb{R}^{n\times p}$ is a basis of M, and $U_2 \in \mathbb{R}^{n\times(n-p)}$ is a basis of $M^{\perp}$. Similarly, let $V = [V_1, V_2] \in \mathbb{R}^{n\times n}$ be an orthogonal matrix such that $V_1 \in \mathbb{R}^{n\times q}$ is a basis of N, and $V_2 \in \mathbb{R}^{n\times(n-q)}$ is a basis of $N^{\perp}$.

Assume now that M and N are complementary. Then $p + q = n$ and $0 < \theta_i \leq \pi/2$ for all $i = 1,\ldots,q$. It is easy to verify that

$$P_{M,N} = U\begin{bmatrix} 0 & \left(V_2^TU_1\right)^{-1} \\ 0 & 0 \end{bmatrix}V^T$$

and

$$P_{N,M} = V\begin{bmatrix} 0 & \left(U_2^TV_1\right)^{-1} \\ 0 & 0 \end{bmatrix}U^T.$$

Let $V_2^TU_1 = X\Sigma Y^T$ be the singular value decomposition of $V_2^TU_1$ with orthogonal matrices $X,Y \in \mathbb{R}^{p\times p}$. The Björck, Golub theorem implies that

$$\Sigma = \operatorname{diag}(\underbrace{1,\ldots,1}_{p-q}, \sin\left(\theta_q\right),\ldots,\sin\left(\theta_1\right)) \in \mathbb{R}^{p\times p}.$$

Then the singular value decomposition of projection $P_{M,N}$ has the form

$$P_{M,N} = [U_1Y, U_2]\begin{bmatrix} \Sigma^{-1} & 0 \\ 0 & 0 \end{bmatrix}\begin{bmatrix} X^TV_2^T \\ V_1^T \end{bmatrix}. \tag{2.43}$$

The Björck, Golub theorem also implies that U_1Y and V_2X are the principal vectors related to the principal angles between M and $N^{\perp}$. Similarly, let

$U_2^T V_1 = \tilde{X}\tilde{\Sigma}\tilde{Y}^T$ be the singular value decomposition of $U_2^T V_1$, where $\tilde{X}, \tilde{Y} \in \mathbb{R}^{q\times q}$ are orthogonal matrices and

$$\tilde{\Sigma} = \operatorname{diag}\left(\sin\left(\theta_q\right), \ldots, \sin\left(\theta_1\right)\right) \in \mathbb{R}^{q\times q}.$$

Then the singular value decomposition of $P_{N,M}$ is given by

$$P_{N,M} = \left[V_1\tilde{Y}, V_2\right]\begin{bmatrix} \tilde{\Sigma}^{-1} & 0 \\ 0 & 0 \end{bmatrix}\begin{bmatrix} \tilde{X}^T U_2^T \\ U_1^T \end{bmatrix}, \tag{2.44}$$

where $V_1\tilde{Y}$ and $U_2\tilde{X}$ are the principal vectors related to the angles between $M^\perp$ and N. The singular value decompositions of $P_{M,N}$ and $P_{N,M}$ and Theorem 1.50 of von Neumann lead to the following result.

PROPOSITION 2.55 *For any projection* $0 \neq P_{M,N} \neq I$,

$$\|P_{M,N}\| = \phi\Big(\frac{1}{\sin\left(\theta_1\right)}, \ldots, \frac{1}{\sin\left(\theta_q\right)}, \overbrace{1, \ldots, 1}^{p-q}, 0, \ldots, 0\Big) \tag{2.45}$$

and

$$\|P_{N,M}\| = \phi\Big(\frac{1}{\sin\left(\theta_1\right)}, \ldots, \frac{1}{\sin\left(\theta_q\right)}, 0, \ldots, 0\Big), \tag{2.46}$$

where ϕ is the symmetric gauge function.

PROPOSITION 2.56 *For any projection $P_{M,N}$ we have*

$$\|P_{M,N}\| \geq \|P_M\| \tag{2.47}$$

in any unitarily invariant matrix norm.

Proof. Since $1/\sin\left(\theta_i\right) \geq 1$, the monotonicity of ϕ implies

$$\|P_{M,N}\| \geq \phi(\overbrace{1, \ldots, 1}^{p}, 0, \ldots, 0).$$

The norm of $P_{M,N}$ is minimal, if $\theta_i = \pi/2$ for all $i = 1, \ldots, q$. This means that $M \perp N$ (or $M = N^\perp$). The minimal value depends only on the dimension p of subspace M. ∎

The result is well known for the spectral norm.

COROLLARY 2.57 *If the symmetric gauge function ϕ satisfies (1.51), then*

$$\frac{1}{\sin\left(\theta_1\right)} \leq \|P_{M,N}\| \leq p - q + \sum_{i=1}^{q} \frac{1}{\sin\left(\theta_i\right)}. \tag{2.48}$$

Proof. The result follows from relations (1.52) and (2.45). ∎

COROLLARY 2.58 *(Ljance).*

$$\|P_{M,N}\|_2 = \frac{1}{\sin(\theta_1)}. \tag{2.49}$$

Proof. Since the spectral norm is defined by $\phi(u) = \max_{1 \le i \le n} |u_i|$, the result clearly follows from Proposition 2.55. ∎

COROLLARY 2.59 *(Kato). For $0 \neq P_{M,N} \neq I$,*

$$\|I - P_{M,N}\|_2 = \|P_{M,N}\|_2 . \tag{2.50}$$

We can prove the following extension of Corollary 2.59.

PROPOSITION 2.60 *Let $P \in \mathbb{R}^{n\times n}$ be an arbitrary projection such that $0 \neq P \neq I$. If P has rank p, $q = n - p$, then we have*

$$\|P\| \le \|I - P\|, \quad p < q, \tag{2.51}$$

$$\|P\| \ge \|I - P\|, \quad p \ge q. \tag{2.52}$$

If in addition the symmetric gauge function ϕ satisfies (1.51), then

$$\|P\| \le \|I - P\| \le \|P\| + q - p, \quad p < q, \tag{2.53}$$

$$\|P\| \ge \|I - P\| \ge \|P\| + q - p, \quad p \ge q. \tag{2.54}$$

Proof. If $p \ge q$, then by the monotonicity of ϕ, we obtain for $P = P_{M,N}$ that

$$\|P_{M,N}\| \ge \|P_{N,M}\| = \|I - P_{M,N}\| .$$

The lower bound follows from

$$\|P_{M,N}\| \le \phi(\frac{1}{\sin(\theta_1)}, \ldots, \frac{1}{\sin(\theta_q)}, 0, \ldots, 0) + \phi(\overbrace{1, \ldots, 1}^{p-q}, 0, \ldots, 0).$$

If $p < q$, we replace $P_{M,N}$ by $P_{N,M}$. ∎

Assume now that $M, N \subset \mathbb{R}^n$ are arbitrary subspaces. For any $x \in \mathbb{R}^n$, the distance between $P_M x \in M$ and $P_N x \in N$ is bounded by

$$\|P_M x - P_N x\| = \|(P_M - P_N)x\| \le \|P_M - P_N\| \|x\| .$$

The quantity $\|P_M - P_N\|$ is the least upper bound for the distance of the projected vectors of unit length. Hence it characterizes the distance of subspaces M and N.

DEFINITION 2.61 *Let $M, N \subset \mathbb{R}^n$ be subspaces. The quantity*

$$\Theta(M, N) = \|P_M - P_N\|_2 \tag{2.55}$$

is called the distance or gap between subspaces M and N.

In Chapter 7 we show that

$$\Theta(M,N) = \max\left\{ \sup_{x\in N,\,\|x\|=1} d(x,M),\ \sup_{x\in M,\,\|x\|=1} d(x,N) \right\}. \tag{2.56}$$

Hence $\Theta(M,N)$ is indeed the distance of M and N.

We now derive the quantity $\|P_M - P_N\|$ for any unitarily invariant matrix norm in terms of principal angles. Since $P_M = U_1U_1^T$ and $P_N = V_1V_1^T$ we have

$$U^T(P_M - P_N)V = \begin{bmatrix} 0 & U_1^TV_2 \\ -U_2^TV_1 & 0 \end{bmatrix}.$$

The unitary invariance of the norm yields

$$\|P_M - P_N\| = \left\| \begin{bmatrix} U_1^TV_2 & 0 \\ 0 & -U_2^TV_1 \end{bmatrix} \right\|.$$

The singular values of the matrix

$$\begin{bmatrix} U_1^TV_2 & 0 \\ 0 & -U_2^TV_1 \end{bmatrix} \quad \left(U_1^TV_2 \in \mathbb{R}^{p\times(n-q)},\ U_2^TV_1 \in \mathbb{R}^{(n-p)\times q}\right)$$

are the singular values of $U_1^TV_2$ and $U_2^TV_1$, and possibly some zeros. Hence Theorem 1.88 implies

$$\begin{aligned}\|P_M - P_N\| &= \phi(\sin(\theta_1),\dots,\sin(\theta_q),\underbrace{1,\dots,1}_{p-q},\\ &\qquad \sin(\theta_1),\dots,\sin(\theta_q),\underbrace{0,\dots,0}_{n-p-q}),\end{aligned}$$

for $p+q \le n$, and

$$\begin{aligned}\|P_M - P_N\| &= \phi(\sin(\theta_1),\dots,\sin(\theta_{n-p}),\underbrace{1,\dots,1}_{p-q},\\ &\qquad \underbrace{0,\dots,0}_{p+q-n},\sin(\theta_1),\dots,\sin(\theta_{n-p})),\end{aligned}$$

for $p+q > n$. As a special case we obtain

PROPOSITION 2.62 *Let $M, N \subset \mathbb{R}^n$ be subspaces. Then*

$$\Theta(M,N) = \|P_M - P_N\|_2 = \begin{cases} \sin(\theta_{\max}), & p = q, \\ 1, & p \ne q. \end{cases} \tag{2.57}$$

It is easy to see that $\Theta(M,N)$ has the following properties for any subspaces M, N and K:

$$0 \leq \Theta(M,N) \leq 1, \quad \Theta(M,N) = 0 \Leftrightarrow M = N, \tag{2.58}$$

$$\Theta(M,N) = \Theta(N,M) = \Theta\left(M^{\perp}, N^{\perp}\right), \tag{2.59}$$

$$\Theta(M,N) \leq \Theta(M,K) + \Theta(K,N). \tag{2.60}$$

The following important result is a simple consequence of (2.57).

COROLLARY 2.63 *If* $\Theta(M,N) < 1$, *then*

$$\dim(M) = \dim(N). \tag{2.61}$$

Hence the subspaces that are close enough have the same dimension. The result can be extended to oblique projections as well (see, e.g., [84]).

THEOREM 2.64 *Let* $\|\cdot\|$ *be any matrix norm and the projections* P *and* Q *be such that* $\|P-Q\| < 1$. *Then rank*(P) =*rank*(Q).

Proof. Let $\{x_i\}_{i=1}^q \subset \mathcal{R}(Q)$ be a basis. We prove that vectors$\{Px_i\}_{i=1}^q \subset \mathcal{R}(P)$ are independent. Assume that $\sum_{i=1}^q \alpha_i P x_i = 0$. Then $y = \sum_{i=1}^q \alpha_i x_i$ is such that $Py = 0$. Thus $(Q-P)y = Qy = y$ and for $y \neq 0$ the inequality $\|y\| = \|(Q-P)y\| \leq \|Q-P\|\,\|y\|$ implies that $\|Q-P\| \geq 1$, which is a contradiction. Hence $y = 0$ is the only possibility implying that $\alpha_i = 0$ $(i = 1,\ldots,q)$ and rank(Q) $\leq$rank(P). We can similarly prove that rank(Q) $\geq$rank(P). ■

PROPOSITION 2.65 *If* $\dim(M) = \dim(N)$, *then*

$$\Theta(M,N) = \|P_M P_{N^\perp}\|_2 = \|P_N P_{M^\perp}\|_2 = \left\|P_{M^\perp P_N}\right\|_2. \tag{2.62}$$

Proof. The symmetry of $\Theta(M,N) = \Theta(N,M)$ implies that $\Theta(M,N) = \|P_M P_{N^\perp}\|$ if and only if $\theta(N,M) = \|P_N P_{M^\perp}\|$. Since $P_M = U_1 U_1^T$ and $P_{N^\perp} = V_2 V_2^T$ we have

$$U^T P_M P_{N^\perp} V = \begin{bmatrix} 0 & U_1^T V_2 \\ 0 & 0 \end{bmatrix}.$$

The singular values of $U_1^T V_2$ are $\sin(\theta_1)$,...,$\sin(\theta_p)$ for $2p \leq n$ $(p = q)$ and $\sin(\theta_1)$,...,$\sin(\theta_{n-p})$ for $2p > n$. Hence $\|P_M P_{N^\perp}\|_2 = \sin(\theta_{\max})$. Similarly, $P_{M^\perp} = U_2 U_2^T$, $P_N = V_1 V_1^T$ and

$$U^T P_{M^\perp P_N} V = \begin{bmatrix} 0 & 0 \\ U_2^T V_1 & 0 \end{bmatrix}.$$

The singular values of $U_2^T V_1$ are also $\sin(\theta_1),\ldots,\sin(\theta_p)$ for $2p \leq n$ $(p = q)$ and $\sin(\theta_1),\ldots,\sin(\theta_{n-p})$ for $2p > n$. Hence $\left\|P_{M^\perp}P_N\right\| = \sin(\theta_{\max})$. ■

We can now reformulate Proposition 1.92 [144].

THEOREM 2.66 *Let $A \in \mathbb{R}^{n\times n}$ be nonsingular and let $A = Q_A R_A$ be its unique QR factorization with positive diagonal entries in R_A. Then A has an LU factorization if and only if*

$$\Theta\left(\mathcal{R}\left(Q_A^{\bar{k}}\right), \mathcal{R}\left(I^{|k}\right)\right) < 1 \tag{2.63}$$

for all $k = 1, \ldots, n-1$.

Proof. Condition $\theta_k\left(\mathcal{R}\left(\left(Q_A^T\right)^{|k}\right), \mathcal{R}\left(I^{|k}\right)\right) < \pi/2$ is equivalent with

$$\Theta\left(\mathcal{R}\left(Q_A^{\bar{k}}\right), \mathcal{R}\left(I^{|k}\right)\right) < 1.$$

■

We can observe that $P_M - P_N$ is nonsingular, if and only if $U_1^T V_2$ and $U_2^T V_1$ are nonsingular matrices. Both matrices are nonsingular if and only if $p + q = n$ and $\theta_1 > 0$. Thus $M \cap N = \{0\}$ and $M + N = \mathbb{R}^n$. For complementary subspaces

$$\left\|(P_M - P_N)^{-1}\right\| = \left\|\begin{bmatrix} \left(U_1^T V_2\right)^{-1} & 0 \\ 0 & -\left(U_2^T V_1\right)^{-1} \end{bmatrix}\right\|$$

$$= \phi\left(\frac{1}{\sin(\theta_1)}, \ldots, \frac{1}{\sin(\theta_q)}, \overbrace{1, \ldots, 1}^{p-q}, \frac{1}{\sin(\theta_1)}, \ldots, \frac{1}{\sin(\theta_q)}\right).$$

This norm is minimal, if and only if $M \perp N$ $(\theta_1 = \cdots = \theta_q = \pi/2)$. The subadditivity of ϕ implies

PROPOSITION 2.67 *If M and N are complementary subspaces, then*

$$\left\|(P_M - P_N)^{-1}\right\| \leq \|P_{M,N}\| + \|P_{N,M}\| \tag{2.64}$$

holds in any unitarily invariant matrix norm.

We can also establish the following result [212].

COROLLARY 2.68 *(Ipsen, Meyer). Subspaces $M, N \subset \mathbb{R}^n$ are complementary, if and only if $P_M - P_N$ is nonsingular. Furthermore*

$$\left\|(P_M - P_N)^{-1}\right\|_2 = \frac{1}{\sin(\theta_1)}. \tag{2.65}$$

A similar result of Buckholtz [71], [72] will be given in Chapter 7. The difference of real oblique projections was investigated by Groß and Trenkler [171].

LEMMA 2.69 *(Groß, Trenkler). Let $P_1, P_2 \in \mathbb{R}^n$ be two oblique projections. Then rank$(P_1 - P_2) = n$ if and only if $\mathcal{R}(P_1) \cap \mathcal{R}(P_2) = \{0\}$ and $\mathcal{N}(P_1) \cap \mathcal{N}(P_2) = \{0\}$.*

We now prove the following complementary result.

PROPOSITION 2.70 *If $M, N \subset \mathbb{R}^n$ are nonzero subspaces, then*

$$\cos(\theta_1) = \|P_M P_N\|_2 = \|P_N P_M\|_2 . \tag{2.66}$$

Proof. Since

$$U^T P_M P_N V = \begin{bmatrix} U_1^T V_1 & 0 \\ 0 & 0 \end{bmatrix}$$

and

$$\|P_M P_N\| = \phi(\cos(\theta_1), \ldots, \cos(\theta_q), 0, \ldots, 0),$$

the statement follows. ∎

Using the last proposition we can state the result of Ljance in a new form.

PROPOSITION 2.71 *(Ljance).*

$$\|P_{M,N}\|_2^2 = \frac{1}{1 - \|P_M P_N\|_2^2} = \frac{1}{1 - \|P_N P_M\|_2}. \tag{2.67}$$

and $\cos(\theta_1) = \|P_M P_N\|_2 = \|P_N P_M\|_2$

Further results on the gap can be found in [301], [165], [161], [331], [354]. Stewart and Sun [331] studied the gap topology induced by unitarily invariant matrix norms. Gohberg, Lancaster and Rodman [161] is the best source on the gap concept and its applications.

6. Bounds for projections

We give bounds for projections and projected vectors. We need the following result.

LEMMA 2.72 *Let $A \in \mathbb{R}^{r \times s}$ and $B \in \mathbb{R}^{r \times k}$ $(1 \leq k, r, s \leq n,\ k + s \leq n)$ be arbitrary matrices. Then, in any unitarily invariant matrix norm of $\mathbb{R}^{n \times n}$,*

$$\|A\| \leq \|[A, B]\| . \tag{2.68}$$

Proof. The singular values of the augmented $n \times n$ matrices

$$\hat{A} = \begin{bmatrix} A & 0 & 0 \\ 0 & 0 & 0 \\ 0 & 0 & 0 \end{bmatrix}, \quad F = \begin{bmatrix} A & B & 0 \\ 0 & 0 & 0 \\ 0 & 0 & 0 \end{bmatrix}$$

are defined by the eigenvalues of

$$C = \hat{A}^T \hat{A} = \begin{bmatrix} A^T A & 0 & 0 \\ 0 & 0 & 0 \\ 0 & 0 & 0 \end{bmatrix}$$

and

$$D = F^T F = \begin{bmatrix} A^T A & A^T B & 0 \\ B^T A & B^T B & 0 \\ 0 & 0 & 0 \end{bmatrix},$$

respectively. The eigenvalues of C are the s eigenvalues of $A^T A$ and $n - s$ zero eigenvalues. The eigenvalues of D are the $k + s$ eigenvalues of

$$\tilde{D} = \begin{bmatrix} A^T A & A^T B \\ B^T A & B^T B \end{bmatrix}$$

and $n - (k + s)$ zeros. Since $\tilde{D}$ is symmetric, the Poincaré separation theorem [209] implies that

$$\sigma_{s-i+1}^2 \left(\hat{A} \right) = \lambda_i \left(A^T A \right) \leq \lambda_{i+k} \left(\tilde{D} \right) = \sigma_{s-i+1}^2 \left(F \right), \quad i = 1, \ldots, s.$$

As

$$\sigma \left(\hat{A} \right) = \left[\sigma_1 \left(A \right), \ldots, \sigma_s \left(A \right), 0, \ldots, 0 \right]^T,$$

$$\sigma \left(F \right) = \left[\sigma_1 \left(F \right), \ldots, \sigma_s \left(F \right), \ldots, \sigma_{k+s} \left(F \right), 0, \ldots, 0 \right]^T$$

we obtain

$$0 \leq \sigma \left(\hat{A} \right) \leq \sigma \left(F \right).$$

Hence

$$\phi \left(\sigma \left(\hat{A} \right) \right) \leq \phi \left(\sigma \left(F \right) \right)$$

follows from the monotonicity of the symmetric gauge function proving the statement of the lemma. ■

Lemma 2.72 was proved in spectral norm by Hanson and Lawson [187] in a different way. It is obvious that

$$\|A^T\| \leq \left\| \begin{bmatrix} A^T \\ B^T \end{bmatrix} \right\|$$

also holds. Lemma 2.72 is not true in general. Bosznay and Garay [51] investigated the induced norms of projections $P : X \to X$ in n-dimensional real vector spaces X. Let $N(X)$ be the set of vector norms defined on X. In any induced operator norm $\|I\| = 1$ and $\|P\| \geq 1$ for $P \neq 0$. Denote by $N_1(X)$ the set of those vector norms for which

$$P : X \to X, \; P^2 = P, \; P \neq I, \; \dim \left(\mathcal{R} \left(P \right) \right) > 1 \Rightarrow \|P\| > 1$$

in the induced operator norm. Bosznay and Garay proved that for $n \geq 3$ the set $N_1(X)$ is open and dense in $N(X)$. Taking such a norm from $N_1(X)$, $X = \mathbb{R}^n$ and

$$P = \begin{bmatrix} 1 & 0 & 0 & 0 \\ 0 & 1 & 0 & 0 \\ 0 & 0 & 0 & 0 \\ 0 & 0 & 0 & 0 \end{bmatrix}$$

we can see that $\|P\| > 1 = \|I\|$.

Using Lemmas 2.72 and 2.29 we can prove the following result [135].

LEMMA 2.73 *Let $A \in \mathbb{R}^{n \times n}$ be an arbitrary nonsingular matrix. Then*

$$\left\|P_{\mathcal{R}(A^{|k}),\mathcal{R}(A^{n-k|})}\right\| \leq cond\,(A), \quad \left\|P_{\mathcal{R}(A^{n-k|}),\mathcal{R}(A^{|k})}\right\| \leq cond\,(A) \tag{2.69}$$

holds in any submultiplicative unitarily invariant matrix norm.

Proof. Lemma 2.29 implies that

$$P_{R(A^{|k}),R(A^{n-k|})} = A^{|k}\left(A^{-1}\right)^{\overline{k}} = \left[A^{|k}, 0\right]\begin{bmatrix} \left(A^{-1}\right)^{\overline{k}} \\ 0 \end{bmatrix}$$

from which

$$\left\|P_{R(A^{|k}),R(A^{n-k|})}\right\| \leq \left\|\left[A^{|k}, 0\right]\right\| \left\|\begin{bmatrix} \left(A^{-1}\right)^{\overline{k}} \\ 0 \end{bmatrix}\right\| \leq \|A\| \left\|A^{-1}\right\| = \text{cond}\,(A)$$

follows. The proof of the other statement is similar. ■

6.1 Estimations for projections

Next we derive different estimations for the length of projected vectors. The first estimate is a lower bound [3].

LEMMA 2.74 *Let $P = I - X\left(Y^TX\right)^{-1}Y^T$ be a given projection. If $z \notin \mathcal{R}(X)$, then*

$$\|Pz\|_2 \geq \left(\left\|X^+z\right\|_2^2 + 1\right)^{1/2} / \left\|[X, z]^+\right\|_2. \tag{2.70}$$

Proof. Let $s = (I - XX^+)\,Pz$. Since s is the orthogonal projection of Pz onto the subspace $\mathcal{R}^\perp(X)$ we have $\|Pz\|_2 \geq \|s\|_2$. But $s = (I - XX^+)\,z$. By Greville's theorem on the pseudoinverse of partitioned matrices (see Section 1.4) we obtain

$$[X, z]^+ = \begin{bmatrix} X^+\left(I - zd^T\right) \\ d^T \end{bmatrix}$$

with $d = (I - XX^+) z / \|(I - XX^+) z\|^2$. This implies

$$[X, z]^+ s = \begin{bmatrix} -X^+ z \\ 1 \end{bmatrix}$$

from which

$$\left(\|X^+ z\|_2^2 + 1 \right)^{1/2} = \|[X, z]^+ s\|_2 \leq \|[X, z]^+\|_2 \|s\|_2$$

and the statement follows. ∎

The geometry behind the lemma is the following. The vector z is orthogonally decomposed into $z = s + z$, where $s \in \mathcal{R}^\perp(X)$ and $z \in \mathcal{R}(X)$. The norm of s is then estimated from below. Thus we have a lower estimation for the norm of z. Since the orthogonal projection of Pz into $\mathcal{R}^\perp(X)$ is also s, we can use the same estimation.

For the next estimate we need the Wielandt, Bauer, Householder inequality [31]. Let $A \in \mathbb{R}^{n \times n}$ be nonsingular, $M = A^T A$ and let $x, y \in \mathbb{R}^n$ be unit vectors such that

$$x^T y = \cos(\phi), \quad 0 \leq \phi \leq \pi/2.$$

We investigate the angle between Ax and Ay. Define the 2×2 matrix

$$G = G(x, y) = [x, y]^T M [x, y] = \begin{bmatrix} x^T M x & x^T M y \\ y^T M x & y^T M y \end{bmatrix}.$$

The squared cosine of the angle between Ax and Ay is

$$\left| \frac{x^T A^T A y}{\|Ax\| \|Ay\|} \right|^2 = \frac{|x^T M y|^2}{x^T M x y^T M y} = 1 - \frac{\det(G)}{x^T M x y^T M y}.$$

Let τ and $\tau\text{cond}(G)$ be the eigenvalues of $G(x, y)$. Then $\det(G) = \tau^2 \text{cond}(G)$ and

$$\left(x^T M x y^T M y \right)^{1/2} \leq \left(x^T M x + y^T M y \right)/2 = \text{tr}(G)/2 = \tau(1 + \text{cond}(G))/2.$$

So we have

$$\left| \frac{x^T A^T A y}{\|Ax\| \|Ay\|} \right|^2 \leq 1 - \frac{4\text{cond}(G)}{(1 + \text{cond}(G))^2} = \left(\frac{\text{cond}(G) - 1}{\text{cond}(G) + 1} \right)^2. \tag{2.71}$$

We now estimate cond(G). Let $z = -x \cot(\phi) + y / \sin(\phi)$. Vector z is of unit length, orthogonal to x and satisfies $[x, z] Q = [x, y]$ with

$$Q = \begin{bmatrix} 1 & \cos(\phi) \\ 0 & \sin(\phi) \end{bmatrix}.$$

Thus we have

$$G(x,y) = Q^T G(x,z)\, Q$$

and

$$\operatorname{cond}(G(x,y)) \leq \operatorname{cond}^2(Q) \operatorname{cond}(G(x,z)).$$

By direct calculation

$$\operatorname{cond}^2(Q) = \operatorname{cond}\left(Q^T Q\right) = \cot^2(\phi/2).$$

Assume that the eigenvalues of $G(x,z)$ and M are numbered in increasing order. Since $x, z \in \mathbb{R}^n$ are orthonormal vectors, the separation theorem of Poincaré implies that

$$\lambda_1(M) \leq \lambda_1(G(x,z)) \leq \lambda_{n-1}(M), \quad \lambda_2(M) \leq \lambda_2(G(x,z)) \leq \lambda_n(M).$$

Hence we obtain the estimations

$$\operatorname{cond}(G(x,z)) = \frac{\lambda_2(G(x,z))}{\lambda_1(G(x,z))} \leq \frac{\lambda_n(M)}{\lambda_1(M)} = \operatorname{cond}(M) = \operatorname{cond}^2(A)$$

and

$$\operatorname{cond}(G(x,y)) \leq \operatorname{cond}^2(A) \cot^2(\phi/2).$$

By the monotonicity of $(x-1)/(x+1)$ for $x \geq 1$, we have

$$\left|\frac{x^T A^T A y}{\|Ax\|\,\|Ay\|}\right|^2 \leq \left(\frac{\operatorname{cond}^2(A)\cot^2(\phi/2) - 1}{\operatorname{cond}^2(A)\cot^2(\phi/2) + 1}\right)^2. \tag{2.72}$$

The last inequality is sharp. If v and u are orthonormal eigenvectors of M with respect to $\lambda_1 = \lambda_1(M)$ and $\lambda_n = \lambda_n(M)$,

$$x = u\cos(\phi/2) - v\sin(\phi/2), \quad y = u\cos(\phi/2) + v\sin(\phi/2),$$

then equality holds in (2.72). If the angle ψ is defined by

$$\left(\operatorname{cond}^2(A)\cot^2(\phi/2) - 1\right) / \left(\operatorname{cond}^2(A)\cot^2(\phi/2) + 1\right) = \cos\psi,$$

we obtain *the generalized inequality of Wielandt,*

$$\frac{\left|x^T M y\right|^2}{x^T M x y^T M y} \leq \cos^2\psi. \tag{2.73}$$

It is noted that the unit vectors x and y can be replaced with arbitrary nonzero vectors. The angle ψ is the minimal angle between Ax and Ay. Elementary calculation yields

$$\operatorname{cond}(A)\cot(\phi/2) = \cot(\psi/2).$$

Thus we proved the following

THEOREM 2.75 *(Bauer, Householder). If $x, y \in \mathbb{R}^n$ $(x, y \neq 0)$ and $|x^T y| \leq \|x\| \|y\| \cos(\phi)$ $(0 \leq \phi \leq \pi/2)$, then*

$$\left|(Ax)^T (Ay)\right| \leq \|Ax\| \|Ay\| \cos(\psi) \quad (0 \leq \psi \leq \pi/2), \tag{2.74}$$

where

$$\cot(\psi/2) = cond(A) \cot(\phi/2). \tag{2.75}$$

An alternative form of the result is the following. If x and y nonzero vectors enclose an angle of not less than ϕ, then Ax and Ay enclose an angle ϕ_A satisfying

$$\cot(\phi_A/2) \leq \operatorname{cond}(A) \cot(\phi/2). \tag{2.76}$$

Another interesting generalization of the Wielandt inequality is given by Yeh [364] (see also [203]).

The following application of the Bauer, Householder theorem gives an upper bound for the orthogonally projected vectors Hs, if the orthogonal projection H has the form

$$H = I - A^T Y \left(Y^T A A^T Y\right)^{-1} Y^T A, \tag{2.77}$$

where $A \in \mathbb{R}^{n \times n}$ is nonsingular and $Y \in \mathbb{R}^{n \times m}$ is of full rank $(m < n)$. As H is idempotent, we can write

$$\|Hs\|^2 = s^T H s = r^T M^{-1} x,$$

where

$$M = AA^T, \quad r = As, \quad x = AHs.$$

It follows from $\mathcal{N}(H) = \mathcal{R}\left(A^T Y\right)$ that $x^T Y = s^T H A^T Y = 0$. Hence $x \in \mathcal{R}^{\perp}(Y)$. As $\|s\|^2 = s^T s = r^T M^{-1} r$ and

$$x^T M^{-1} x = s^T H A^T M^{-1} A H s = s^T H s,$$

we have

$$\frac{\|Hs\|^2}{\|s\|^2} = \frac{s^T H s}{s^T s} = \frac{\left|r^T M^{-1} x\right|^2}{r^T M^{-1} r x^T M^{-1} x}.$$

Here we apply the Bauer, Householder theorem by noting that

$$\operatorname{cond}\left(M^{-1}\right) = \operatorname{cond}(M) = \operatorname{cond}^2(A).$$

If the angle between x and r is at least ϕ, then

$$\frac{\|Hs\|}{\|s\|} \leq \cos\psi = \frac{\operatorname{cond}^2(A) \cot^2(\phi/2) - 1}{\operatorname{cond}^2(A) \cot^2(\phi/2) + 1} \tag{2.78}$$

with

$$\cot(\psi/2) = \mathrm{cond}(A)\cot(\phi/2). \tag{2.79}$$

The vector x is orthogonal to $\mathcal{R}(Y)$. So $\pi/2 - \phi$ is the angle between r and $\mathcal{R}(Y)$, or more precisely between r and $P_{R(Y)}r$. Thus

$$\sin(\phi) = \cos\left(\frac{\pi}{2} - \phi\right) = \frac{r^T P_{R(Y)} r}{\|r\| \left\|P_{R(Y)} r\right\|} = \frac{\left\|P_{R(Y)} r\right\|}{\|r\|} = \frac{\left\|Y\left(Y^T Y\right)^{-1} Y^T r\right\|}{\|r\|}.$$

If $r \in \mathcal{R}(Y)$, then $\phi = \pi/2$ and $\cos(\psi)$ is minimal. The minimal angle is given by

$$\cos(\psi) = \frac{\mathrm{cond}^2(A) - 1}{\mathrm{cond}^2(A) + 1}.$$

We can establish the following theorem of Bauer and Householder [31].

THEOREM 2.76 *If ψ is defined as above,*

$$H = I - A^T Y\left(Y^T A A^T Y\right)^{-1} Y^T A,$$

and $s = A^{-1}r$, then

$$\left\|Y\left(Y^T Y\right)^{-1} Y^T r\right\| \geq \|r\| \sin(\phi) \quad (0 \leq \phi \leq \pi/2)$$

implies $\|Hs\| \leq \|s\| \cos(\psi)$.

It is obvious, that under the conditions of the theorem, $\|(I - H)s\| \geq \|s\| \sin(\psi)$ and $\|Hs\| / \|(I - H)s\| \leq \cot(\psi)$. The theorem can be written in the following form. If $d(r, \mathcal{R}(Y)) \leq \|r\| \cos(\phi)$, $As = r$, then

$$d\left(s, \mathcal{R}\left(A^T Y\right)\right) \leq \|s\| \cos(\psi).$$

The *Kantorovich inequality*, which can also be derived from the generalized Wielandt inequality (2.73) of Bauer and Householder (see, e.g., [206]) has the following form:

$$\sin^2(\psi)\left(x^T M x\right)\left(x^T M^{-1} x\right) \leq \|x\|_2^4 \quad \left(x \in \mathbb{R}^n\right). \tag{2.80}$$

Equality holds, if $x = u + v$, where v and u are orthonormal eigenvectors of M corresponding to its smallest and largest eigenvalues. An equivalent formulation of the Kantorovich inequality is

$$\sin(\psi)\|Ax\|_2 \left\|A^{-1}x\right\|_2 \leq \|x\|_2 \quad \left(x \in \mathbb{R}^n\right). \tag{2.81}$$

The Kantorovich inequality will be used in the convergence analysis of certain projection methods. Generalizations of the Kantorovich inequality and related matters can be found e.g., in Mitrinovic [257], Horn and Johnson [203]. Gastinel and Joly [156] gave an extension of the Wielandt, Bauer, Householder and Kantorovich inequalities to abstract spaces and an application to a general projection method. Gustafson [174] describes the relationship between the angle ψ and his operator angle $\phi(A)$.

Meany [254] proved the following result on the norm of a product of orthogonal projections of rank $n-1$.

THEOREM 2.77 *(Meany). Let $x_i \in \mathbb{R}^n$ have unit length for $i = 1, \ldots, k$, $k \leq n$, $X_k = [x_1, \ldots, x_k]$ and $Q_k = \left(I - x_k x_k^T\right) \cdots \left(I - x_2 x_2^T\right)\left(I - x_1 x_1^T\right)$. Then*

$$\max_{y \in \mathcal{R}(X_k),\ \|y\|_2 = 1} \|Q_k y\|_2 \leq \left(1 - \det\left(X_k^T X_k\right)\right)^{1/2}. \tag{2.82}$$

Proof. The result will be proved in three steps. We call a vector z normalized if $\|z\|_2 = 1$. Let the normalized vectors $y_1, \ldots, y_{n-k}$ be orthogonal to $x_1, \ldots, x_k$ and to each other and let $Y = [y_1, \ldots, y_{n-k}]$. Then

$$\begin{aligned} |\det([X_k, Y])| &= \left(\det\left([X_k, Y]^T [X_k, Y]\right)\right)^{1/2} \\ &= \left(\det\left(\begin{bmatrix} X_k^T X_k & 0 \\ 0 & I \end{bmatrix}\right)\right)^{1/2} \\ &= \left(\det\left(X_k^T X_k\right)\right)^{1/2}. \end{aligned} \tag{2.83}$$

Write $x_k = x_S + x_N$, where $x_S \in \mathcal{R}(X_{k-1})$ and $x_N \in \mathcal{R}^{\perp}(X_{k-1})$. We show that

$$\det\left([X_{k-1}, x_k]^T [X_{k-1}, x_k]\right) = \|x_N\|^2 \det\left(X_{k-1}^T X_{k-1}\right). \tag{2.84}$$

If $x_N = 0$, then $x_1, \ldots, x_k$ are linearly dependent and both sides of (2.84) are 0. Otherwise let $y_0 = x_N / \|x_N\|$ and choose arbitrarily normalized vectors $y_1, \ldots, y_{n-k}$ orthogonal to $x_1, \ldots, x_k$ and to each other. Then $x_k = \|x_N\| y_0 + c_1 x_1 + \cdots + c_{k-1} x_{k-1}$, where $c_1, \ldots, c_{k-1}$ are scalars, and we have, with the use of (2.83),

$$\begin{aligned} \det\left([X_{k-1}, x_k]^T [X_{k-1}, x_k]\right) &= \left(\det\left(X_{k-1}, x_k, y_1, \ldots, y_{n-k}\right)\right)^2 \\ &= \|x_N\|^2 \left(\det\left(X_{k-1}, y_0, y_1, \ldots, y_{n-k}\right)\right)^2 \\ &= \|x_N\|^2 \det\left(X_{k-1}^T X_{k-1}\right). \end{aligned}$$

Finally we use mathematical induction on k. The theorem is clearly true when $k = 1$, for in this case both sides of (2.82) are 0. Suppose that

$$\max_{y \in \mathcal{R}(X_{k-1}), \|y\| = 1} \|Q_{k-1} y\| \leq \left(1 - \det\left(X_{k-1}^T X_{k-1}\right)\right)^{1/2} \tag{2.85}$$

is true. Write $x_k = x_N + x_S$, where $x_S \in \mathcal{R}(X_{k-1})$ and $x_N \in \mathcal{R}^{\perp}(X_{k-1})$. Let y be an arbitrary normalized vector in the space $\mathcal{R}(X_k)$, and write $y = y_S + y_N$, where $y_S \in \mathcal{R}(X_{k-1})$ and $y_N \in \mathcal{R}^{\perp}(X_{k-1})$. Finally, let $z = \left(I - x_{k-1}x_{k-1}^T\right)\cdots\left(I - x_1 x_1^T\right) y_S$. Using the facts that $\left(I - x_k x_k^T\right)^2 = \left(I - x_k x_k^T\right)$, $y_N^T x_N = \pm \|y_N\| \|x_N\|$ (as x_N and y_N are parallel or $y_N = 0$) and z is orthogonal to x_N and y_N (because z belongs to $\mathcal{R}(X_{k-1})$), we then have

$$\begin{aligned}
\|Q_k y\|^2 &= \left\|\left(I - x_k x_k^T\right)(y_N + z)\right\|^2 \\
&= \left(y_N^T + z^T\right)\left(I - x_k x_k^T\right)(y_N + z) \\
&= \|y_N\|^2 + \|z\|^2 - \left(y_N^T x_k\right)^2 - 2y_N^T x_k z^T x_k - \left(z^T x_k\right)^2 \\
&= \|y_N\|^2 + \|z\|^2 - \left(y_N^T x_N\right) - 2y_N^T x_N z^T x_S - \left(z^T x_S\right)^2 \\
&= \|y_N\|^2 + \|z\|^2 - \|y_N\|^2 \|x_N\|^2 \pm 2\|y_N\| \|x_N\| \left|z^T x_S\right| - \left|z^T x_s\right|^2 \\
&\le \|y_N\|^2 + \|z\|^2 - \|y_N\|^2 \|x_N\|^2 + 2\|y_N\| \|x_N\| \left|z^T x_S\right| - \left|z^T x_s\right|^2 \\
&= \|y_N\|^2 + \|z\|^2 - \left(\|y_N\| \|x_N\| - \left|z^T x_S\right|\right)^2 .
\end{aligned} \tag{2.86}$$

There are now two cases to consider.

Case 1. $\|x_N\| \le \|y_S\|$. With use of (2.86), the induction hypothesis (2.85), which says that

$$\|z\| \le \|y_S\| \left(1 - \det\left(X_{k-1}^T X_{k-1}\right)\right)^{1/2},$$

and relation (2.84), we then have that

$$\begin{aligned}
\|Q_k y\|^2 &\le \|y_N\|^2 + \|y_S\|^2 \left(1 - \det\left(X_{k-1}^T X_{k-1}\right)\right) \\
&\le 1 - \|x_N\|^2 \det\left(X_{k-1}^T X_{k-1}\right) \\
&= 1 - \det\left(\left[X_{k-1}, x_k\right]^T \left[X_{k-1}, x_k\right]\right).
\end{aligned}$$

Case 2. $\|x_N\| > \|y_S\|$. In this case

$$\|y_N\| = \left(1 - \|y_S\|^2\right)^{1/2} > \left(1 - \|x_N\|^2\right)^{1/2} = \|x_S\|,$$

so that

$$\|y_N\| \|x_N\| > \|y_S\| \|x_S\| \ge \|z\| \|x_S\| \ge \left|z^T x_S\right|$$

and hence

$$\|y_N\| \|x_N\| - \left|z^T x_S\right| \ge \|y_N\| \|x_N\| - \|z\| \|x_S\| . \tag{2.87}$$

With the use of (2.86), (2.87), the induction hypothesis (2.85), Cauchy's inequality ($2\sqrt{ab} \leq a + b$), and relation (2.84), in that order, we then have

$$\begin{aligned}
\|Q_k y\|^2 &\leq \|y_N\|^2 + \|z\|^2 - (\|y_N\| \|x_N\| - \|z\| \|x_S\|)^2 \\
&= \|y_N\|^2 \left(1 - \|x_N\|^2\right) + \|z\|^2 \left(1 - \|x_S\|^2\right) \\
&\qquad + 2 \|y_N\| \|x_N\| \|z\| \|x_S\| \\
&= \|y_N\|^2 \|x_S\|^2 + \|z\|^2 \|x_N\|^2 + 2 \|y_N\| \|x_N\| \|z\| \|x_S\| \\
&= (\|y_N\| \|x_S\| + \|z\| \|x_N\|)^2 \\
&\leq \left(\|y_N\| \|x_S\| + \|y_S\| \|x_N\| \left(1 - \det\left(X_{k-1}^T X_{k-1}\right)\right)^{1/2}\right)^2 \\
&\leq \left(\|y_N\|^2 + \|y_S\|^2\right) \left(\|x_S\|^2 + \|x_N\|^2 \left(1 - \det\left(X_{k-1}^T X_{k-1}\right)\right)\right) \\
&= 1 - \|x_N\|^2 \left(1 - \det\left(X_{k-1}^T X_{k-1}\right)\right) \\
&= 1 - \det\left(\left[X_{k-1}, x_k\right]^T \left[X_{k-1}, x_k\right]\right).
\end{aligned}$$

Hence we proved the theorem. ■

The matrix Q_k arose in a relaxed Kaczmarz type algorithm of Whitney and Meany [357]. In fact,

$$Q_k = \left(I - P_{\mathcal{R}(x_k)}\right) \cdots \left(I - P_{\mathcal{R}(x_1)}\right) = P_{\mathcal{R}^\perp(x_k)} \cdots P_{\mathcal{R}^\perp(x_1)} \tag{2.88}$$

and Meany's result is related to the method of alternating projections (see, Chapter 7).

If $\mathcal{R}(X_n) = \mathbb{R}^n$, then $\max_{y \in \mathcal{R}(X_n), \|y\|=1} \|Q_n y\|_2 = \|Q_n\|_2$ and

$$\|Q_n\|_2 \leq \left(1 - \det\left(X_n^T X_n\right)\right)^{1/2}. \tag{2.89}$$

Note that $X_k^T X_k = \left[x_i^T x_j\right]_{i,j=1}^k$ is the Gram matrix. Hence

$$0 \leq \det\left(X_k^T X_k\right) \leq \|x_1\|^2 \|x_2\|^2 \cdots \|x_k\|^2. \tag{2.90}$$

The lower bound occurs if and only if the vectors x_i are linearly dependent. The upper bound occurs if and only if the vectors x_i are pairwise orthogonal.

If $\|x_i\| = 1$ for all i, then the sequence $\mu_k = \det\left(X_k^T X_k\right)$ is decreasing, that is $\mu_k \leq \mu_{k-1} \leq 1$. Hence the bound (2.82) is increasing along with k.

If, in addition, the vectors $x_1, \ldots, x_n$ are linearly independent, then $\mu_n = \det\left(X_n^T X_n\right) > 0$ and

$$\|Q_n\|_2 \leq \gamma = \sqrt{1 - \mu_n} < 1. \tag{2.91}$$

If the normalized vectors x_i are pairwise orthogonal, then $\mu_n = 1$. This implies $\|Q_n\|_2 = 0$ and $Q_n = 0$.

7. Perturbations of projections

We investigate the perturbations of projections. We give bounds for the change of spectrum and norm of projections. We need the following result (see e.g., [165]).

THEOREM 2.78 *(Bauer-Fike). If μ is an eigenvalue of $A+E \in \mathbb{C}^{n\times n}$ and $X^{-1}AX = D = diag(\lambda_1, \ldots, \lambda_n)$, then*

$$\min_{1\le i\le n} |\lambda_i - \mu| \le cond_p(X)\, \|E\|_2 , \tag{2.92}$$

where $\|\cdot\|_p$ denotes any of the Hölder norms.

An extension of the Bauer-Fike theorem to cases when A is nondiagonalizable is given by Chu [86].

THEOREM 2.79 *Let $P \in \mathbb{R}^{n\times n}$ be any projection of rank k. Assume that P has the form*

$$P = V \begin{bmatrix} I_k & B \\ 0 & 0_{n-k} \end{bmatrix} V^T, \tag{2.93}$$

where V is orthogonal. Let μ be an eigenvalue of the perturbed matrix $\widehat{P} = P + \delta P$. Then

$$\min_{1\le i\le n} |\lambda_i - \mu| \le \left(n + \|B\|_F^2\right) \|\delta P\|_2 . \tag{2.94}$$

Proof. Let

$$X = \begin{bmatrix} I_k & -B \\ 0 & I_{n-k} \end{bmatrix}.$$

Then the Jordan form of P is given by

$$P = VX \begin{bmatrix} I_k & 0 \\ 0 & 0 \end{bmatrix} X^{-1} V^T,$$

where

$$X^{-1} = \begin{bmatrix} I_k & B \\ 0 & I_{n-k} \end{bmatrix}.$$

The Bauer-Fike theorem gives the estimate

$$\min_{1\le i\le n} |\lambda_i - \mu| \le \mathrm{cond}_2(VX)\, \|\delta P\|_2 . \tag{2.95}$$

Since $\|A\|_2 \le \|A\|_F$ we have $\mathrm{cond}_2(VX) \le \mathrm{cond}_2(X) \le \mathrm{cond}_F(X)$. On the other hand

$$\|X\|_F = \left\|X^{-1}\right\|_F = \left(n + \|B\|_F^2\right)^{1/2}$$

and

$$\mathrm{cond}_F(X) = n + \|B\|_F^2 .$$

■

Since the two distinct eigenvalues of a projection are 0 and 1, the eigenvalues of $P + \delta P$ remain in two clusters if δP is small enough. If P is an orthogonal projection, then $B = 0$, $X = I$ and bound (2.95) becomes

$$\min_{1 \leq i \leq n} |\lambda_i - \mu| \leq \|\delta P\|_2 . \tag{2.96}$$

Let $M, N \subset \mathbb{C}^n$ be complementary subspaces so that $M \oplus N = \mathbb{C}^n$ and consider the oblique projection $P_{M,N}$. Let $M_1, N_1 \subset \mathbb{C}^n$ be also complementary subspaces (i.e., $M_1 \oplus N_1 = \mathbb{C}^n$) that are close, in some sense, to M and N, respectively. Then we consider the oblique projection P_{M_1,N_1} as a perturbation of $P_{M,N}$ and investigate $\|P_{M_1,N_1} - P_{M,N}\|$ as a function of the subspace perturbations. The subspace perturbation can be measured either by the gap metric of subspaces or by the norm of the changes in the corresponding subspace bases.

For example, let us consider P_M under the assumption that the columns of $U \in \mathbb{C}^{n \times k}$ are orthonormal and $\mathcal{R}(U) = M$. Let $U_1 = U + \delta_U \in \mathbb{C}^{n \times k}$ have orthonormal columns and $\mathcal{R}(U_1) = M_1$. Consider now $P_{M_1} - P_M$. By definition

$$P_{M_1} - P_M = U_1 U_1^H - U U^H = U \delta_U^H + \delta_U U^H + \delta_U \delta_U^H$$

and

$$\|P_{M_1} - P_M\| \leq 2\|\delta_U\| + \|\delta_U\|^2 . \tag{2.97}$$

The bound $2\|\delta_U\| + \|\delta_U\|^2$ is less than 1 if $\|\delta_U\| < \sqrt{2} - 1$. If $\|\delta_U\| \to 0$, then $P_{M_1} \to P_M$. We also obtained that if the subspaces M_1 and M have orthonormal bases $U_1 \subset M_1$ and $U \subset M$ such that $\|U_1 - U\| < \sqrt{2} - 1$, then $\Theta(M_1, M) < 1$. In Chapter 7 we quote a kind of reverse result. The perturbation of orthogonal projections is studied by Stewart [326] (see, also [331]). For oblique projections we can easily develop similar, but more complicated estimates.

Next we show the Lipschitz continuity of oblique projections in the gap metric. This result of Wimmer [360] improves a similar theorem of Gohberg, Lancaster and Rodman [161] (Theorem 13.1.3, pp. 390, 391). Here we use the spectral norm.

Let $M \oplus N = \mathbb{C}^n$ be a nontrivial direct sum decomposition. The restriction of $P_{M,N}$ on a subspace X will be denoted by $P_{M,N}|_X$.

LEMMA 2.80 *Assume that* $M \oplus N = \mathbb{C}^n$.
(i) For a subspace X *of* $\mathbb{C}^n$ *we have*

$$\left\| P_{M,N}|_X \right\| = \|P_{M,N} P_X\| . \tag{2.98}$$

(ii) if $\dim(X) = \dim(N)$, *then*

$$\left\| P_{M,N}|_X \right\| \leq \|P_{M,N}\| \, \Theta(N, X) . \tag{2.99}$$

Proof. (i) For $y \in X$, $X \neq \{0\}$, we have $P_{M,N}|_X\, y = P_{M,N}P_X y$. Therefore

$$\left\| P_{M,N}|_X \right\| = \max_{y\in X,\, y\neq 0} \left\{ \frac{\|P_{M,N}P_X y\|}{\|y\|} \right\} \leq \|P_{M,N}P_X\|$$
$$= \max_{y\in \mathbb{C}^n,\, y\neq 0} \left\{ \frac{\|P_{M,N}P_X y\|}{\|y\|} \right\}$$
$$\leq \max_{y\in \mathbb{C}^n,\, P_X y\neq 0} \left\{ \frac{\|P_{M,N}P_X y\|}{\|P_X y\|} \right\} = \left\| P_{M,N}|_X \right\|.$$

(ii) From $P_{M,N} = P_{M,N}\,(I - P_N)$ and (i) it follows that

$$\left\| P_{M,N}|_X \right\| = \|P_{M,N}\,(I - P_N)\,P_X\|.$$

Hence (2.62) yields (ii). ■

LEMMA 2.81 *Assume that* $M \oplus N = \mathbb{C}^n$.
(i) If $M \neq \{0\}$, *then*

$$\max_{x\in M} \frac{1}{d\,(x, N)} = \|P_{M,N}\|. \tag{2.100}$$

(ii) If $M \neq \{0\}$ *and* $N \neq \{0\}$, *then*

$$\|P_{M,N}\| = \|P_{N,M}\|. \tag{2.101}$$

Proof. (i) We show that (2.100) is equivalent to the identity

$$\frac{1}{1 - \|P_M P_N\|^2} = \|P_{M,N}\|^2 \tag{2.102}$$

of Ljance. Set

$$\tau = \min_{x\in M,\, \|x\|=1} d\,(x, N) = \min_{x\in M,\, \|x\|=1} \|(I - P_N)\,P_M x\|.$$

Then the left-hand side of (2.100) is equal to $1/\tau$. If $x \in M$ and $\|x\| = 1$, then

$$\|(I - P_N)\,P_M x\|^2 + \|P_N P_M x\|^2 = 1.$$

Hence

$$\tau^2 = 1 - \max_{x\in M,\, \|x\|=1} \|P_N P_M x\|^2 = 1 - \|P_N P_M\|^2.$$

Therefore

$$\frac{1}{\tau^2} = \frac{1}{1 - \|P_N P_M\|^2},$$

and (2.100) follows from (2.102).

(ii) Since $P_N P_M = 0$ implies $P_M P_N = 0$, we note that either $P_N P_M = P_M P_N = 0$ or both $P_N P_M \neq 0$ and $P_M P_N \neq 0$. In each case we have $\|P_N P_M\| = \|P_M P_N\|$. Hence (2.102) implies (2.101). ∎

The identity of Ljance is proved in Section 2.5 and Chapter 7. Relation (2.101) is contained in Proposition 2.55.

THEOREM 2.82 *(Wimmer). Assume that $M \oplus N = \mathbb{C}^n$, $M \neq \{0\}$, $N \neq \{0\}$. (i) Let N_1 be a subspace of $\mathbb{C}^n$ with $\dim(N_1) = \dim(N)$ and*

$$\mu = \left\| P_{M,N}|_{N_1} \right\| < 1. \tag{2.103}$$

Then

$$M \oplus N_1 = \mathbb{C}^n \tag{2.104}$$

and

$$\|P_{M,N_1} - P_{M,N}\| \leq \frac{\mu}{1-\mu} \|P_{M,N}\|. \tag{2.105}$$

(ii) If a subspace N_1 satisfies

$$\Theta(N, N_1) \leq \frac{1-c}{\|P_{M,N}\|} \quad (0 < c < 1), \tag{2.106}$$

then (2.104) holds and

$$\|P_{M,N_1} - P_{M,N}\| \leq \frac{1}{c} \|P_{M,N}\|^2 \Theta(N, N_1). \tag{2.107}$$

Proof. (i) Suppose that $0 \neq x \in M \cap N_1$. Then $\left(P_{M,N}|_{N_1}\right) x = x$. Hence $\left\|P_{M,N}|_{N_1}\right\| \geq 1$, which contradicts (2.103). Therefore, we have $M \cap N_1 = \{0\}$ and (2.104). We need the following identities: $P_X P_{X,Y} = P_{X,Y}$, $P_{X,Y} P_X = P_X$, $P_{X,Y} P_{Y,Z} = 0$, and $P_{X,Y} = P_{X,Y} P_{Z,Y}$. The first three relations are clear by definition. The last relation is obtained from $0 = P_{X,Y} P_{Y,Z} = P_{X,Y}(I - P_{Z,Y})$. Using $P_{X,Y} = I - P_{Y,X}$ and (2.101) we obtain

$$\begin{aligned} \|P_{M,N_1} - P_{M,N}\| &= \|P_{N_1,M} - P_{N,M}\| = \|P_{N_1,M} P_{N,M} - P_N P_{N,M}\| \\ &= \|(P_{N_1,M} P_N - P_N) P_{N,M}\| \\ &\leq \|P_{N_1,M} P_N - P_N\| \|P_{N,M}\|. \end{aligned}$$

The identity

$$\begin{aligned} P_{N_1,M} P_N - P_N &= P_{N_1,M} P_N - P_{N,M} P_N = P_{N_1,M} P_N - P_{N,M} P_{N_1,M} P_N \\ &= (I - P_{N,M}) P_{N_1,M} P_N = P_{M,N} P_{N_1} P_{N_1,M} P_N, \end{aligned}$$

yields the bound

$$\|P_{N_1,M} P_N - P_N\| \leq \|P_{M,N} P_{N_1}\| \|P_{N_1,M} P_N\| \leq \mu \|P_{N_1,M} P_N\|.$$

Since

$$\|P_{N_1,M}P_N\| - \|P_N\| \leq \|P_{N_1,M}P_N - P_N\| \leq \mu \|P_{N_1,M}P_N\|,$$

we easily obtain that $\|P_{N_1,M}P_N\| \leq 1/(1-\mu)$ and $\|P_{N_1 M}P_N - P_N\| \leq \mu/(1-\mu)$. Thus we proved inequality (2.105).

(ii) From (2.106) and (2.99) we obtain that

$$\mu \leq \|P_{M,N}\| \Theta(N, N_1) \leq 1 - c < 1.$$

Then $\|P_{M,N}\| \geq 1$ implies that $\Theta(N, N_1) < 1$ and $\dim(N_1) = \dim(N)$. Because of $\mu < 1$ we can use (i) and conclude that P_{M,N_1} exists. Since $\mu \leq 1 - c$ is equivalent to

$$0 < \frac{1}{1-\mu} \leq \frac{1}{c},$$

the estimate (2.107) follows immediately from (2.105). ∎

The proof of statement (i) is simpler and shorter than that of Wimmer [360]. The theorem includes the stability of the direct sum decomposition in a special case. For a more general result we refer to Wimmer [359].

In Wimmer's theorem the range M was fixed and the null space N changed to N_1. Since $P_{N,M} = I - P_{M,N}$ and $\|P_{N,M}\|_2 = \|P_{M,N}\|_2$ we have the alternate result for fixed N and varying M.

THEOREM 2.83 *Assume that $M \oplus N = \mathbb{C}^n$, $M \neq \{0\}$, $N \neq \{0\}$.*
(i) Let M_1 be a subspace of $\mathbb{C}^n$ with $\dim(M_1) = \dim(M)$ *and*

$$\mu = \left\| P_{N,M}|_{M_1} \right\| < 1. \tag{2.108}$$

Then $M_1 \oplus N = \mathbb{C}^n$ and

$$\|P_{M_1,N} - P_{M,N}\| \leq \frac{\mu}{1-\mu} \|P_{M,N}\|. \tag{2.109}$$

(ii) If a subspace M_1 satisfies

$$\Theta(M, M_1) \leq \frac{1-c}{\|P_{M,N}\|} \quad (0 < c < 1), \tag{2.110}$$

then $M_1 \oplus N = \mathbb{C}^n$ and

$$\|P_{M_1,N} - P_{M,N}\| \leq \frac{1}{c} \|P_{M,N}\|^2 \Theta(M, M_1). \tag{2.111}$$

THEOREM 2.84 *Let $M \oplus N = \mathbb{C}^n$, $M \neq \{0\}$, $N \neq \{0\}$. If M_1 and N_1 are sufficiently close to M and N, respectively, then*

$$\|P_{M_1,N_1} - P_{M,N}\| \leq \frac{1}{c} \|P_{M,N}\|^2 \Gamma(\Theta(M, M_1), \Theta(N, N_1)), \tag{2.112}$$

where

$$\Gamma(\alpha,\beta)=\alpha+\left(1+\frac{1}{c}\left\|P_{M,N}\right\|\alpha\right)^{2}\beta \quad \left(\alpha,\beta\in\mathbb{R}^{+}\right). \tag{2.113}$$

Proof. The bound follows from the estimates

$$\begin{aligned}\left\|P_{M_1,N_1}-P_{M,N}\right\| &\le \left\|P_{M_1,N_1}-P_{M_1,N}\right\|+\left\|P_{M_1,N}-P_{M,N}\right\|\\ &\le \frac{1}{c}\left\|P_{M_1,N}\right\|^{2}\Theta(N,N_1)+\frac{1}{c}\left\|P_{M,N}\right\|^{2}\Theta(M,M_1)\end{aligned}$$

and

$$\begin{aligned}\left\|P_{M_1,N}\right\| &\le \left\|P_{M_1,N}-P_{M,N}\right\|+\left\|P_{M,N}\right\|\\ &\le \frac{1}{c}\left\|P_{M,N}\right\|^{2}\Theta(M,M_1)+\left\|P_{M,N}\right\|.\end{aligned}$$

■

It is also possible to estimate the perturbation of an oblique projection by using group inverse representation. Suppose that $A\in\mathbb{C}^{n\times n}$ has a group inverse $A^{\#}$ and it is perturbed such that $B=A+E$ also has a group inverse. Since $AA^{\#}=P_{\mathcal{R}(A),\mathcal{N}(A)}$ and $BB^{\#}=P_{\mathcal{R}(B),\mathcal{N}(B)}$, the quantity $\left\|AA^{\#}-BB^{\#}\right\|$ is the measure of the projection perturbation. Wei [356] proved the following result.

THEOREM 2.85 *(Wei). Assume that $B=A+E$ and Ind(A) =Ind$(B)=1$. Then*

$$\left\|B^{\#}-A^{\#}\right\|\le\left(1+2\max\left\{\left\|AA^{\#}\right\|,\left\|BB^{\#}\right\|\right\}\right)\max\left\{\left\|A^{\#}\right\|^{2},\left\|B^{\#}\right\|^{2}\right\}\left\|E\right\| \tag{2.114}$$

and

$$\left\|BB^{\#}-AA^{\#}\right\|\le 2\max\left\{\left\|AA^{\#}\right\|,\left\|BB^{\#}\right\|\right\}\max\left\{\left\|A^{\#}\right\|,\left\|B^{\#}\right\|\right\}\left\|E\right\|. \tag{2.115}$$

If in addition

$$\left\|A^{\#}\right\|\left\|E\right\|<\frac{1}{1+\left\|AA^{\#}\right\|}<1/2, \tag{2.116}$$

then

$$\frac{\left\|BB^{\#}-A^{\#}A\right\|}{\left\|A^{\#}A\right\|}\le\frac{\left\|A^{\#}\right\|\left\|E\right\|\left(1-\left\|A^{\#}\right\|\left\|E\right\|\right)\left(2-\left\|A^{\#}\right\|\left\|E\right\|\right)}{\left(1-\left\|A^{\#}\right\|\left\|E\right\|-\left\|A^{\#}\right\|\left\|E\right\|\left\|AA^{\#}\right\|\right)^{2}}. \tag{2.117}$$

Improvements of the above results can be found in Li and Wei [234]. Note that the perturbation E is subject to the constraint Ind(A) =Ind$(A+E)=1$.

Finally we mention that approximate perturbation bounds can also be obtained by having the derivative of the projection matrix. Derivatives of certain projection matrices were first given by Golub and Pereyra [164]. Harville [191] is a good source of related results. Here we show only the result of Hanke [186], which is related to the group inverse representation of orthogonal projections.

Let $B, B_\omega \in \mathbb{C}^{n\times n}$ be such that $\text{Ind}(B) = \text{Ind}(B_\omega) = 1$ and let B_ω be a continuous matrix function of $\omega \in \mathbb{C}^{r\times r}$ in a neighborhood of $\omega = 0$ with $B_0 = B$.

THEOREM 2.86 *(Hanke). Let $P = P_{\mathcal{R}(B),\mathcal{N}(B)}$, $P_\omega = P_{\mathcal{R}(B_\omega),\mathcal{N}(B_\omega)}$. Then*

$$P_\omega - P = B_\omega^{\#}(B_\omega - B)(I - P) + (I - P_\omega)(B_\omega - B)B^{\#}. \tag{2.118}$$

If B_ω has a Fréchet derivative $\mathbf{D}B$ in $\omega = 0$ and if the rank of B_ω is constant in a neighborhood of $\omega = 0$, then P_ω is Fréchet differentiable in $\omega = 0$ with derivative

$$\mathbf{D}P = B^{\#}\mathbf{D}B\,(I - P) + (I - P)\,\mathbf{D}BB^{\#}. \tag{2.119}$$

Proof. Equation (2.118) can be easily checked using the identities

$$P = B^{\#}B = BB^{\#}, \quad P_\omega = B_\omega^{\#}B_\omega = B_\omega^{\#}B_\omega.$$

Under the given assumptions, $B_\omega^{\#} \to B^{\#}$ as $\omega \to 0$ [74] and the derivative $\mathbf{D}P$ can be obtained from (2.118) by letting $\omega \to 0$. ■

8. Further results

It is almost impossible to perceive all the results published on projections in the last two decades. However Trenkler [344] collected and organized much of the material into more than one hundred equivalent characterizations of oblique and orthogonal projections. These characterizations are given in sixteen groups which relate to rank, eigenvalue, norm and other properties. Here we recall some characterizations by means of matrix ordering.

For any given real matrices A and B these partial orderings are given by the following conditions.

Löwner ordering.

$$A \overset{L}{\leq} B \Leftrightarrow B - A = CC^T \text{ for some matrix } C. \tag{2.120}$$

Star ordering (Drazin).

$$A \leq_* B \Leftrightarrow A^TA = A^TB \text{ and } AA^T = BA^T. \tag{2.121}$$

Rank subtractive ordering (Hartwig).

$$A \overset{rs}{\leq} B \Leftrightarrow \text{rank}\,(B - A) = \text{rank}\,(B) - \text{rank}\,(A)\,. \tag{2.122}$$

The rank subtractivity ordering is also called the minus partial order. The following results are due to Hartwig and Styan [190].

THEOREM 2.87 *The following statements are equivalent.*
(i) P *is a projection;*
(ii)$P \overset{rs}{\leq} I$*;*
(iii) $P \leq_* Q$ *for some projection* Q*;*
(iv) $P \overset{rs}{\leq} Q$ *for some projection* Q*;*
(v) $P \overset{L}{\leq} Q$ *for some projector* Q *such that* $PQ = QP$ *and all eigenvalues of* P *are* 0 *or* 1.

THEOREM 2.88 *The following statements are equivalent.*
(i) P *is an orthogonal projection;*
(ii)$P \overset{rs}{\leq} P^T P$*;*
(iii) $P \leq_* Q$ *for some orthogonal projection* Q*;*
(iv) $P \overset{L}{\leq} Q$ *for some orthogonal projection* Q *and all eigenvalues of* P *are* 0 *or* 1*;*
(v) $0 \overset{L}{\leq} P \overset{L}{\leq} Q$ *and* $(Q-P)^2 = Q - P$ *for some orthogonal projection* Q.

For other partial ordering characterizations, see [344].

Finally we mention the following two generalizations of projections. Groß and Trenkler [169] call the matrix $A \in \mathbb{C}^{n \times n}$ generalized projector , if $A^H = A^2$. Groß, Trenkler and Troschke [172] call the matrix $A \in \mathbb{C}^{n \times n}$ a semi-orthogonal projector if $A^H A = H(A) = \frac{1}{2}\left(A + A^H\right)$.

Chapter 3

FINITE PROJECTION METHODS FOR LINEAR SYSTEMS

We can define projection methods in several ways. Consider the equation

$$Ax = y, \tag{3.1}$$

where $A \in \mathbb{R}^{m\times n}$. The *projection method* can be defined as follows. Let $\{X_k\} \subset \mathbb{R}^n$ and $\{Y_k\} \subset \mathbb{R}^m$ be sequences of linear subspaces, and let $P_k : \mathbb{R}^m \to Y_k$ be projections. Now replace equation $Ax = y$ by the approximation

$$P_k (Ax_k - y) = 0 \quad (x_k \in X_k). \tag{3.2}$$

If $m = n$ and $X_k = Y_k$ ($k = 1, 2, \ldots$), the projection method is called the Galerkin method. Assume that $\{\phi_j\}_{j=1}^n \subset \mathbb{R}^n$ and $\{\psi_j\}_{j=1}^m \subset \mathbb{R}^m$ are such that $\mathcal{R}(\phi_1, \ldots, \phi_n) = \mathbb{R}^n$ and $\mathcal{R}(\psi_1, \ldots, \psi_m) = \mathbb{R}^m$. Let $X_k = \mathcal{R}(\phi_1, \ldots, \phi_{i_k})$ and $Y_k = \mathcal{R}(\psi_1, \ldots, \psi_{i_k})$, $i_k < i_{k+1}$.

LEMMA 3.1 *Let $P_k : \mathbb{R}^m \to Y_k$ be the orthogonal projection. Then equations*

$$P_k z = 0 \tag{3.3}$$

and

$$\langle z, \psi_j \rangle = 0 \quad (j = 1, \ldots, i_k) \tag{3.4}$$

are equivalent.

Proof. $\mathcal{N}(P_k) = \mathcal{R}^{\perp}(P_k)$ implies that $z \in \mathcal{N}(P_k) \Leftrightarrow z \perp \mathcal{R}(P_k) = Y_k = \mathcal{R}(\psi_1, \ldots, \psi_{i_k})$. ∎

Assume that $P_k : \mathbb{R}^m \to Y_k$ is the orthogonal projection. Then by the previous lemma $P_k (Ax_k - y) = 0$ holds if and only if

$$\langle Ax_k - y, \psi_j \rangle = 0 \quad (j = 1, \ldots, i_k). \tag{3.5}$$

Taking

$$x_k = \sum_{j=1}^{i_k} \xi_j \phi_j, \tag{3.6}$$

we can write this condition as

$$\sum_{j=1}^{i_k} \langle A\phi_j, \psi_i \rangle \xi_j = \langle y, \psi_i \rangle \quad (i = 1, \ldots, i_k), \tag{3.7}$$

which is a linear system for the coefficients ξ_j. Hence x_k is defined by the orthogonality condition

$$Ax_k - y \perp \psi_j \quad (j = 1, \ldots, i_k). \tag{3.8}$$

This version is called the Galerkin-Petrov projection method . If

$$\psi_i = A\phi_i \quad (i = 1, 2, \ldots), \tag{3.9}$$

the Galerkin-Petrov method is known as the *least-squares method* . If $m = n$ and $\psi_i = \phi_i$ $(i = 1, 2, \ldots, m)$, the projection method is called the Bubnov-Galerkin method .

1. The Galerkin-Petrov projection method

We consider the linear systems of the form

$$Ax = b \quad (A \in \mathbb{R}^{m \times m}), \tag{3.10}$$

where A is nonsingular and the exact solution is $\omega = A^{-1}b$. Let $\phi_k = U_k \in \mathbb{R}^{m \times m_k}$ and $\psi_k = V_k \in \mathbb{R}^{m \times m_k}$ $(k = 1, \ldots, r)$ be such that $U = [U_1, \ldots, U_r] \in \mathbb{R}^{m \times m}$ and $V = [V_1, \ldots, V_r] \in \mathbb{R}^{m \times m}$ are nonsingular. Furthermore let

$$X_k = \mathcal{R}\left(U^{|k}\right) = \mathcal{R}\left([U_1, \ldots, U_k]\right), \tag{3.11}$$

$$Y_k = \mathcal{R}\left(V^{|k}\right) = \mathcal{R}\left([V_1, \ldots, V_k]\right) \tag{3.12}$$

and $\dim(X_k) = \dim(Y_k) = n_k$. If $x_k \in X_k$, then there exists a unique $a_k \in \mathbb{R}^{n_k}$ such that $x_k = U^{|k} a_k$. From the orthogonality condition $r_k = Ax_k - b \perp Y_k = \mathcal{R}\left(V^{|k}\right)$ it follows that

$$V^{|kT}(Ax_k - b) = 0.$$

Thus we obtain $V^{|kT} A U^{|k} a_k = V^{|kT} b$. If $V^{|kT} A U^{|k}$ is nonsingular, then $a_k = \left(V^{|kT} A U^{|k}\right)^{-1} V^{|kT} b$. Hence

$$x_k = U^{|k}\left(V^{|kT} A U^{|k}\right)^{-1} V^{|kT} b \tag{3.13}$$

$$= U^{|k}\left(V^{|kT} A U^{|k}\right)^{-1} V^{|kT} A\omega, \tag{3.14}$$

where the matrix $U^{|k}\left(V^{|kT}AU^{|k}\right)^{-1}V^{|kT}A$ is a projector onto $\mathcal{R}\left(U^{|k}\right)$ along $\mathcal{R}^{\perp}\left(A^TV^{|k}\right)$. Any of the expressions yields

PROPOSITION 3.2 *For $k = r$ the Galerkin-Petrov projection method gives the exact solution, that is $x_r = \omega$.*

The above form of the Galerkin-Petrov method is not suitable for practical computations. Therefore, we seek for a class of projection methods, where x_k is recursively calculated from x_{k-1} and the initial point x_0 is arbitrary. Let $x_k = \Delta_k + x_0$ and consider the solution of

$$P_k\left(A\left(\Delta_k + x_0\right) - b\right) = P_k\left(A\Delta_k - \left(b - Ax_0\right)\right) = 0.$$

Here Δ_k plays the role of x_k and $-r_0 = b - Ax_0$ plays the role of b. Hence the above formula becomes

$$x_k = x_0 - U^{|k}\left(V^{|kT}AU^{|k}\right)^{-1}V^{|kT}r_0. \tag{3.15}$$

The residual error $r_k = Ax_k - b$ has the form

$$r_k = \left(I - AU^{|k}\left(V^{|kT}AU^{|k}\right)^{-1}V^{|kT}\right)r_0, \tag{3.16}$$

where the matrix $I - AU^{|k}\left(V^{|kT}AU^{|k}\right)^{-1}V^{|kT}$ is a projector onto $\mathcal{R}^{\perp}\left(V^{|k}\right)$ along $\mathcal{R}\left(AU^{|k}\right)$. The absolute error $s_k = x_k - \omega$ can also be written as

$$s_k = \left(I - U^{|k}\left(V^{|kT}AU^{|k}\right)^{-1}V^{|kT}A\right)s_0, \tag{3.17}$$

where $I - U^{|k}\left(V^{|kT}AU^{|k}\right)^{-1}V^{|kT}A$ is a projector onto $\mathcal{R}^{\perp}\left(A^TV^{|k}\right)$ along $\mathcal{R}\left(U^{|k}\right)$.

Let us consider the difference between x_{k-1} and x_k. By definition

$$x_k - x_{k-1} = -\left(Z_k - Z_{k-1}\right)s_0, \tag{3.18}$$

where the matrix $Z_k = U^{|k}\left(V^{|kT}AU^{|k}\right)^{-1}V^{|kT}A$ is a projector onto $\mathcal{R}\left(U^{|k}\right)$ along $\mathcal{R}^{\perp}\left(A^TV^{|k}\right)$ for $k = 1, \ldots, r$.

LEMMA 3.3 *$Z_k - Z_{k-1}$ is a projector onto $\mathcal{R}\left(U^{|k}\right) \cap \mathcal{R}^{\perp}\left(A^TV^{|k-1}\right)$ along $\mathcal{R}^{\perp}\left(A^TV^{|k}\right) \oplus \mathcal{R}\left(U^{|k-1}\right)$.*

Proof. We exploit the Banachiewicz inversion formula (1.12). Let $B_k = V^{|kT}AU^{|k}$. Then

$$B_k = \begin{bmatrix} B_{k-1} & \left(V^{|k-1}\right)^TAU_k \\ V_k^TAU^{|k-1} & V_k^TAU_k \end{bmatrix}$$

and

$$B_k^{-1} = \begin{bmatrix} B_{k-1}^{-1} & 0 \\ 0 & 0 \end{bmatrix} + \begin{bmatrix} B_{k-1}^{-1} \left(V^{|k-1}\right)^T AU_k \\ -I \end{bmatrix} S^{-1} \left[V_k^T AU^{|k-1} B_{k-1}^{-1}, -I\right].$$

By simple calculations we can easily demonstrate that $Z_k Z_{k-1} = Z_{k-1} Z_k = Z_{k-1}$. Hence by Theorem 2.13, $Z_k - Z_{k-1}$ is a projector onto $\mathcal{R}\left(U^{|k}\right) \cap \mathcal{R}^{\perp}\left(A^T V^{|k-1}\right)$ along $\mathcal{R}^{\perp}\left(A^T V^{|k}\right) \oplus \mathcal{R}\left(U^{|k-1}\right)$. ■

We restrict projectors $Z_k - Z_{k-1}$ such that the equality

$$\mathcal{R}\left(U^{|k}\right) \cap \mathcal{R}^{\perp}\left(A^T V^{|k-1}\right) = \mathcal{R}\left(U_k\right) \tag{3.19}$$

should hold. This means that $x_k - x_{k-1} \in \mathcal{R}(U_k)$. The elements of the left-hand side can be written as $U^{|k} y$, where $y \in \mathbb{R}^{m_k}$ satisfies $\left(V^{|k-1}\right)^T AU^{|k} y = 0$. The elements of $\mathcal{R}(U_k)$ are of the form $U_k z$ ($z \in \mathbb{R}^{m_k}$). Relation (3.19) holds if and only if for any $U^{|k} y$ there is a corresponding z such that $U^{|k} y = U_k z$ and vice versa. However, $\left(V^{|k-1}\right)^T AU^{|k} y = \left(\left(V^{|k-1}\right)^T AU_k\right) z = 0$ ($\forall z \in \mathbb{R}^{m_k}$) implies that $\left(V^{|k-1}\right)^T AU_k = 0$. This holds for all $k = 2, \ldots, r$ if and only if $V^T AU$ is block lower triangular.

Assume that $V^T AU = L$ is a nonsingular block lower triangular matrix and let $V^{|kT} AU^{|k} = L_k$ for $k = 1, \ldots, r$. Then

$$\begin{aligned} \left(V^{|kT} AU^{|k}\right)^{-1} &= \begin{bmatrix} L_{k-1} & 0 \\ V_k^T AU^{|k-1} & V_k^T AU_k \end{bmatrix}^{-1} \\ &= \begin{bmatrix} L_{k-1}^{-1} & 0 \\ -\left(V_k^T AU_k\right)^{-1} V_k^T AU^{|k-1} L_{k-1}^{-1} & \left(V_k^T AU_k\right)^{-1} \end{bmatrix} \end{aligned}$$

and

$$Z_k - Z_{k-1} = U_k \left(V_k^T AU_k\right)^{-1} V_k^T A \left(I - U^{|k-1} L_{k-1}^{-1} \left(V^{|k-1}\right)^T A\right).$$

As $As_k = r_k$ we obtain that

$$\left(Z_k - Z_{k-1}\right) s_0 = U_k \left(V_k^T AU_k\right)^{-1} V_k^T As_{k-1} = U_k \left(V_k^T AU_k\right)^{-1} V_k^T r_{k-1}$$

and

$$x_k - x_{k-1} = -U_k \left(V_k^T AU_k\right)^{-1} V_k^T r_{k-1} \quad (k = 1, \ldots, r). \tag{3.20}$$

THEOREM 3.4 *If $V^T AU$ is a nonsingular block lower triangular matrix, the Galerkin-Petrov method (3.15) has the finitely terminating recursive form (3.20).*

The algorithm can also be interpreted as a sequence of projection methods, where $X_k = \mathcal{R}(U_k)$, $Y_k = \mathcal{R}(V_k)$, $P_k : X \to X_k$ being orthogonal, and $x_k = x_{k-1} + \Delta_k$.

In the next section we show that the obtained subclass of the Galerkin-Petrov method is identical with the conjugate direction methods of Stewart [325].

1.1 The conjugate direction methods

We define the concept of conjugate direction methods due to Stewart [325], who obtained it as a common generalization of different conjugate gradient methods known at that time.

DEFINITION 3.5 *Let $A, U, V \in \mathbb{R}^{m \times m}$ be nonsingular and assume that matrices U and V are partitioned in the form $U = [U_1, U_2, \ldots, U_r]$ and $V = [V_1, V_2, \ldots, V_r]$, where $U_k, V_k \in \mathbb{R}^{m \times m_k}$ ($k = 1, \ldots, r$). The pair (U, V) is said to be block A-conjugate (with respect to the partition $\{m_1, m_2, \ldots, m_r\}$), if the matrix $L = V^T A U = \left[V_i^T A U_j\right]_{i,j=1}^{r}$ is block lower triangular, that is $V_k^T A U_j = 0$ for $k < j$. The pair (U, V) is said to be block A-biorthogonal (biconjugate), if the matrix $L = V^T A U$ is block diagonal.*

If $r = m$, then $m_i = 1$ for all i. In this case we simply speak of A-conjugacy and A-biorthogonality (A-biconjugacy). We also say that P is A-orthogonal if $P^T A P = D$ is diagonal nonsingular. If $A = I$, then P is orthogonal.

Assume that $A, U, V \in R^{m \times m}$ are nonsingular, and the pair (U, V) is block A-conjugate with respect to the partition $\{m_1, m_2, \ldots, m_r\}$ $(1 \le r \le m)$.

Algorithm CDM (Conjugate Direction Method)

$x_0 \in \mathbb{R}^m$

for $k = 1, \ldots, r$

$\quad r_{k-1} = A x_{k-1} - b$

$\quad d_k = (V_k^T A U_k)^{-1} V_k^T r_{k-1}$

$\quad x_k = x_{k-1} - U_k d_k$

end

The conjugate direction methods are exactly the finitely terminating recursive projection methods (3.20) we specified just before. If V equals to a permutation matrix, then the algorithm becomes a row-action method [77] Although we proved the finite termination property we quote the following theorem of Stewart [325] for its proof technique and intrinsic characterization of the algorithm.

THEOREM 3.6 *(Stewart). For any initial point x_0, Algorithm CDM terminates in at most r iterations, that is*

$$A x_r = b. \tag{3.21}$$

Let $s_k = x_k - \omega$. Then

$$s_k = x_{k-1} - U_k d_k - \omega = s_{k-1} - U_k \left(V_k^T A U_k\right)^{-1} V_k^T r_{k-1}.$$

Now $r_k = A s_k$ implies

$$s_k = (I - R_k)\, s_{k-1}, \quad k = 1, \ldots, r,$$

where

$$R_k = U_k \left(V_k^T A U_k\right)^{-1} V_k^T A$$

is a projector of rank m_k with

$$\mathcal{N}(R_k) = \mathcal{R}^{\perp}\left(A^T V_k\right), \quad \mathcal{R}(R_k) = \mathcal{R}(U_k).$$

Generally we have

$$s_k = (I - R_k)(I - R_{k-1}) \ldots (I - R_1)\, s_0, \quad k = 1, \ldots, r. \tag{3.22}$$

As the pair (U, V) is block A-conjugate, projectors $R_1, \ldots, R_r$ satisfy $R_i R_j = 0$ for $i < j$. Hence it is natural to introduce the concept of *conjugate projectors* [325].

DEFINITION 3.7 *Let $R_1, \ldots, R_k$ be projectors. Then $R_1, \ldots, R_k$ are conjugate projectors, if*

$$i < j \Rightarrow R_i R_j = 0. \tag{3.23}$$

We note that

$$\mathcal{R}(R_i) \cap \mathcal{R}(R_j) = \{0\}, \quad i \neq j$$

holds for conjugate projectors $\{R_i\}_{i=1}^k$. Assume that there is a vector $x \neq 0$ such that $x \in \mathcal{R}(R_i) \cap \mathcal{R}(R_j)$ holds for some $i < j$. Definition 3.7 then implies that $x = R_i x = R_i R_j x = 0$, which is a contradiction. The following result of Stewart [325] will be exploited in several ways.

THEOREM 3.8 *If $R_1, \ldots, R_k$ are conjugate projectors, then*

$$Q_k = (I - R_k)(I - R_{k-1}) \ldots (I - R_1) \tag{3.24}$$

is a projector with

$$\mathcal{R}(Q_k) = \cap_{j=1}^k \mathcal{N}(R_j) \tag{3.25}$$

and

$$\mathcal{N}(Q_k) = \sum_{j=1}^k \mathcal{R}(R_j). \tag{3.26}$$

Proof. For $i \leq k$ it follows from Definition 3.7 that $R_i(I - R_k) \ldots (I - R_1) = 0$ and $Q_k^2 = Q_k$ proving that Q_k is a projector. We recall that

$\mathcal{R}(H) = \{x|Hx = x\}$ holds for any projector H. If $x \in \mathcal{R}(Q_k)$, then $R_j x = R_j(Q_k x) = (R_j Q_k)x = 0x = 0$ implies that $x \in \mathcal{N}(R_j)$ for all $1 \leq j \leq k$. Hence $\mathcal{R}(Q_k) \subseteq \cap_{j=1}^k \mathcal{N}(R_j)$. If $x \in \cap_{j=1}^k \mathcal{N}(R_j)$, then

$$Q_k x = (I - R_k) \dots (I - R_1)x = x \in \mathcal{R}(Q_k),$$

which proves the relation $\mathcal{R}(Q_k) = \cap_{j=1}^k \mathcal{N}(R_j)$. As the transposed projectors $R_k^T, \dots, R_1^T$ are also conjugate $\mathcal{R}(Q_k^T) = \cap_{j=1}^k \mathcal{N}(R_j^T)$. The relation $\mathcal{N}(R_j^T) = \mathcal{R}^{\perp}(R_j)$ implies $\mathcal{R}(Q_k^T) = [\sum_{j=1}^k \mathcal{R}(R_j)]^{\perp}$, from which $\mathcal{N}(Q_k) = \sum_{j=1}^k \mathcal{R}(R_j)$ readily follows. ■

REMARK 3.9 *$\mathcal{N}(Q_k)$ is the direct sum of subspaces $\mathcal{R}(R_j)$.*

The proof of Theorem 3.6. If we apply Theorem 3.8 to (3.22) we obtain that

$$\begin{aligned} \mathcal{R}(Q_k) &= \cap_{j=1}^k \mathcal{R}^{\perp}\left(A^T V_j\right) \\ &= \left[\sum_{j=1}^k \mathcal{R}\left(A^T V_j\right)\right]^{\perp} \\ &= \mathcal{R}^{\perp}\left(A^T V^{|k}\right) \\ &= \mathcal{R}\left[U_{k+1}, \dots, U_r\right] = \mathcal{R}\left(U^{|r-k}\right) \end{aligned}$$

and

$$\mathcal{N}(Q_k) = \sum_{j=1}^k \mathcal{R}(U_j) = \mathcal{R}\left([U_1, \dots, U_k]\right) = \mathcal{R}\left(U^{|k}\right).$$

For $k = r$ the matrix Q_r becomes the zero projector with

$$\mathcal{N}(Q_r) = \mathcal{R}\left([U_1, \dots, U_r]\right) = \mathbb{R}^m, \quad \mathcal{R}(Q_r) = \{0\}.$$

Hence $s_r = Q_r s_0 = 0$ proving Theorem 3.6.

The connection of the conjugate direction methods with projections is evident from the proof of Theorem 3.6. It is also clear that

$$Q_k = P_{\mathcal{R}(U^{r-k|}), \mathcal{R}(U^{|k})},$$

where the Householder notation is used blockwise.

REMARK 3.10 *Under the assumption of A-conjugacy*

$$I - U^{|k}\left(V^{|kT} A U^{|k}\right)^{-1} V^{|kT} A = Q_k = (I - R_k)(I - R_{k-1}) \dots (I - R_1).$$

This means that the oblique projector, which determines s_k is a product of k projectors.

We prove now some simple results on the absolute and residual error of the iterates x_k.

PROPOSITION 3.11 *The absolute errors s_k of Algorithm CDM satisfy the inequality*

$$\|s_k\| \leq \|Q_k\| \, \|s_0\| \quad (k = 1, \ldots, r). \tag{3.27}$$

In any unitarily invariant matrix norm, $\|Q_k\|$ is minimal for all k if and only if U is orthogonal. In any submultiplicative unitarily invariant matrix norm

$$\|Q_k\| \leq cond\,(U) \quad (k = 1, \ldots, r)\,. \tag{3.28}$$

Proof. The claim follows from Proposition 2.56 and Lemma 2.73. The projectors Q_k have minimal norm, if $Q_k = Q_k^T$ for all k. The matrix Q_k is symmetric if and only if

$$\mathcal{R}\left(U^{|k}\right) = \mathcal{R}^{\perp}\left(U^{r-k|}\right) \quad (k = 1, \ldots, r).$$

In other words, all Q_k are symmetric if and only if U is orthogonal. ∎

We consider now the residual errors of Algorithm CDM. Since $r_k = As_k$, we have

$$r_k = As_k = AQ_k s_0 = \widetilde{Q}_k r_0, \tag{3.29}$$

where

$$\widetilde{Q}_k = AQ_k A^{-1} = A\left\{U^{r-k|}\left(U^{-1}\right)^{\underline{r-k}}\right\}A^{-1} = (AU)^{r-k|}\left(U^{-1}A^{-1}\right)^{\underline{r-k}} \tag{3.30}$$

is a projector with $\mathcal{R}\left(\widetilde{Q}_k\right) = \mathcal{R}\left((AU)^{r-k|}\right) = \mathcal{R}\left(AU^{r-k|}\right)$ and $\mathcal{N}\left(\widetilde{Q}_k\right) = \mathcal{R}\left((AU)^{|k}\right) = \mathcal{R}\left(AU^{|k}\right)$. From the definition of block A-conjugacy it follows that $\mathcal{R}\left(\widetilde{Q}_k\right) = \mathcal{R}^{\perp}\left(V^{|k}\right)$.

COROLLARY 3.12 *The residual errors r_k of Algorithm CDM satisfy*

$$r_k = \widetilde{Q}_k r_0, \tag{3.31}$$

where

$$\mathcal{R}\left(\widetilde{Q}_k\right) = \mathcal{R}\left((AU)^{r-k|}\right), \quad \mathcal{N}\left(\widetilde{Q}_k\right) = \mathcal{R}\left((AU)^{|k}\right). \tag{3.32}$$

PROPOSITION 3.13 *The residual errors r_k of Algorithm CDM satisfy the inequality*

$$\|r_k\| \leq \left\|\widetilde{Q}_k\right\| \|r_0\| \quad (k = 1, \ldots, r)\,.$$

In any unitarily invariant matrix norm, $\left\|\widetilde{Q}_k\right\|$ *is minimal for all k if and only if U is* A^TA*-orthogonal. In any submultiplicative unitarily invariant matrix norm*

$$\left\|\widetilde{Q}_k\right\| \leq cond(AU) \quad (k = 1, \dots, r). \tag{3.33}$$

Proof. It is the same as in the last proposition. Projector $\widetilde{Q}_k$ is symmetric if and only if

$$\mathcal{R}\left(AU^{r-k|}\right) = \mathcal{R}^{\perp}\left(AU^{k|}\right) \quad (k = 1, \dots, r).$$

All $\widetilde{Q}_k$ are symmetric if and only if $U^TA^TAU = I$. ■

We can give similar estimates for the iterates x_k of Algorithm CDM.

COROLLARY 3.14 *The iterates of Algorithm CDM satisfy*

$$\|x_k\| \leq \|I - Q_k\| \|\omega\| + \|Q_k\| \|x_0\| \quad (k = 1, \dots, r). \tag{3.34}$$

If the matrix norm is unitarily invariant and submultiplicative, then

$$\|x_k\| \leq cond(U)\left(\|\omega\| + \|x_0\|\right) \quad (k = 1, \dots, r). \tag{3.35}$$

Proof. It is obvious that

$$x_k = (I - Q_k)\,\omega + Q_k x_0, \quad k = 1, \dots, r,$$

which implies the first bound. Lemma 2.73 implies

$$\|Q_k\| \leq \mathrm{cond}\,(U), \quad \|I - Q_k\| \leq \mathrm{cond}\,(U)$$

and the second inequality. ■

REMARK 3.15 *If all* Q_k *are symmetric, that is U is orthogonal, then*

$$\|x_k\|_2 \leq \|\omega\|_2 + \|x_0\|_2 .$$

Using a result of Broyden [66] we can show that A-conjugacy is not only a sufficient, but also a necessary condition for the finite termination property of the conjugate direction methods or iteration methods of the form

$$x_k - x_{k-1} = -U_k\left(V_k^TAU_k\right)^{-1}V_k^Tr_{k-1} \quad (r_{k-1} = Ax_{k-1} - b,\ k = 1, \dots, r).$$

THEOREM 3.16 *(Broyden). Assume that r steps are performed with Algorithm CDM, where the nonsingular matrices*

$$V = [V_1, \dots, V_r], \quad U = [U_1, \dots, U_r]$$

are such that $V_i^T AU_i$ is nonsingular for $i = 1, \ldots, r$. Then a necessary and sufficient condition for the vector x_r to solve $Ax = b$ for any starting point x_0 is that the matrix $L = V^T AU$ be nonsingular block lower triangular.

Proof. We prove only the necessity as the sufficiency follows from Stewart's result. Define the vector $\alpha = \left[d_1^T, \ldots, d_r^T\right]^T \in \mathbb{R}^m$. For $1 \leq j \leq r$ the vector x_j computed by Algorithm CDM has the form

$$x_j = x_0 - \sum_{k=1}^{j} U_k d_k \tag{3.36}$$

and the residual is

$$r_j = r_0 - A \sum_{k=1}^{j} U_k d_k. \tag{3.37}$$

By the definition of d_i the residual r_i is orthogonal to V_i. Thus multiplying equation (3.37) on the left by V_j we have

$$V_j^T r_j = V_j^T r_0 - V_j^T A \sum_{k=1}^{j} U_k d_k = 0, \quad j = 1, \ldots, r. \tag{3.38}$$

Let L be the block lower triangular matrix with the elements

$$L_{ij} = V_i^T AU_j, \quad j \leq i, \ i = 1, \ldots, r.$$

Observe that by assumption L is nonsingular. The linear system (3.38) can be written into the form $Ld = V^T r_0$ from which $\alpha = L^{-1} V^T r_0$ follows. For $j = r$, equation (3.36) can be written as

$$x_r = x_0 - UL^{-1} V^T r_0. \tag{3.39}$$

Assume that $x_r = \omega$ holds for arbitrary x_0. Then (3.39) becomes

$$\omega = x_0 - UL^{-1} V^T A \left(x_0 - \omega\right),$$

or,

$$z = UL^{-1} V^T Az \quad \left(z = \omega - x_0\right).$$

Since z is arbitrary, it follows that $I = UL^{-1} V^T A$. Multiplying this on the left by U^{-1} and on the right by U, we obtain $I = L^{-1} V^T AU$ from which the condition $L = V^T AU$ easily follows. Thus we proved the necessity. ∎

We obtained Algorithm CDM in various ways. The recursive Galerkin-Petrov algorithms are identified with the class of conjugate direction methods. It was also shown that A-conjugacy is the key of the finite termination property and recursivity.

2. Conjugation procedures

We deal with conjugation procedures. There are many such procedures. Here we prefer the factorization based conjugation procedures which are obtained through the rank reduction procedure. The material of the whole section is based on [142] and [147].

2.1 The rank reduction procedure

The rank reduction procedure is a simple but a very general technique. It determines the rank of a matrix, provides a related full rank factorization and has a built in conjugation property. The rank reduction algorithm is based on the following result.

THEOREM 3.17 *(The rank reduction theorem). Let $UR^{-1}V^T$ ($U \in \mathbb{R}^{m\times k}$, $R \in \mathbb{R}^{k\times k}$, $V \in \mathbb{R}^{n\times k}$) be a full rank decomposition of S. Then the necessary and sufficient conditions for*

$$rank\,(H - S) = rank\,(H) - rank\,(S)\,, \tag{3.40}$$

are

$$U = HX, \quad V^T = Y^T H, \quad Y^T HX = R \tag{3.41}$$

for some matrices $X \in \mathbb{R}^{n\times k}$ and $Y \in \mathbb{R}^{m\times k}$.

Proof. Let

$$B = \begin{bmatrix} Y^T HX & Y^T H \\ HX & H \end{bmatrix} = \begin{bmatrix} Y^T & I_k \\ I_m & 0 \end{bmatrix} \begin{bmatrix} H & 0 \\ 0 & 0 \end{bmatrix} \begin{bmatrix} X & I_n \\ I_k & 0 \end{bmatrix}. \tag{3.42}$$

The first and last matrices of the right-hand side are nonsingular. Hence $\operatorname{rank}(H) = \operatorname{rank}(B)$. The Guttman lemma then implies

$$\operatorname{rank}\left(H - HX\left(Y^T HX\right)^{-1} Y^T H\right) = \operatorname{rank}(H) - \operatorname{rank}(S)\,.$$

Assume that $\operatorname{rank}(H - S) = \operatorname{rank}(H) - k$ and define

$$B = \begin{bmatrix} R & V^T \\ U & H \end{bmatrix}.$$

The Guttman lemma implies

$$\operatorname{rank}(H) \le \operatorname{rank}(B) = \operatorname{rank}(R) + \operatorname{rank}\left(H - UR^{-1}V^T\right) = \operatorname{rank}(H)\,.$$

Hence $\operatorname{rank}(H) = \operatorname{rank}(B)$ and there exist matrices X and Y such that

$$\left[R, V^T\right] = Y^T\left[U, H\right], \quad \begin{bmatrix} R \\ U \end{bmatrix} = \begin{bmatrix} V^T \\ H \end{bmatrix} X.$$

Thus we obtain $V^T = Y^T H$, $U = HX$ and $R = V^T X = Y^T U = Y^T H X$. ■

The rank reduction theorem is constructive. If $\operatorname{rank}(H) \geq k$, then H has a $k \times k$ nonsingular submatrix $H[\alpha, \beta]$. Letting $Y = \Pi_\alpha$ and $X = \Pi_\beta$, $S = H\Pi_\beta \left(\Pi_\alpha^T H \Pi_\beta\right)^{-1} \Pi_\alpha^T H$ and $\operatorname{rank}(H - S) = \operatorname{rank}(H) - k$. The rank reduction operation is defined by

$$\widehat{H} = H - HX\left(Y^T H X\right)^{-1} Y^T H, \tag{3.43}$$

where $H \in \mathbb{R}^{m \times n}$, $X \in \mathbb{R}^{n \times k}$ and $Y \in \mathbb{R}^{m \times k}$.

Proposition 3.18 *The rank reduction operation has the following properties:*
(i) $\widehat{H}$ can be written in the forms

$$\widehat{H} = (I - R) H = H (I - S), \tag{3.44}$$

where $R = HX\left(Y^T H X\right)^{-1} Y^T$ and $S = X\left(Y^T H X\right)^{-1} Y^T H$ are oblique projectors with

$$\mathcal{N}(R) = \mathcal{R}^\perp(Y), \quad \mathcal{R}(R) = \mathcal{R}(HX) \tag{3.45}$$

and

$$\mathcal{N}(S) = \mathcal{R}^\perp\left(H^T Y\right), \quad \mathcal{R}(S) = \mathcal{R}(X); \tag{3.46}$$

(ii) $\widehat{H}X = 0$ and $Y^T \widehat{H} = 0$;
(iii) Let $M = X\left(Y^T H X\right)^{-1} Y^T$. Then $H = M^-$. If $k = \operatorname{rank}(M) < \operatorname{rank}(H)$, then $M \neq H^-$. If $k = \operatorname{rank}(H)$, then $M \in H^{\{1,2\}}$.
(iv) Let $H = FG^T$ be a full rank factorization. If $X = G\left(G^T G\right)^{-1} = G^+$ and $Y = F\left(F^T F\right)^{-1} = \left(F^+\right)^T$ ($k = \operatorname{rank}(H)$), then $M = H^+$. If $k < \operatorname{rank}(H)$, $X = G\left(G^T G\right)^{-1} I^{|k}$ and $Y = F\left(F^T F\right)^{-1} I^{|k}$, then $\widehat{H}$ has the form $\widehat{H} = F\left(I - I^{|k} I^{|kT}\right) G^T$.

Proof. Properties (i) and (iii) are obvious. (i) implies (ii) as $\widehat{H}X = H(I - S)X = 0$ and $Y^T \widehat{H} = Y^T (I - R) H = \left(H^T (I - R)^T Y\right)^T = 0$. If $H = FG^T$, $X = G\left(G^T G\right)^{-1}$ and $Y = F\left(F^T F\right)^{-1}$, then by substitution $M = G\left(G^T G\right)^{-1}\left(F^T F\right)^{-1} F^T$, which is the Moore-Penrose inverse of H. In the case $X = G\left(G^T G\right)^{-1} I^{|k}$ and $Y = F\left(F^T F\right)^{-1} I^{|k}$ matrix $\widehat{H}$ has the form $\widehat{H} = FG^T - FI^{|k} I^{|kT} G^T$. ■

We note that for any H^-, $\widehat{H} H^- \widehat{H} = \widehat{H}$, that is $H^- \in \widehat{H}^{\{1\}}$ ($H^- = \widehat{H}^-$).

Proposition 3.19 *The rank reduction operation does not change the zero columns or rows. If X or Y contains unit vectors, the rank reduction operation makes the corresponding columns or rows zero.*

Proof. Assume that $He_j = 0$ or $e_i^T H = 0$. Then by definition $\widehat{H}e_j = 0$ or $e_i^T\widehat{H} = 0$ also holds. If $e_i = Ye_j$ for some $1 \le j \le k$, then by the identity $e_i^T\widehat{H} = e_j^T Y^T \widehat{H} = 0$ the ith row of $\widehat{H}$ is zero. Similarly, if $e_p = Xe_q$ for some $1 \le q \le k$, then relation $\widehat{H}e_p = \widehat{H}Xe_q = 0$ implies that the pth column of $\widehat{H}$ is zero. ■

LEMMA 3.20 *Let $X = \Pi_\beta$ and $Y = \Pi_\alpha$. Then*

$$\widehat{H}\left[\alpha', \beta'\right] = H\left[\alpha', \beta'\right] - H\left[\alpha', \beta\right]\left(H\left[\alpha, \beta\right]\right)^{-1} H\left[\alpha, \beta'\right], \tag{3.47}$$

while the other elements of $\widehat{H}$ are zero.

Proof. By definition $\widehat{H}\Pi_\beta = 0$, $\Pi_\alpha^T\widehat{H} = 0$ and

$$\widehat{H}\left[\alpha', \beta'\right] = \Pi_{\alpha'}^T\widehat{H}\Pi_{\beta'} = \Pi_{\alpha'}^T H\Pi_{\beta'} - \Pi_{\alpha'}^T H\Pi_\beta\left(\Pi_\alpha^T H\Pi_\beta\right)^{-1}\Pi_\alpha^T H\Pi_{\beta'}.$$

■

Expression (3.47) is also called the generalized Schur complement. For particular choices we have the following consequences.

COROLLARY 3.21 *Assume that $H \in \mathbb{R}^{n\times n}$ is partitioned according to*

$$H = \begin{bmatrix} H_{11} & H_{12} \\ H_{21} & H_{22} \end{bmatrix} \quad \left(H_{11} \in \mathbb{R}^{k\times k}\right) \tag{3.48}$$

(1.10) and let $J = I^{n-k|} \in \mathbb{R}^{n\times(n-k)}$, $\widehat{J} = I^{|k} \in \mathbb{R}^{n\times k}$. If H is nonsingular, then

$$H - HJ\left(J^T HJ\right)^{-1} J^T H = \begin{bmatrix} (H/H_{22}) & 0 \\ 0 & 0 \end{bmatrix}. \tag{3.49}$$

If H_{11} is nonsingular, then

$$H - H\widehat{J}\left(\widehat{J}^T H\widehat{J}\right)^{-1}\widehat{J}^T H = \begin{bmatrix} 0 & 0 \\ 0 & (H/H_{11}) \end{bmatrix}. \tag{3.50}$$

REMARK 3.22 *If H is replaced by H^{-1}, then $\left(H^{-1}/S^{-1}\right) = H_{11}^{-1}$ and*

$$H^{-1} - H^{-1}J\left(J^T H^{-1}J\right)^{-1} J^T H^{-1} = \begin{bmatrix} H_{11}^{-1} & 0 \\ 0 & 0 \end{bmatrix}. \tag{3.51}$$

Thus one rank reduction step on matrix H^{-1} results the inverse of the leading principal submatrix H_{11}. Relation (3.51) gives a direct connection between the rank reduction and the bordered inversion method based on formula (1.11). This connection was first exploited by Egerváry [114].

The rank reduction procedure is the repeated use of the rank reduction operation.

Let $H_1 = H$, $X_i \in \mathbb{R}^{n\times l_i}$, $Y_i \in \mathbb{R}^{m\times l_i}$, $l_i \geq 1$ and $Y_i^T H_i X_i$ be nonsingular for $i = 1, 2, \ldots, k$.

The rank reduction procedure

$$H_{i+1} = H_i - H_i X_i \left(Y_i^T H_i X_i\right)^{-1} Y_i^T H_i \quad (i = 1, 2, \ldots, k), \tag{3.52}$$

where $\operatorname{rank}(H) \geq \sum_{i=1}^k l_i$.

It is clear that

$$\operatorname{rank}(H_{i+1}) = \operatorname{rank}(H_i) - l_i = \operatorname{rank}(H) - \sum_{j=1}^{i} l_j.$$

The algorithm stops, if $H_{k+1} = 0$. If $H_{k+1} = 0$, then $\operatorname{rank}(H) = \sum_{i=1}^k l_i$.

We can write H_{k+1} in the forms

$$H_{k+1} = H_1 - \sum_{i=1}^{k} H_i X_i \left(Y_i^T H_i X_i\right)^{-1} Y_i^T H_i$$

and

$$H_{k+1} = H_1 - QD^{-1}P^T,$$

where

$$P = \left[H_1^T Y_1, \ldots, H_k^T Y_k\right], \quad Q = \left[H_1 X_1, \ldots, H_k X_k\right] \tag{3.53}$$

and

$$D = \operatorname{diag}\left(Y_1^T H_1 X_1, \ldots, Y_k^T H_k X_k\right). \tag{3.54}$$

If $H_{k+1} = 0$, then we obtain the rank additive decomposition

$$H = \sum_{i=1}^{k} H_i X_i \left(Y_i^T H_i X_i\right)^{-1} Y_i^T H_i \tag{3.55}$$

and the related full rank factorization

$$H = QD^{-1}P^T. \tag{3.56}$$

For the general properties and characterizations of rank additive decompositions we refer to Carlson [76], and Marsaglia and Styan [244].

The rank reduction procedure is said to be breakdown free, if $Y_i^T H_i X_i$ is nonsingular for all $i = 1, 2, \ldots, k$.

THEOREM 3.23 *Let $X = [X_1, \ldots, X_k]$ and $Y = [Y_1, \ldots, Y_k]$. Then the rank reduction procedure can be carried out breakdown free if and only if $Y^T HX = \left[Y_i^T H X_j\right]_{i,j=1}^k$ is block strongly nonsingular. In this event the rank reduction procedure has the canonical form*

$$H_{k+1} = H - HX\left(Y^T HX\right)^{-1} Y^T H. \tag{3.57}$$

Proof. We first assume that k successive steps were performed, which means that $Y_i^T H_i X_i$ is nonsingular for $i = 1, \dots, k$. Then

$$H_{k+1} = H_1 - QD^{-1}P^T,$$

where

$$P = \left[H_1^T Y_1, \dots, H_k^T Y_k\right], \quad Q = \left[H_1 X_1, \dots, H_k X_k\right] \tag{3.58}$$

and

$$D = \operatorname{diag}\left(Y_1^T H_1 X_1, \dots, Y_k^T H_k X_k\right). \tag{3.59}$$

Let us observe that $Y_i^T H_j = 0$ $(i < j)$ and $H_i X_j = 0$ $(i > j)$. Therefore $Y^T Q = \left[Y_i^T H_j X_j\right]_{i,j=1}^k = LD$ and $P^T X = \left[Y_i^T H_i X_j\right]_{i,j=1}^k = DU$, where the nonsingular L and U are unit block lower and upper triangular, respectively. We can also observe that

$$H_1 X = (H_1 - H_{k+1}) X = QD^{-1}P^T X = QU.$$

This implies that $Q = H_1 X U^{-1}$. Similarly we obtain

$$Y^T H_1 = Y^T (H_1 - H_{k+1}) = Y^T Q D^{-1} P^T = LP^T$$

and $P^T = L^{-1} Y^T H_1$. Hence $H_{k+1} = H_1 - QD^{-1}P^T$ implies

$$H_{k+1} = H_1 - H_1 X U^{-1} D^{-1} L^{-1} Y^T H_1 = H - HX (LDU)^{-1} Y^T H.$$

As

$$Y^T H_1 X = Y^T Q U = LDU \tag{3.60}$$

we proved that $Y^T H X$ is block strongly nonsingular. By substitution we obtain formula (3.57). Let us suppose that $Y^T H X$ is block strongly nonsingular. We must prove that $Y_i^T H_i X_i$ is nonsingular for $i = 1, \dots, k$. By the initial assumption $Y_1^T H_1 X_1$ is nonsingular. Hence matrix $H_2 = H_1 - H_1 X_1 \left(Y_1^T H_1 X_1\right)^{-1} Y_1^T H_1$ is defined. Let us assume that H_i is defined for $i \le k$, $X^{|i} = [X_1, \dots, X_i]$ and $Y^{|i} = [Y_1, \dots, Y_i]$. Then

$$H_i = H - HX^{|i-1} \left[\left(Y^{|i-1}\right)^T HX^{|i-1}\right]^{-1} \left(Y^{|i-1}\right)^T H$$

and

$$Y_i^T H_i X_i = Y_i^T H X_i - Y_i^T H X^{|i-1} \left[\left(Y^{|i-1}\right)^T H X^{|i-1}\right]^{-1} \left(Y^{|i-1}\right)^T H X_i.$$

As $Y_i^T H_i X_i$ is the Schur complement of the block bordered matrix

$$Y^{|iT} H X^{|i} = \begin{bmatrix} \left(Y^{|i-1}\right)^T H X^{|i-1} & \left(Y^{|i-1}\right)^T H X_i \\ Y_i^T H X^{|i-1} & Y_i^T H X_i \end{bmatrix},$$

rank $\left(Y^{|iT}HX^{|i}\right) = \sum_{j=1}^{i} l_j$ and rank $\left(\left(Y^{|i-1}\right)^T HX^{|i-1}\right) = \sum_{j=1}^{i-1} l_j$, the Guttman lemma implies that rank $\left(Y_i^T H_i X_i\right) = l_i$. Hence $Y_i^T H_i X_i$ is nonsingular and H_{i+1} is defined. ■

We emphasize that k successive rank reduction operations can be replaced by one block rank reduction. Thus the characterizations of one rank reduction remain valid for the rank reduction algorithm as well. In addition we give the following

PROPOSITION 3.24 *Let H^- be any (1)-inverse of H. Then*

$$H_k H^- H_j = H_j H^- H_k = H_k \quad (j \leq k). \tag{3.61}$$

provided that $Y^T HX$ is block strongly nonsingular.

Observe that $H^- H_k$'s are projectors. If $H_1 = I \in \mathbb{R}^{n\times n}$, then all H_k's are projectors. Another proof of this fact follows from the rank subtractivity partial ordering

$$H_{i+1} \overset{rs}{\leq} H_i, \quad H_i X_i \left(Y_i^T H_i X_i\right)^{-1} Y_i^T H_i \overset{rs}{\leq} H_i \tag{3.62}$$

produced by the rank reduction algorithm. Hartwig and Styan [190] proved that Π is a projector if and only if $\Pi \overset{rs}{\leq} \Omega$ for some projector Ω. Hence $H_1 = I$ implies that H_i and $\Delta_i = H_i X_i \left(Y_i^T H_i X_i\right)^{-1} Y_i^T H_i$ are projectors for $i = 1, \ldots, k$. For $H_{k+1} = 0$ we obtain the rank additive decomposition $\sum_{i=1}^{k} \Delta_i = I$, where all Δ_i's are oblique projectors. This special case of decomposition (3.55) is related to generalizations of the Cochran theorem (cf. Carlson [76], Marsaglia and Styan [244]).

We assume that $H_{k+1} = 0$ (rank $(H) = \sum_{i=1}^{k} l_i$) and $Y^T HX$ is block strongly nonsingular. Let $X^{(i)} = [X_1, \ldots, X_i]$ and $Y^{(i)} = [Y_1, \ldots, Y_i]$.

THEOREM 3.25 *If $Y^T HX$ is block strongly nonsingular and $H_{k+1} = 0$, then H_{i+1} has the canonical form*

$$H_{i+1} = HX\left(\left(Y^T HX\right)^{-1} - \begin{bmatrix} \left(Y^{(i)T} HX^{(i)}\right)^{-1} & 0 \\ 0 & 0 \end{bmatrix}\right) Y^T H \quad (i = 1, \ldots, k). \tag{3.63}$$

Proof. Formula (3.57) gives

$$H_{i+1} = H - HX^{(i)} \left(Y^{(i)T} HX^{(i)}\right)^{-1} Y^{(i)T} H \quad (i = 1, \ldots, k).$$

Let $\omega = \sum_{j=1}^{l} l_j$. Relations $H = HX\left(Y^T HX\right)^{-1} Y^T H$, $X^{(i)} = XI^{|\omega}$ and $Y^{(i)} = YI^{|\omega}$ imply

$$H_{i+1} = HX\left[\left(Y^T HX\right)^{-1} - I^{|\omega} \left(Y^{(i)T} HX^{(i)}\right)^{-1} \left(I^{|\omega}\right)^T\right] Y^T H,$$

which clearly has the form in question. ■

The canonical form (3.63) indicates that the rank reduction procedure implicitly approximates the inverse of $Y^T HX$ by the inverses of the leading principal submatrices bordered with zero matrices of appropriate size. Thus another connection with the bordered inversion method is clarified.

2.2 Rank reduction and factorizations

For the rest of this chapter we assume that $\sum_{i=1}^{k} l_i = \text{rank}(H)$. Let $B = L_B D_B U_B$ be the unique block LDU-decomposition of matrix $B \in \mathbb{R}^{m\times m}$ with unit block lower triangular L_B, block diagonal D_B and unit block upper triangular U_B.

The following consequence of Theorem 3.23 gives the components of the full rank factorization (3.56) in terms of the parameters X and Y. The result will play a key role in the conjugation via the rank reduction.

THEOREM 3.26 *Let $Y^T HX$ be block strongly nonsingular. The components of the full rank factorization (3.56) are*

$$P = H^T Y L_{Y^T HX}^{-T}, \quad Q = HXU_{Y^T HX}^{-1}, \quad D = D_{Y^T HX}. \tag{3.64}$$

Proof. From the proof of Theorem 3.23 (equation (3.60)) it follows that $L = L_{Y^T HX}$, $D = D_{Y^T HX}$, $U = U_{Y^T HX}$, $P^T = L_{Y^T HX}^{-1} Y^T H$ and $Q = HXU_{Y^T HX}^{-1}$. ■

Here Z^{-T} stands for $\left(Z^{-1}\right)^T$. Matrices P and Q can also be written in the form $P = H^T Y U_{X^T H^T Y}^{-1}$ and $Q = HXL_{X^T H^T Y}^{-T}$. The role of P and Q can be changed by transposing the sequence (3.52). Hence the results on P can be given for Q, as well. It follows from Theorem 3.26, that all full rank factorizations can be generated by the rank reduction procedure.

PROPOSITION 3.27 *Let $H = FG$ be any full rank factorization and let $X = G^T \left(GG^T\right)^{-1}$ and $Y = F\left(F^T F\right)^{-1}$. Then $Q = F$, $D = I$ and $P^T = G$.*

Proof. As $Y^T HX = I$ is strongly nonsingular, the rank reduction procedure is breakdown free. Formula (3.64) yields $Q = F$ and $P = G^T$. ■

In the rest of this section we assume that all matrices $H, X, Y \in \mathbb{R}^{m\times m}$ are nonsingular and $Y^T HX = \left[Y_i^T HX_j\right]_{i,j=1}^{k}$ is block strongly nonsingular. Then Q and P are also nonsingular and can be written in the form

$$P^T = D_{Y^T HX} U_{Y^T HX} X^{-1}, \quad Q = Y^{-T} L_{Y^T HX} D_{Y^T HX}. \tag{3.65}$$

Using (3.65) one can easily prove the following results.

PROPOSITION 3.28 *P (Q) is block lower triangular, if and only if X (Y) is block upper triangular.*

Proof. The matrix $P^T = (D_{Y^THX}U_{Y^THX})X^{-1} = \widetilde{U}$ is block upper triangular if and only if $X = \widetilde{U}^{-1}U_{Y^THX}^{-1}D_{Y^THX}^{-1}$ is also block upper triangular. Similarly, $Q = Y^{-T}L_{Y^THX}D_{Y^THX} = \widetilde{L}$ is block lower triangular if and only if Y^T is lower triangular. ■

COROLLARY 3.29 *Q is block lower triangular and P^T is block upper triangular, if and only if X and Y are block upper triangular.*

If both Q and P are block lower triangular, then factorization (3.56) simplifies to the block LDU type factorization

$$H = (L_H D_H D_X)\left(D_X^{-1} D_H^{-1} D_{Y^T}^{-1}\right)(D_{Y^T} D_H U_H).$$

We recall that P is B-orthogonal if and only if the pair (P, P) is B-biconjugate.

PROPOSITION 3.30 *Let $B \in \mathbb{R}^{m\times m}$ be a symmetric and positive definite matrix. P is B-orthogonal, if and only if $X = BH^TYU$ holds with any nonsingular Y and a nonsingular upper triangular U.*

Proof. P is B-orthogonal, if and only if

$$P^TBP = L_{Y^THX}^{-1}Y^THBH^TYL_{Y^THX}^{-T} = D,$$

where D is diagonal. This holds exactly if $Y^THBH^TY = L_{Y^THX}DL_{Y^THX}^T$. This implies $L_{Y^THBH^TY} = L_{Y^THX}$. Hence a nonsingular upper triangular U exists such that $Y^THBH^TYU = Y^THX$, from which $X = BH^TYU$ follows. As B is symmetric and positive definite, the nonsingularity of Y and U is the condition for the strong nonsingularity of Y^THX. ■

COROLLARY 3.31 *P is orthogonal up to a diagonal scaling and Q is lower triangular, if and only if $X = H^TYU$ holds with any nonsingular upper triangular U and Y. In such a case, factorization (3.56) is the LQ-factorization of H (QR factorization of H^T).*

PROPOSITION 3.32 *Let $B \in \mathbb{R}^{m\times m}$ be a symmetric and positive definite matrix. Q is B-orthogonal, if and only if $Y = BHXL^T$ holds with any nonsingular X and a nonsingular lower triangular L.*

Proof. Q is B-orthogonal, if and only if

$$Q^TBQ = U_{Y^THX}^{-T}X^TH^TBHXU_{Y^THX}^{-1} = D,$$

where D is diagonal. This holds exactly if $X^TH^TBHX = U_{Y^THX}^TDU_{Y^THX}$ from which $U_{X^TH^TBHX} = U_{Y^THX}$ follows. Hence a nonsingular lower triangular matrix L exists such that $LX^TH^TBHX = Y^THX$. This implies that

$Y = BHXL^T$. As B is symmetric and positive definite, the nonsingularity of X and L is the condition for the strong nonsingularity of Y^THX. ■

COROLLARY 3.33 *Q is orthogonal up to a diagonal scaling and P_m^T is upper triangular, if and only if $Y = HXL^T$ holds with any nonsingular lower triangular L and upper triangular X. In such a case, factorization (3.56) is the QR-factorization of H.*

PROPOSITION 3.34 *P is block upper Hessenberg , if and only if H^TY is block upper Hessenberg. Q is block upper Hessenberg, if and only if HX is block upper Hessenberg.*

Proof. The block upper Hessenberg form is invariant under multiplication by block upper triangular matrices from both sides. By definition $P = H^TYL_{Y^THX}^{-T} = F$ is block upper Hessenberg, if and only if $H^TY = FL_{Y^THX}^T$ is block upper Hessenberg. Similarly, $Q = HXU_{Y^THX}^{-1} = F$ is upper Hessenberg, if and only if $HX = FU_{Y^THX}$ is block upper Hessenberg. ■

If $H = I$, then the reciprocal relation $I = QD^{-1}P^T$, so that

$$Q^T = \left(PD^{-1}\right)^{-1}. \tag{3.66}$$

In this case the rank reduction algorithm produces both P and its inverse P^{-1} apart from a diagonal scaling. We exploit this property in the following observation on the Lánczos reduction to tridiagonal form (see, e.g., [206]).

THEOREM 3.35 *Let $Y_1 \in \mathbb{R}^m$ be such that $Y = \left[Y_1, BY_1, \dots, B^{m-1}Y_1\right]$ is nonsingular. If $H = I$ and X is such that Y^TX is strongly nonsingular, then $Q^TB\left(PD^{-1}\right)$ is an upper Hessenberg matrix similar to B. If, in addition $X = \left[X_1, B^TX_1, \dots, \left(B^T\right)^{m-1}X_1\right]$, then $Q^TB\left(PD^{-1}\right)$ is a tridiagonal matrix, which is similar to B.*

Proof. As $B^mY_1 = \sum_{i=1}^m \gamma_i B^{i-1}Y_1$, the relation $BY = YF$ holds, where the companion matrix $F = [f_{ij}]_{i,j=1}^m$ is defined by

$$f_{ij} = \begin{cases} 1, & i = j+1, \\ \gamma_i, & j = n, \\ 0, & \text{otherwise.} \end{cases}$$

By the reciprocal relation (3.66) the matrix

$$\begin{aligned} Q^TB\left(PD^{-1}\right) &= D_{Y^TX}L_{Y^TX}^T\left(Y^{-1}BY\right)L_{Y^TX}^{-T}D_{Y^TX}^{-1} \\ &= D_{Y^TX}L_{Y^TX}^TFL_{Y^TX}^{-T}D_{Y^TX}^{-1}, \end{aligned} \tag{3.67}$$

is a similarity transformation of B. As any upper Hessenberg matrix F multiplied by two upper triangular matrices keeps the upper Hessenberg form, the rank reduction algorithm defines a similarity transformation of matrix B to upper Hessenberg form. If $X = \left[X_1, B^T X_1, \ldots, \left(B^T\right)^{m-1} X_1\right]$ for a suitable vector $X_1 \in \mathbb{R}^m$, then $B^T X = X F_1$ holds with another companion matrix F_1. So relation (3.67) becomes

$$Q^T B \left(PD^{-1}\right) = U_{Y^T X}^{-T} \left(X^T B X^{-T}\right) U_{Y^T X}^T = U_{Y^T X}^{-T} F_1^T U_{Y^T X}^T,$$

where the lower Hessenberg matrix F_1^T is multiplied by two lower triangular matrices. This multiplication keeps the lower Hessenberg form. Thus $Q^T B \left(PD^{-1}\right)$ is of both upper and lower Hessenberg form. Hence the matrix must be tridiagonal. ■

The rank reduction algorithm defines a similarity transformation of matrix B to tridiagonal form giving the transformation matrices Q^T and Q^{-T} simultaneously.

2.3 Rank reduction and conjugation

Let us assume that $H \in \mathbb{R}^{m \times n}$, X and Y are such that $Y^T H X$ is block strongly nonsingular. The following observations show the inherent conjugation properties of the rank reduction algorithm (3.52).

PROPOSITION 3.36 *If $X = A^T V$ and $Y = B^T W$, then the pair (P, V) is block A-conjugate and the pair (Q, W) is block B-conjugate.*

Proof. It follows from Theorem 3.26, that

$$V^T A P = X^T H^T Y U_{X^T H^T Y}^{-1} = L_{X^T H^T Y} D_{X^T H^T Y}.$$

and

$$W^T B Q = Y^T H X U_{Y^T H X}^{-1} = L_{Y^T H X} D_{Y^T H X}.$$

■

PROPOSITION 3.37 *For any block A-conjugate pair (P, V) there exists a matrix Y with orthogonal columns such that the rank reduction algorithm (3.52) with $H = I$, $X = A^T V$ and Y generates P.*

Proof. Let $P = \widetilde{Q} D R_1$ be a QR-factorization of P with diagonal D and unit upper triangular R_1. As $V^T A P = X^T P = X^T \widetilde{Q} D R_1 = L$ implies $X^T \widetilde{Q} D = L R_1^{-1}$, the matrix $X^T \widetilde{Q} D$ is strongly nonsingular. Let $H = I$, $X = A^T V$ and $Y = \widetilde{Q} D$. Hence Theorem 3.26 implies $P = \widetilde{Q} D U_{X^T \widetilde{Q} D}^{-1} = \widetilde{Q} D R_1$. ■

Propositions 3.36 and 3.37 indicate how the rank reduction process can be used directly for A-conjugation. We can write the rank reduction process in the following form:

Conjugation via rank reduction
$H_1 = H$
for $i = 1, \dots, k$
$\quad P_i = H_i^T Y_i$
$\quad H_{i+1} = H_i - H_i A^T V_i \left(Y_i^T H_i A^T V_i\right)^{-1} Y_i^T H_i$
end

In general, $P = [P_1, \dots, P_r] = H^T Y L_{Y^T H A^T V}^{-T} = H^T Y U_{V^T A H^T Y}^{-1}$. By Proposition 3.37 we can suppose that $H = I$.

PROPOSITION 3.38 *Let H^- denote any (1)-inverse of H. Then the pair (Q, P) is block H^--biconjugate.*

Proof. By definition

$$P^T H^- Q = L_{Y^T H X}^{-1} Y^T \left(H H^- H\right) X U_{Y^T H X}^{-1} = D_{Y^T H X}.$$

■

Using Proposition 3.38 we can specialize the rank reduction algorithm to biconjugation. If H is a reflexive inverse, that is $H = B^{(1,2)}$ for some B, then the pair (Q, P) is block B-biconjugate. Let $B \in \mathbb{R}^{m \times n}$, X and Y be such that $Y^T B X$ is block strongly nonsingular. Let $H = X \left(Y^T B X\right)^{-1} Y^T$ be the reflexive inverse of B,

$$H_1 = H, \tag{3.68}$$

$$H_{i+1} = H_i - H_i F_i \left(G_i^T H_i F_i\right)^{-1} G_i^T H_i \quad (i = 1, \dots, k) \tag{3.69}$$

and

$$P_i = H_i^T G_i, \quad Q_i = H_i F_i \quad (i = 1, \dots, k). \tag{3.70}$$

As $H = B^{(1,2)}$ holds, the pair (Q, P) is block B-biconjugate provided that $G^T H F$ is block strongly nonsingular, where $F = [F_1, \dots, F_k]$ and $G = [G_1, \dots, G_k]$. Theorem 3.26 implies $P = H^T G L_{G^T H F}^{-T}$ and $Q = H F U_{G^T H F}^{-1}$. If $F = BX$ and $G = B^T Y$, then $G^T H F = Y^T B X$ is block strongly nonsingular and

$$P = Y L_{Y^T B X}^{-T}, \quad Q = X U_{Y^T B X}^{-1}. \tag{3.71}$$

Hence (3.69) implies

$$H_{i+1} = H_1 - \sum_{j=1}^{i} H_j F_j \left(G_j^T H_j F_j\right)^{-1} G_j^T H_j$$

from which

$$P_{i+1} = H_1^T G_{i+1} - \sum_{j=1}^{i} P_j \left(F_j^T H_j^T G_j\right)^{-1} Q_j^T G_{i+1}$$

and

$$Q_{i+1} = H_1 F_{i+1} - \sum_{j=1}^{i} Q_j \left(G_j^T H_j F_j\right)^{-1} P_j^T F_{i+1}$$

follow. As $H_j = H_j H^- H_j = H_j B H_j$ we can write $G_j^T H_j F_j = P_j^T B Q_j$. Also, by the special choice of H, we have $H_1^T G_i = Y_i$ and $H_1 F_i = X_i$. Thus we obtained the following algorithm.

The block two-sided Gram-Schmidt algorithm

$P_1 = Y_1$, $Q_1 = X_1$
for $i = 1, \ldots, k-1$
$\quad P_{i+1} = Y_{i+1} - \sum_{j=1}^{i} P_j \left(Q_j^T B^T P_j\right)^{-1} Q_j^T B^T Y_{i+1}$
$\quad Q_{i+1} = X_{i+1} - \sum_{j=1}^{i} Q_j \left(P_j^T B Q_j\right)^{-1} P_j^T B X_{i+1}$
end

PROPOSITION 3.39 *The pair (Q, P) is block B-biconjugate with P and Q having the form (3.71).*

For more on the Gram-Schmidt method we refer to [193], [87] and [69].

2.4 Other conjugation procedures

There are several conjugation procedures, especially for biconjugation. Most of them are based on Krylov sequences and go with the calculation of the iterates $\{x_k\}$. Concerning these we refer to the cited literature (see, e.g., [352], [193], [306]). Here we recall those algorithms which produce general conjugate directions.

2.4.1 Stewart's conjugation algorithm

Stewart's A-conjugation process is based on the LU factorization $V^T A Q = LS$, where $Q = [q_1, \ldots, q_m]$ is the parameter [325]. Clearly, $\widetilde{P} = QS^{-1}$ defines the A-conjugate pair $\left(\widetilde{P}, V\right)$.

Stewart's conjugation algorithm

$\tilde{p}_1 = q_1$
for $k = 1, \ldots, m$
$\quad$**if** $k > 1$
$\qquad s_k = \left(L^{|\overline{k-1}}\right)^{-1} \left(V^{|k-1}\right)^T A q_k$
$\qquad \tilde{p}_k = \sigma_{kk}^{-1} \left(q_k - \tilde{P}^{|k-1} s_k\right), \quad \sigma_{kk} \neq 0$
$\quad$**end**
$\quad L^{|\overline{k}} = \left(V^{|k}\right)^T A \tilde{P}^{|k}$
end

It is easy to see that

$$\tilde{p}_{k+1} = \sigma_{k+1,k+1}^{-1}\left(I - \tilde{P}^{|k}\left(V^{|kT}A\tilde{P}^{|k}\right)^{-1}V^{|kT}A\right)q_{k+1}.$$

As $\tilde{P} = QS^{-1}$ we can write $\tilde{P}^{|k} = Q^{|k}\left(S^{-1}\right)^{|\overline{k}}$ and

$$\tilde{P}^{|k}\left(V^{|kT}A\tilde{P}^{|k}\right)^{-1}V^{|kT}A = Q^{|k}\left\{V^{|kT}AQ^{|k}\right\}^{-1}V^{|kT}A.$$

Substituting $y_k = \sigma_{kk}^{-1}q_k$ we obtain $\tilde{p}_k = H_k^T y_k$ for $k = 1,\ldots,m$. Thus Stewart's conjugation procedure is algebraically equivalent with the rank reduction conjugation algorithm in the special case $H_1 = I$ and $X = A^T V$. In fact, Stewart's algorithm implicitly uses the canonical form (3.57).

2.4.2 The ABS conjugation algorithm

The *ABS* conjugation procedure of Abaffy, Broyden and Spedicato [1] (see also [4]) explicitly uses the rank reduction procedure.

The ABS conjugation algorithm
$H_1 \in \mathbb{R}^{m\times m}$ is arbitrary nonsingular
for $k = 1,\ldots,r$
  $U_k = H_k^T Z_k$
  $H_{k+1} = H_k - H_k A^T V_k\left(Y_k^T H_k A^T V_k\right)^{-1} Y_k^T H_k$
end

The parameters $Z_k, Y_k \in \mathbb{R}^{m\times m_k}$ must satisfy $\det\left(X_k^T A U_k\right) \neq 0$ and $\det\left(Y_k^T H_k A^T V_k\right) \neq 0$. The calculation of matrices H_k is just the rank reduction procedure with the choice $X_k = A^T V_k$ $(k = 1,\ldots,r)$. If $Z_k = Y_k$ for all $k = 1,\ldots,r$, then the ABS conjugation algorithm is the same as the rank reduction conjugation algorithm with $X_k = A^T V_k$.

PROPOSITION 3.40 *The ABS conjugation algorithm produces A-conjugate pairs (U,V).*

Proof. Let $X = A^T V$. For $i < j$ the relation $X_i \in \mathcal{N}(H_j)$ implies $X_i^T H_j^T = X_i^T H_j^T Z_j = X_i^T U_j = V_i^T A U_j = 0$. Hence the matrix $L = X^T U = V^T A U$ is nonsingular block lower triangular. ■

We seek for conditions under which the above algorithm produces nonsingular A-conjugate pairs (U,V). This happens if the matrices $V_k^T A U_k$ and $Y_k^T H_k A^T V_k \neq 0$ are all nonsingular for $i = 1,\ldots,r$. For simplicity, we use the notation $X = A^T V$. Hence we require $X_k^T U_k$ and $Y_k^T H_k X_k$ to be nonsingular for all i. It follows from Theorem 3.23 that the latter condition holds, if and only if $Y^T H_1 X$ is strongly nonsingular. By definition $X_1^T U_1 = X_1^T H_1^T Z_1$ and

$$X_{k+1}^T U_{k+1} = X_{k+1}^T H_1^T Z_{k+1} - X_{k+1}^T H_1^T Y^{|k}\left(X^{|kT}H_1^T Y^{|k}\right)^{-1}X^{|kT}H_1^T Z_{k+1},$$

which is the Schur complement of the bordered matrix

$$\left(X^{|k+1}\right)^T H_1^T \left[Y^{|k}, Z_{k+1}\right] = \begin{bmatrix} X^{|kT} H_1^T Y^{|k} & X^{|kT} H_1^T Z_{k+1} \\ X_{k+1}^T H_1^T Y^{|k} & X_{k+1}^T H_1^T Z_{k+1} \end{bmatrix}.$$

By the Guttman lemma

$$\operatorname{rank}\left(X_{k+1}^T U_{k+1}\right) = \operatorname{rank}\left(\left(X^{|k+1}\right)^T H_1^T \left[Y^{|k}, Z_{k+1}\right]\right) - \operatorname{rank}\left(X^{|kT} H_1^T Y^{|k}\right).$$

The matrix $X_{k+1}^T U_{k+1}$ is nonsingular, if its rank is equal to its size m_{k+1}. This happens exactly, if the matrix $\left(X^{|k+1}\right)^T H_1^T \left[Y^{|k}, Z_{k+1}\right]$ is nonsingular. Hence we obtained the following result [142].

THEOREM 3.41 *The ABS conjugation algorithm produces a nonsingular A-conjugate pair (U, V), if and only if the matrix $Y^T H_1 A^T V$ is block strongly nonsingular, and the matrices $V_1^T A H_1^T Z_1 \neq 0$, $\left(V^{|k+1}\right)^T A H_1^T \left[Y^{|k}, Z_{k+1}\right]$ $(k = 1, \ldots, r-1)$ are nonsingular.*

3. Perturbation analysis

We study the stability of the finitely terminating Galerkin-Petrov projection (or conjugate direction) methods. We also investigate the stability of conjugation via the rank reduction. The latter is related to the stability of certain triangular factorizations.

3.1 The stability of conjugate direction methods

Concerning the numerical stability of Algorithm CDM we follow Broyden's backward error analysis technique [66]. The material of this section is based on [135] (see also [4]). The basic idea of the stability analysis is the following. For the solution of some problem we consider any finite algorithm of the form

$$X_k = \Psi_k(X_{k-1}) \quad (k = 1, \ldots, r), \tag{3.72}$$

where X_r is the solution of the problem. Assume that an error ε_j occurs at step j and this error propagates further. It is also assumed that no other source of error occurs. The exact solution X_r is given by

$$X_r = \Psi_r \circ \Psi_{r-1} \circ \ldots \circ \Psi_j (X_j) = \Omega_{rj} (X_j), \tag{3.73}$$

while the perturbed solution X_r' is given by

$$X_r' = \Omega_{rj}(X_j + \varepsilon_j). \tag{3.74}$$

If the quantity $\|X_r - X_r'\|$ is large, algorithm (3.72) is considered as unstable.

We use the projector technique of Stewart [325] shown in Section 3.1.1. Let again

$$R_k = U_k \left(V_k^T A U_k\right)^{-1} V_k^T A, \tag{3.75}$$

where the matrix R_k is a projector of rank m_k with $\mathcal{R}(R_k) = \mathcal{R}(U_k)$ and $\mathcal{N}(R_k) = \mathcal{R}^{\perp}(A^T V_k)$. Using notation (3.75) we can rewrite Algorithm CDM in the form

$$x_k = (I - R_k)x_{k-1} + d_k \quad \left(d_k = U_k \left(V_k^T A U_k\right)^{-1} V_k^T b,\ k = 1, \dots, r\right). \tag{3.76}$$

Furthermore let

$$Q_{k,j} = \begin{cases} (I - R_k) \cdots (I - R_j) & (k \geq j), \\ I \quad (k < j). \end{cases} \tag{3.77}$$

The solution of recursion (3.76) is then given by

$$x_r = Q_{r,j+1} x_j + \sum_{i=j+1}^{r} Q_{r,i+1} d_i. \tag{3.78}$$

Suppose that an error occurs at step j $(0 \leq j \leq r-1)$ resulting in x'_j instead of x_j. If this error propagates further, the modified sequence

$$x'_k = (I - R_k)\, x'_{k-1} + d_k \quad (k = j+1, \dots, r) \tag{3.79}$$

is calculated instead of (3.76). Hence we obtain

$$x'_r = Q_{k,j+1} x'_j + \sum_{i=j+1}^{r} Q_{k,i+1} d_i \quad (k = j+1, \dots, r). \tag{3.80}$$

The error occurring in the final step is thus given by

$$\omega - x'_r = x_r - x'_r = Q_{r,j+1}(x_j - x'_j).$$

The matrix $Q_{r,j+1}$ can be considered as the error matrix. Hence we have the error bound

$$\left\| \omega - x'_r \right\| \leq \left\| Q_{r,j+1} \right\| \left\| x_j - x'_j \right\|. \tag{3.81}$$

DEFINITION 3.42 *(Broyden). A particular method of Algorithm CDM is said to be optimally stable, if $\|Q_{r,j+1}\|$ is minimal for all j.*

Here we assume submultiplicative unitarily invariant matrix norm. As the projectors R_k are conjugate in the sense of Definition 3.7, Theorem 3.8 of Stewart implies that $Q_{r,j+1}$ is a projector with

$$\mathcal{R}(Q_{r,j+1}) = \cap_{i=j+1}^{r} \mathcal{N}(R_i), \quad \mathcal{N}(Q_{r,j+1}) = \sum_{i=j+1}^{r} \mathcal{R}(R_i). \tag{3.82}$$

It is easy to see that

$$\mathcal{R}(Q_{r,j+1}) = \cap_{i=j+1}^{r} \mathcal{R}^{\perp}(A^T V_i) = \mathcal{R}^{\perp}(A^T V^{r-j|}) \tag{3.83}$$

and

$$\mathcal{N}(Q_{r,j+1}) = \sum_{i=j+1}^{r} \mathcal{R}(U_i) = \mathcal{R}(U^{r-j|}) = \mathcal{R}^{\perp}(A^T V^{|j}). \tag{3.84}$$

REMARK 3.43 *$Q_{r,j+1} = 0$ for $j = 0$ in agreement with Theorem 3.8. Thus the error propagation can influence the final result only for $j \geq 1$.*

The projector $Q_{r,j+1}$ has minimal norm if and only if it is symmetric . A projector P is symmetric, if and only if $\mathcal{R}(P) = \mathcal{N}^{\perp}(P)$. Thus $Q_{r,j+1}$ is symmetric, if and only if

$$\mathcal{R}(A^T V^{|j}) = \mathcal{R}^{\perp}(A^T V^{r-j|}). \tag{3.85}$$

A method is optimally stable, if and only if (3.85) is satisfied for all j. The latter condition is equivalent to the orthogonality condition

$$A^T V_i \perp A^T V_j \quad (i \neq j). \tag{3.86}$$

In matrix formulation it means that $V^T A A^T V = D$ holds, where D is block diagonal. Thus we proved

THEOREM 3.44 *A method of the Algorithm CDM class is optimally stable, if and only if (3.86), or equivalently $V^T A A^T V = D$ holds with a block diagonal matrix D.*

The result was originally obtained by Broyden [66] in a different way. The projector technique however gives the structure of the error matrix $Q_{r,j+1}$.

An optimally stable method always exists for a given A. Let $A = V \sum U^T$ be the singular value decomposition of A. Then $V^T A A^T V = \sum^2$. However the requirement for the optimal stability is too strong from a practical point of view.

By Theorem 2.9

$$Q_{r,j+1}^T = P_{\mathcal{R}(A^T V^{|j}),\mathcal{R}(A^T V^{r-j|})}.$$

Hence by Lemma 2.29 it can be represented in the form

$$Q_{r,j+1}^T = \left(A^T V\right)^{|j} \left(\left(A^T V\right)^{-1}\right)^{\bar{j}},$$

which clearly gives the representation

$$Q_{r,j+1} = (A^{-1} V^{-T})^{|j} (V^T A)^{\bar{j}}. \tag{3.87}$$

By Lemma 2.73 the bound

$$\|Q_{r,j+1}\| \leq \left\|A^{-1}V^{-T}\right\| \left\|V^T A\right\| = \mathrm{cond}(V^T A) \leq \mathrm{cond}(V)\mathrm{cond}(A) \quad (3.88)$$

holds in any submultiplicative unitarily invariant matrix norm.

Using the inequality (3.81) we can establish

THEOREM 3.45 *For the error propagation model (3.79) the bound*

$$\left\|\omega - x_r'\right\| \leq cond(V^T A) \left\|x_j - x_j'\right\| \leq cond(V) cond(A) \left\|x_j - x_j'\right\| \quad (3.89)$$

holds in any submultiplicative unitarily invariant matrix norm.

REMARK 3.46 *If V is a unitary matrix, then $1 \leq cond(V) \leq m$ holds for unitarily invariant matrix norms generated by normalized symmetric gauge functions (cf. inequality (1.52). Thus the error bound (3.89) is proportional to $cond(A)$. Particularly, $cond(V) = 1$ in the spectral norm.*

Next we define the residual perturbation as $r_k' = A(x_k - x_k')$. Then for the error propagation model (3.79) we have

$$r_{r+1}' = AQ_{r,j+1}A^{-1}r_j'. \quad (3.90)$$

Using the relation $(AB)^{|k}(CD)^{\overline{k}} = A\left(B^{|k}C^{\overline{k}}\right)D$ and (3.87) we can show that

$$AQ_{r,j+1}A^{-1} = \left(V^{-T}\right)^{|j}\left(V^T\right)^{\overline{j}} \quad (3.91)$$

is a projector onto $\mathcal{R}((V^{-T})^{|j})$ along $\mathcal{R}((V^{-T})^{r-j|})$ (cf. Lemma 2.29). Again by Lemma 2.73, $\left\|AQ_{r,j+1}A^{-1}\right\| \leq$cond$(V)$ holds in any submultiplicative unitarily invariant matrix norm. The quantity $\left\|AQ_{r,j+1}A^{-1}\right\|$ is minimal, if and only if $AQ_{r,j+1}A^{-1}$ is symmetric, that is

$$\mathcal{R}\left(\left(V^{-T}\right)^{|j}\right) = \mathcal{R}^{\perp}\left(\left(V^{-T}\right)^{r-j|}\right). \quad (3.92)$$

Relation (3.92) holds for all j if and only if $V^T V = D$ holds with a block diagonal matrix D. So we have

THEOREM 3.47 *For the residual error the inequality*

$$\left\|r_r'\right\| \leq \left\|AQ_{r,j+1}A^{-1}\right\| \left\|r_j'\right\| \leq cond(V) \left\|r_j'\right\| \quad (3.93)$$

holds for all j. The error constant $\left\|AQ_{r,j+1}A^{-1}\right\|$ is minimal for all j, if $V^T V = D$ holds with a suitable block diagonal matrix D.

The structure of Algorithm CDM yields the following simple extension of the error propagation model (3.79). Assume that an ε_k error occurs at each step k and the perturbed recursion (3.76) can be written in the form

$$x_k' = (I - R_k)\left(x_{k-1}' + \varepsilon_{k-1}\right) + d_k \quad (k = 1, \ldots, r). \quad (3.94)$$

Here we assume that the errors ε_k occur independently of each other. Writing (3.94) in the form

$$x'_k = (I - R_k)\, x'_{k-1} + [(I - R_k)\, \varepsilon_{k-1} + d_k] \quad (k = 1, \ldots, r)$$

we get the solution

$$x'_r = Q_{r,1} x'_0 + \sum_{i=1}^{r} Q_{r,i+1} \left[(I - R_i)\, \varepsilon_{i-1} + d_i\right].$$

A comparison with the solution of recursion (3.76) immediately gives the error term

$$\omega - x'_r = \sum_{i=1}^{r} Q_{r,i} \varepsilon_{i-1} \tag{3.95}$$

from which the bound

$$\left\| \omega - x'_r \right\| \leq \operatorname{cond}\left(V^T A\right) \sum_{i=1}^{r} \|\varepsilon_{i-1}\| \leq \operatorname{cond}(V) \operatorname{cond}(A) \sum_{i=1}^{r} \|\varepsilon_{i-1}\| \tag{3.96}$$

follows by (3.88).

THEOREM 3.48 *For the extended error propagation model (3.94) the inequality (3.96) holds.*

We remark that $\|\omega - x'_r\| \leq \sum_{i=1}^{r} \|\varepsilon_{i-1}\|_2$ holds for the optimally stable method.

3.2 The stability of conjugation

We now investigate the stability of the conjugation via rank reduction, which in fact means the stability of the full rank factorization (3.56). The results are based on [146], [147], [148] and [133].

Let $B = L_B \widehat{U}_B$ and $B = \widehat{L}_B U_B$ be the unique $L_1 U$ and $L U_1$ factorizations of B, respectively. Then the components (3.65) can be written as $P^T = \widehat{U}_{Y^T H X} X^{-1}$ and $Q = Y^{-T} \widehat{L}_{Y^T H X}$.

We assume that $H, X, Y \in \mathbb{R}^{m \times m}$ and $Y^T H X$ is strongly nonsingular. If Y, H and X are subject of perturbations $\widetilde{\delta}_Y$, $\widetilde{\delta}_H$ and $\widetilde{\delta}_X$, respectively, then we can write

$$\left(Y^T + \widetilde{\delta}_Y^T\right)\left(H_1 + \widetilde{\delta}_H\right)\left(X + \widetilde{\delta}_X\right) = Y^T (H + \delta_H)\, X.$$

Hence formally we can assume that only H_1 is subject to perturbation and the parameter matrices X and Y are exact or unchanged. Let δ_H be the perturbation of H. Thus $\widetilde{H} = H + \delta_H$,

$$Y^T \widetilde{H} X = Y^T H X + Y^T \delta_H X = Y^T H X + \delta_{Y^T H X}$$

and the perturbed full rank factorization $\widetilde{H} = \widetilde{Q}\widetilde{D}^{-1}\widetilde{P}^T$ reads as

$$\widetilde{P}^T = \widehat{U}_{Y^T\widetilde{H}X}X^{-1},\ \widetilde{D} = D_{Y^T\widetilde{H}X},\ \widetilde{Q} = Y^{-T}\widehat{L}_{Y^T\widetilde{H}X}.$$

The initial assumption implies that only the triangular and the diagonal factors of the LU and LDU decompositions of Y^TH_1X are subject to change in the above expression.

THEOREM 3.49 *Let $B = DQ^{-1}\delta_H P^{-T}$ and $\widetilde{B} = Q^{-1}\delta_H P^{-T}D$. If the perturbation δ_H is such that both Y^THX and $Y^THX + Y^T\delta_H X$ are nonsingular and have LU factorizations, then*

$$\delta P^T = triu(G)\,P^T,\quad \delta D = diag(G)\,D,\quad \delta Q = Q tril\left(\widetilde{G}\right), \tag{3.97}$$

where G and $\widetilde{G}$ are the unique solutions of equations $G = B - tril^*(G)\,B$ and $\widetilde{G} = \widetilde{B} - \widetilde{B}triu^*\left(\widetilde{G}\right)$, respectively. Hence

$$\left|\delta P^T\right| \le triu(|G|)\left|P^T\right|,\quad |\delta D| \le diag(|G|)\,|D|,\quad |\delta Q| \le |Q|\,tril\left(\left|\widetilde{G}\right|\right). \tag{3.98}$$

If $\rho(|B|) < 1$ and $\rho\left(\left|\widetilde{B}\right|\right) < 1$, then

$$\left|\delta P^T\right| \le triu\left(G^{b,1}\right)\left|P^T\right|,\ |\delta D| \le diag\left(G^{b,1}\right)|D|,\ |\delta Q| \le |Q|\,tril\left(\widetilde{G}^{b,1}\right), \tag{3.99}$$

where $G^{b,1}$ and $\widetilde{G}^{b,1}$ are the unique solutions of equations

$$G = |B| + tril^*(G)\,|B| \quad and \quad \widetilde{G} = \left|\widetilde{B}\right| + \left|\widetilde{B}\right| triu^*\left(\widetilde{G}\right), \tag{3.100}$$

respectively.

Proof. Let $Z = Y^THX$, $\delta_Z = Y^T\delta_H X$, $Z = L_Z\widehat{U}_Z$ and

$$Z + \delta_Z = (L_Z + \delta_{L_1})\left(\widehat{U}_Z + \delta_U\right).$$

Then $P^T = \widehat{U}_Z X^{-1}$ and $\widetilde{P}^T = \left(\widehat{U}_Z + \delta_U\right)X^{-1}$, where by Theorem 1.95, $\delta_U = \mathrm{triu}(G)\,\widehat{U}_Z$, $G = B - \mathrm{tril}^*(G)\,B$ and $B = L_Z^{-1}\left(Y^T\delta_H X\right)\widehat{U}_Z^{-1}$. Hence

$$\delta P^T = \widetilde{P}^T - P^T = \delta_U X^{-1} = \mathrm{triu}(G)\,\widehat{U}_Z X^{-1} = \mathrm{triu}(G)\,P^T.$$

Matrix B can also be written as

$$B = L_Z^{-1}Y^T\delta_H X\widehat{U}_Z^{-1} = D_Z\left(D_Z^{-1}L_Z^{-1}Y^T\right)\delta_H\left(X\widehat{U}_Z^{-1}\right).$$

Let $Z = L_Z D_Z U_Z$ and $Z + \delta_Z = (L_Z + \delta_{L_1})(D_Z + \delta_D)(U_Z + \delta_{U_1})$ be the unique LDU factorizations. Then by definition $D = D_Z$, $\widetilde{D} = D_Z + \delta_D$ and $\delta D = \delta_D$. Hence by Theorem 1.97

$$\delta D = \operatorname{diag}(G)\, D,$$

where G is the unique solution of equation $G = B - \operatorname{tril}^*(G)\, B$ and $B = L_Z^{-1}(Y^T \delta_H X)\, \widehat{U}_Z^{-1}$. Finally, let $Z = \widehat{L}_Z U_Z$ and

$$Z + \delta_Z = \left(\widehat{L}_Z + \delta_L\right)(U_Z + \delta_{U_1})$$

be the unique LU_1 factorizations. By definition $Q = Y^{-T}\widehat{L}_Z$ and $\widetilde{Q} = Y^{-T}\left(\widehat{L}_Z + \delta_L\right)$. By noticing that $Z^T = U_Z^T \widehat{L}_Z^T$ is an $L_1 U$ factorization of Z^T we can easily obtain that $\delta_L = \widehat{L}_Z \operatorname{tril}\left(\widetilde{G}\right)$, where $\widetilde{G}$ is the unique solution of equation $G = \widetilde{B} - \widetilde{B}\operatorname{triu}^*(G)$ with $\widetilde{B} = \widehat{L}_Z^{-1}(Y^T \delta_H X)\, U_Z^{-1}$. Hence

$$\delta Q = \widetilde{Q} - Q = Y^{-T}\delta_L = Y^{-T}\widehat{L}_Z \operatorname{tril}(G) = Q \operatorname{tril}(G).$$

We can also write $\widetilde{B}$ in the form

$$\widetilde{B} = \widehat{L}_Z^{-1} Y^T \delta_H X U_Z^{-1} = \left(\widehat{L}_Z^{-1} Y^T\right) \delta_H \left(X U_Z^{-1} D_Z^{-1}\right) D_Z.$$

The rest of the proof follows from Theorems 1.98 and 1.100. ∎

REMARK 3.50 *For $\rho(|B|) < 1$ and $\rho\left(\left|\widetilde{B}\right|\right) < 1$ the upper estimates $G^{b,1} \leq |B|(I - |B|)^{-1}$ and $\widetilde{G}^{b,1} \leq \left(I - \left|\widetilde{B}\right|\right)^{-1}\left|\widetilde{B}\right|$ hold.*

We can observe the strong similarity between the perturbations of full rank factorization (3.56) and the triangular factorizations. Having Theorem 3.26 this comes as no surprise. However we can conclude that the full rank factorization and conjugation algorithm of Egerváry is stable, whenever the corresponding triangular factorization is stable. For norm estimates, see [148].

4. A survey of particular methods

There are plenty of conjugate direction methods and other related methods that can be classified as projection methods. These various conjugate gradient or Krylov subspace methods either minimize an error or project an operator or error. For the history and development of such methods we refer to [126], [163] [7] and [307].

First we classify some cases of the conjugate direction methods in the context of rank reduction conjugation or ABS methods. Then we quote the basic formulas of the Hestenes-Stiefel conjugate gradient method, which initiated the study of the Krylov subspace methods. Since the basic Krylov methods can be found essentially in every numerical linear algebra textbook we refer to the literature for further references. We also mention that papers [352], [193], [67] and [68] try to unify the various Krylov methods.

The ABS method of Abaffy, Broyden and Spedicato [1], [4] is a combination of the conjugate direction method (Algorithm CDM) and the rank reduction based ABS conjugation procedure in the following form:

Algorithm ABS (ABS method)
$x_1 \in \mathbb{R}^m$, $H_1 = I$
for $k = 1, \ldots, m$

$$p_k = H_k^T z_k \quad (z_k \in \mathbb{R}^m,\ p_k^T A^T v_k \neq 0)$$
$$x_{k+1} = x_k - p_k \left(v_k^T A p_k\right)^{-1} v_k^T (A x_k - b)$$
$$H_{k+1} = H_k - H_k A^T v_k \left(w_k^T H_k A^T v_k\right)^{-1} w_k^T H_k \quad (w_k \in \mathbb{R}^m)$$

end

The conjugation procedure becomes the rank reduction conjugation procedure if $w_k = z_k = q_k$ for all $k = 1, \ldots, m$. In that case $P = [p_1, \ldots, p_m] = QU_{V^TAQ}^{-1}$, where $Q = [q_1, \ldots, q_m]$. We now identify three special cases of the algorithm.

Case A: The implicit LU algorithm

The implicit LU algorithm is defined by the parameters $W = Z = I$, when $P = U_A^{-1}$. Thus the algorithm works with the inverse of the upper triangular factor of the LU decomposition of matrix A. It is clear from Theorem 3.23 that A must be strongly nonsingular (see, also [4]). For the special properties of the implicit LU method we refer to Abaffy and Spedicato [4] and Fletcher [124]. Fletcher [124] also gives a historical survey on the implicit LU method and points out that the seemingly different methods of Purcell [118], Pietrzykowski [282], Sloboda [319], Benzi and Meyer [41], and many others are essentially the implicit LU method.

Case B: The Huang method

The Huang method is defined by the parameters $w_i = z_i = A^T v_i$ ($i = 1, \ldots, m$). Hence $P = A^T V U_{V^TAA^TV}^{-1}$ is an orthogonal matrix. The matrix $V^T A A^T V$ is strongly nonsingular if V is nonsingular. The very useful special properties of the Huang method are given in [4]. The Huang method is related to the Gram-Schmidt orthogonalization procedure. This and other connections are studied in [28].

Case C: Symmetrized conjugate direction methods

This subclass of conjugate direction methods is defined by the parameters $W = Z = Q$ and $V = FP$, where the matrix $G = F^T A$ is symmetric and positive definite. For a given A it is always possible to find an orthogonal matrix F such that $G = F^T A$ is symmetric and positive definite. We prove the following result [137].

THEOREM 3.51 *The symmetrized conjugate direction methods can be performed without breakdown if and only if Q is nonsingular. Furthermore*

$$P = QU_{Q^TGQ}^{-1}, \quad V = FQU_{Q^TGQ}^{-1}. \tag{3.101}$$

Proof. Theorem 3.23 implies that the conjugation (and the algorithm) can be performed without breakdown if and only if V^TAQ has the LU factorization $V^TAQ = (L_{V^TAQ}D_{V^TAQ})\, U_{V^TAQ}$. Then $P = QU_{V^TAQ}^{-1}$ and

$$V^TAQ = U_{V^TAQ}^{-T}Q^TF^TAQU_{V^TAQ}^{-1} = U_{V^TAQ}^{-T}Q^TGQU_{V^TAQ}^{-1} = L_{V^TAQ}D_{V^TAQ}.$$

Since Q^TGQ is symmetric and positive definite it follows that

$$L_{V^TAQ}D_{V^TAQ} = D$$

and D is diagonal with positive main diagonal entries. Thus we obtain $Q^TGQ = U_{V^TAQ}^TDU_{V^TAQ}$ which means that Q^TGQ is strongly nonsingular. As G is positive definite this holds if and only if Q is nonsingular. ■

REMARK 3.52 *The symmetrized conjugate direction methods produce A-biorthogonal pairs (P, V) such that $V^TAP = P^TGP = D$, where D is a diagonal matrix.*

The idea of symmetrized conjugate direction methods came from the subclass S3 of the ABS methods [4]. It can be easily seen [137] that the subclasses S2, S3, S4 and S5 of the ABS methods [4] are special cases of the symmetrized conjugate direction methods.

Next we show the conjugate gradient methods of Hestenes and Stiefel [196], [197].

Assume that $A \in \mathbb{R}^{m\times m}$ is a symmetric positive definite matrix, $x_0 \in \mathbb{R}^m$ and $r_0 = b - Ax_0$. The conjugate gradient method of Hestenes and Stiefel [197] computes the iterate x_k $(k \geq 1)$ with the minimization property

$$\|x_k - \omega\|_A = \min_{x\in x_0+K_k(r_0;A)} \|x - \omega\|_A, \tag{3.102}$$

where

$$K_k(r_0; A) = \mathcal{R}\left(\left[r_0, Ar_0, \ldots, A^{k-1}r_0\right]\right) \tag{3.103}$$

is the kth Krylov subspace of $\mathbb{R}^m$ generated by r_0 and $\|\cdot\|_A$ is the weighted Euclidean norm.

Since $\|x - \omega\|_A^2 = x^TAx - 2b^Tx + b^T\omega$, vector x_k also minimizes the strictly convex function $F(x) = \frac{1}{2}x^TAx - b^Tx$ on $x_0 + K_k(r_0; A)$.

The Hestenes-Stiefel conjugate gradient method has the following form: $x_0 \in \mathbb{R}^m$, $r_0 = b - Ax_0$

for $k = 0, 1, \ldots$
$x_{k+1} = x_k + \sigma_k d_k, \quad \sigma_k = r_k^T r_k / d_k^T A d_k$
$r_{k+1} = x_k - \sigma_k A d_k$
$d_{k+1} = r_{k+1} + \beta_k d_k, \quad \beta_k = r_{k+1}^T r_{k+1} / r_k^T r_k$
end

The main properties of the Hestenes-Stiefel algorithm are formulated in

THEOREM 3.53 *(i) There is a first index* m *with* $0 \leq m \leq n$ *such that* $d_m = 0$*. The vectors* $d_0, \ldots, d_{m-1}$ *are linearly independent and* A*-conjugate:*

$$d_i^T A d_j = 0 \quad (0 \leq i < j < m);$$

(ii) $r_i^T r_j = 0$ *for* $0 \leq i < j \leq m$*,* $r_i^T r_i > 0$ *for* $0 \leq i < m$*;*
(iii)

$$d_i^T r_j = \begin{cases} 0, & 0 \leq i < j \leq m, \\ r_i^T r_i, & 0 \leq j \leq i \leq m; \end{cases}$$

(iv) $\mathcal{R}([d_0, d_1, \ldots, d_{k-1}]) = K_k(r_0; A) = \mathcal{R}([r_0, r_1, \ldots, r_{k-1}])$ *for* $1 \leq k \leq m$*;*
(v) $x_m = \omega$*.*

There are many different but mathematically equivalent formulations of the Hestenes-Stiefel method [197]. It is easy to see that the conjugate gradient method is indeed a conjugate direction method.

Finally, we note that the Hestenes-Stiefel and several other conjugate gradient methods have equivalent representations in the framework of the ABS conjugate direction methods (for details, see [4]).

Chapter 4

ITERATIVE PROJECTION METHODS FOR LINEAR ALGEBRAIC SYSTEMS

We consider linear algebraic systems of the form

$$Ax = b \quad \left(A \in \mathbb{R}^{n\times n},\ b \in \mathbb{R}^n\right) \tag{4.1}$$

provided that A is nonsingular. We use the following notation: ω is the exact solution ($\omega = A^{-1}b$), $s_i = x_i - \omega$, $r_i = Ax_i - b = As_i$ and $A^T = [a_1, \ldots, a_n]$. The ith equation of the linear system is given by $a_i^T x = b_i$ or $e_i^T Ax = e_i^T b$.

DEFINITION 4.1 *A sequence $\{x_i\}_{i=0}^{\infty} \subset \mathbb{R}^n$ is said to converge linearly to its limit x (with rate β), if $\beta \in [0,1)$ and there is some $\alpha \geq 0$ such that*

$$\|x_i - x\| \leq \alpha\beta^i \quad (i \geq 0)\,.$$

We use the following technical result (see, e.g., [32]).

THEOREM 4.2 *Suppose $\{x_i\}_{i=1}^{\infty} \subset \mathbb{R}^n$, $x \in \mathbb{R}^n$, and p is some positive integer. If the subsequence $\{x_{pi}\}_{i=1}^{\infty}$ converges linearly to x with rate β and $\{\|x_i - x\|\}_{i=0}^{\infty}$ is decreasing, then the entire sequence $\{x_i\}_{i=1}^{\infty}$ converges linearly to x with rate $\beta^{1/p}$.*

Proof. There exist $\alpha > 0$ and $0 \leq \beta < 1$ such that

$$\|x_{pi} - x\| \leq \alpha\beta^i \quad (i \geq 0)\,.$$

Now fix an arbitrary positive integer m and divide by p with remainder r, i.e.,

$$m = pi + r \quad (r \in \{0, 1, \ldots, p-1\})\,.$$

The monotonicity assumption implies

$$\begin{aligned}\|x_m - x\| \le \|x_{pi} - x\| &\le \alpha\left(\beta^{\frac{1}{p}}\right)^{ip} \\ &= \frac{\alpha\left(\beta^{\frac{1}{p}}\right)^{ip+r}}{\left(\beta^{\frac{1}{p}}\right)^{r}} \le \left(\frac{\alpha}{\left(\beta^{\frac{1}{p}}\right)^{p-1}}\right)\left(\beta^{\frac{1}{p}}\right)^{m},\end{aligned}$$

and the result follows. ■

1. Construction principles

The construction principles of iterative projection methods are quite simple. There are essentially three variations:

1. Iterated projection of a point onto a sequence of hyperplanes (Kaczmarz).
2. Iterated projection of a point onto a sequence of subspaces (Altman).
3. Convex combination of the reflections of the current iteration on a set of hyperplanes (Cimmino).

These principles are also mixed with the introduction of a relaxation parameter to accelerate the convergence of the methods.

The Cimmino and Kaczmarz methods have variants for the convex feasibility problem. For the unification of the construction principles we refer to Censor [77], Bauschke and Borwein [32].

There are various techniques to prove and analyze the convergence of these methods. They are usually based on the observation that certain subsequences of the projection methods satisfy a stationary iteration of the form

$$x_{k+1} = Hx_k + f \quad (k \ge 0), \tag{4.2}$$

where $f \in \mathbb{R}^n$, $H \in \mathbb{R}^{n\times n}$, and $x_0 \in \mathbb{R}^n$ is arbitrary. Hence the rich theory of iterative methods can be applied (see, e.g., [366], [176], or [96]). The matrix H is said to be "weakly convergent" if $\lim_{k\to\infty} H^k$ exists. Let λ_i for $i = 1, \dots, n$, denote the eigenvalues of H. An inspection of the Jordan canonical form of H leads to the following conclusion (see, e.g., [366]). H is weakly convergent whenever the following property holds: $|\lambda_i| \le 1$ for $i = 1, \dots, n$, $|\lambda_i| = 1$ implies $\lambda_i = 1$, and all the elementary divisors that correspond to $\lambda_i = 1$ are linear.

THEOREM 4.3 *The following conditions are equivalent:*
(a) The sequence $\{x_k\}_{k=0}^{\infty}$ *converges for any choice of* x_0.
(b) $\lim_{k\to\infty} \|x_{k+1} - x_k\| = 0$ *for any choice of* x_0.
(c) H *is weakly convergent and the linear system*

$$(I - H)\,x = f \tag{4.3}$$

is solvable.

Condition (b) is due to Dax [96]. For the rest, see [96]. The matrix is said to be "convergent" if $\lim_{k\to\infty} H^k = 0$. The matrix H is convergent if and only if $\rho(H) < 1$. If H is convergent, then the sequence $\{x_k\}_{k=0}^{\infty}$ converges to the same limit $\left(\omega = (I-H)^{-1} f\right)$ for any choice of x_0. In this case the convergence is clearly linear (see, e.g., [366]).

Most of the iterative projection algorithms are related to the alternating projection method studied in Chapter 7. This connection provides a way to prove convergence. It is also pointed out that the Kaczmarz and Cimmino algorithms, the two archetypes of iterative projection methods, have versions developed for the convex feasibility problem studied in Chapter 6. Those proof techniques also apply here.

For rectangular or singular linear systems of the form $Ax = b$ the projection methods of the chapter produce linear least squares solutions. Concerning this and the solution concepts we refer to [338], [96], [186] or [255].

2. Classical projection methods and their extensions

The first known iterative projection method is due to Kaczmarz [215], who published it in 1937. This was followed by a large number of iterative projection methods many of which are variants or extensions of the Kaczmarz method. The growing interest is due to certain applications, where these methods proved to be quite useful (see, e.g., [77]). Next we deal with the most important methods and related theory.

2.1 The method of Kaczmarz

The Kaczmarz algorithm [215] starts from an arbitrary point $x_0 \in \mathbb{R}^n$. The $(j+1)^{\text{th}}$ iteration x_{j+1} is the orthogonal projection of x_j onto the hyperplane $a_i^T x = b_i$. Hence

$$x_{j+1} = x_j + \frac{b_i - a_i^T x_j}{a_i^T a_i} a_i = x_j - \frac{e_i^T r(x_j)}{\|A^T e_i\|_2^2} A^T e_i \quad (j \geq 0).$$

The ith hyperplane is selected cyclically usually starting with the first one. If so, the Kaczmarz method takes the form

$$x_{j+1} = x_j - \frac{e_i^T r(x_j)}{\|A^T e_i\|_2^2} A^T e_i \quad (i \equiv j \pmod{n} + 1). \tag{4.4}$$

The method can also be written in the form

$$x_{j+1} = \left(I - \frac{A^T e_i e_i^T A}{\|A^T e_i\|_2^2}\right) x_j + \frac{b_i}{\|A^T e_i\|_2^2} A^T e_i \quad (i \equiv j \pmod{n} + 1). \tag{4.5}$$

The Kaczmarz method for solving $Ax = b$ is equivalent to the Gauss-Seidel method for the normal equations (see, e.g., [57]). Kaczmarz [215] assumed that A is normalized such that $\left\|A^T e_i\right\|_2 = 1$ for all i. This clearly makes no restriction of any kind, but the formula becomes somewhat simpler. Kaczmarz [215], Bodewig [49], Gastinel [155] and Tanabe [337] proved the convergence of the method without rate. The next result establishes the convergence with rate.

THEOREM 4.4 *If A is nonsingular, the Kaczmarz method is linearly convergent for any initial vector x_0 with rate $(\gamma_{Kac}(A))^{1/n}$, where $0 \leq \gamma_{Kac}(A) < 1$ is a constant.*

Proof. The exact solution ω satisfies the equation

$$\omega = \left(I - \frac{A^T e_i e_i^T A}{\|A^T e_i\|_2^2}\right)\omega + \frac{b_i}{\|A^T e_i\|_2^2} A^T e_i.$$

Hence

$$x_{j+1} - \omega = \left(I - \frac{A^T e_i e_i^T A}{\|A^T e_i\|_2^2}\right)(x_j - \omega)$$

and $\|x_{j+1} - \omega\|_2 \leq \|x_j - \omega\|_2$ for all j. For $k \geq 1$,

$$\begin{aligned} x_{kn} - \omega &= \left(I - \frac{A^T e_n e_n^T A}{\|A^T e_n\|_2^2}\right) \cdots \left(I - \frac{A^T e_1 e_1^T A}{\|A^T e_1\|_2^2}\right)\left(x_{(k-1)n} - \omega\right) \\ &= Q^k (x_0 - \omega), \end{aligned}$$

where

$$Q = \left(I - \frac{A^T e_n e_n^T A}{\|A^T e_n\|_2^2}\right) \cdots \left(I - \frac{A^T e_1 e_1^T A}{\|A^T e_1\|_2^2}\right)$$

is a product of orthogonal projectors. Letting $y_i = A^T e_i / \left\|A^T e_i\right\|_2$ we can apply Meany's result (Theorem 2.77 of Chapter 2). Hence

$$\|Q\|_2 \leq \sqrt{1 - \det\left(Y_n^T Y_n\right)} = \gamma_{Kac}(A) < 1$$

and

$$\|x_{kn} - \omega\|_2 \leq \|Q\|_2^k \|x_0 - \omega\|_2 \leq (\gamma_{Kac}(A))^k \|x_0 - \omega\|_2 .$$

The rest follows from Theorem 4.2. ■

The convergence behavior of the method is obviously independent of the ordering of equations. We note that McCormick [252] also gave a convergence proof for the Kaczmarz method with rate (see, also [21]). His result and proof are essentially the same as those of Gastinel [152] (see Theorem 4.22 or 4.27).

The Kaczmarz method converges even if the linear system is overdetermined and inconsistent. The first result of this kind was obtained by Whitney and Meany [357] for a relaxed version of the Kaczmarz method called single pattern least mean squares algorithm. Later Tanabe [337] proved the similar result for the Kaczmarz method with a similar but refined technique. Here we present Tanabe's result and leave the Whitney and Meany theorem for Section 4.3.1 on the relaxed Kaczmarz methods.

Consider the linear system

$$Ax = b \quad \left(A \in \mathbb{R}^{m\times n},\ m \geq n\right), \tag{4.6}$$

where $\|a_i\| > 0$ is assumed for all i. Let

$$P_i = \frac{a_i a_i^T}{\|a_i\|_2^2} = \frac{a_i a_i^T}{\alpha_i} \quad (i = 1, \ldots, m), \tag{4.7}$$

$Q_{k,j} = (I - P_k)\cdots(I - P_j)$ $(k \geq j)$ and $Q_{k,j} = I$ $(k < j)$. Then recursion (4.5) can be written in the form

$$x_{km+i} = (I - P_i)\, x_{km+i-1} + \frac{b_i}{\alpha_i} a_i \quad (i = 1, \ldots, m,\ k = 0, 1, \ldots).$$

It can be shown (Chapter 3, (3.76), (3.78)) that

$$x_{km} = Q_{m,1} x_{(k-1)m} + \sum_{i=1}^{m} Q_{m,i+1} \frac{b_i}{\alpha_i} a_i.$$

Let the $n \times m$ matrix

$$R = \left[Q_{m,2}\frac{a_1}{\alpha_1}, \ldots, Q_{m,m+1}\frac{a_m}{\alpha_m}\right].$$

Hence the subsequence $\{x_{km}\}_{k=0}^{\infty}$ is produced by the stationary iteration

$$x_{km} = Q_{m,1} x_{(k-1)m} + Rb \quad (k \geq 0). \tag{4.8}$$

LEMMA 4.5 $Q_{m,1} + RA = I$.

Proof. An easy calculation shows that

$$\begin{aligned} RA &= \sum_{i=1}^{m} R e_i e_i^T A = \sum_{i=1}^{n} Q_{m,i+1} \frac{a_i}{\alpha_i} a_i^T = \sum_{i=1}^{m} Q_{m,i+1} P_i \\ &= \sum_{i=1}^{m} \left(Q_{m,i+1} - Q_{m,i}\right) = Q_{m,m+1} - Q_{m,1} = I - Q_{m,1}. \end{aligned}$$

■

It is easy to see that

$$\mathcal{N}(A) = \cap_{i=1}^{m} \mathcal{N}(P_i), \quad \mathcal{R}\left(A^T\right) = \sum_{i=1}^{m} \mathcal{R}(P_i).$$

LEMMA 4.6 $\|Q_{m,1}x\|_2 = \|x\|_2$ *if and only if* $x \in \mathcal{N}(A)$.

Proof. If $x \notin \mathcal{N}(A)$, then there is a number $i_0 \leq m$ such that $P_{i_0}x \neq 0$ and i_0 is the smallest of them. Thus $(I - P_{i_0})x \neq x$ and $\|(I - P_{i_0})x\|_2 < \|x\|_2$. Since each projector has unit norm, we have

$$\|Q_{m,1}x\|_2 = \|Q_{m,i_0+1}Q_{i_0,1}x\|_2 \leq \|Q_{i_0,1}x\|_2 = \|(I - P_{i_0})x_0\|_2 < \|x\|_2.$$

Conversely, if $x \in \mathcal{N}(A)$, then $(I - P_i)x = x$ implies that $Q_{m,1}x = x$. ∎

COROLLARY 4.7 $\|Q_{m,1}\|_2 \leq 1$. *If* $\operatorname{rank}(A) < n$, *then* $\|Q_{m,1}\|_2 = 1$.

Proof. It is obvious that $\|Q_{m,1}\|_2 \leq 1$. If $\dim(\mathcal{N}(A)) = n - \operatorname{rank}(A) > 0$, then there exists an $x \in \mathcal{N}(A)$ such that $\|x\|_2 = 1$ and $Q_{m,1}x = x$. Hence $\|Q_{m,1}\|_2 = 1$. ∎

COROLLARY 4.8 $Q_{m,1}x = x$ *if and only if* $x \in \mathcal{N}(A)$.

Proof. If $x \in \mathcal{N}(A)$, then $Q_{m,1}x = x$. Conversely, if $Q_{m,1}x = x$, then $\|Q_{m,1}x\|_2 = \|x\|_2$ and by Lemma 4.6, $x \in \mathcal{N}(A)$. ∎

THEOREM 4.9 $Q_{m,1}$ *can be decomposed in the form*

$$Q_{m,1} = P_{\mathcal{N}(A)} + \widetilde{Q},$$

where $\widetilde{Q} = Q_{m,1}P_{\mathcal{N}^\perp(A)}$, $\widetilde{Q}P_{\mathcal{N}(A)} = P_{\mathcal{N}(A)}\widetilde{Q} = 0$ *and* $\left\|\widetilde{Q}\right\|_2 < 1$.

Proof. It follows from $I = P_{\mathcal{N}(A)} + P_{\mathcal{N}^\perp(A)}$ that

$$Q_{m,1} = Q_{m,1}P_{\mathcal{N}(A)} + Q_{m,1}P_{\mathcal{N}^\perp(A)}.$$

For any $x \in \mathbb{R}^n$, $Q_{m,1}P_{\mathcal{N}(A)}x = P_{\mathcal{N}(A)}x$. Hence $Q_{m,1}P_{\mathcal{N}(A)} = P_{\mathcal{N}(A)}$. Similarly,

$$\widetilde{Q}P_{\mathcal{N}(A)} = Q_{m,1}P_{\mathcal{N}^\perp(A)}P_{\mathcal{N}(A)} = Q_{m,1}\left(I - P_{\mathcal{N}(A)}\right)P_{\mathcal{N}(A)} = 0.$$

Since $Q_{m,1}^T P_{\mathcal{N}(A)} = P_{\mathcal{N}(A)}$ also holds,

$$\begin{aligned} P_{\mathcal{N}(A)}\widetilde{Q} &= P_{\mathcal{N}(A)}Q_{m,1}P_{\mathcal{N}^\perp(A)} = \left(Q_{m,1}^T P_{\mathcal{N}(A)}\right)^T P_{\mathcal{N}^\perp(A)} \\ &= P_{\mathcal{N}(A)}P_{\mathcal{N}^\perp(A)} = 0. \end{aligned}$$

It is clear that $\left\|\widetilde{Q}\right\|_2 \leq \|Q_{m,1}\|_2 \leq 1$. If $\left\|\widetilde{Q}\right\|_2 = 1$, then there exists $x_0 \neq 0$ such that

$$\left\|\widetilde{Q}x_0\right\|_2 = \left\|Q_{m,1}P_{\mathcal{N}^\perp(A)}x_0\right\|_2 = \|Q_{m,1}z_0\|_2 = \|x_0\|_2 .$$

This x_0 must be in $\mathcal{N}^\perp(A) = \mathcal{R}\left(A^T\right)$. If not, $\left\|P_{\mathcal{N}^\perp(A)}x_0\right\|_2 < \|x_0\|_2$ and $\left\|\widetilde{Q}x_0\right\|_2 < \|x_0\|_2$. Hence $z_0 = x_0$ and $\|Q_{m,1}x_0\|_2 = \|x_0\|_2$. Lemma 4.6 implies that $x_0 \in \mathcal{N}(A)$. Thus $x_0 = 0$, which is a contradiction. ∎

REMARK 4.10 *If* $\operatorname{rank}(A) = n$, *then* $\widetilde{Q} = Q_{m,1}$ *and* $\|Q_{m,1}\|_2 < 1$.

COROLLARY 4.11 $\lim_{i\to\infty} Q^i_{m,1} = P_{\mathcal{N}(A)}$.

Proof. It follows from Theorem 4.9 that $Q^i_{m,1} = P_{\mathcal{N}(A)} + \widetilde{Q}^i$. As $\widetilde{Q}^i \to 0$ we have the requested result. ∎

From relation (4.8) it follows that

$$x_{km} = Q^k_{m,1}x_0 + \left(\sum_{i=0}^{k-1} Q^i_{m,1}R\right) b. \tag{4.9}$$

The behavior of the sum is characterized in

THEOREM 4.12 *The following statements hold.*
(i) $\lim_{k\to\infty} \sum_{i=0}^{k-1} Q^i_{m,1}R$ *exists and* $\left(I - \widetilde{Q}\right)^{-1} R = \sum_{i=0}^{\infty} Q^i_{m,1}R$, *where* R *may not be removed as a factor from the series.*
(ii) The $n \times m$ *matrix* $\left(I - \widetilde{Q}\right)^{-1} R$, *which will be denoted by* G, *is a* $(1,2,4)$*-inverse of* A *such that*

$$AGA = A, \quad GAG = G, \quad (GA)^T = GA.$$

Proof. For any $x \in \mathcal{N}(A)$, we have, for $i = 1, \ldots, m$,

$$x^T Q_{m,i+1}a_i = \left(Q^T_{m,i+1}x\right)^T a_i = x^T a_i = 0.$$

Hence $Re_i \in \mathcal{N}^\perp(A)$. Since $Q^i_{m,1} = P_{\mathcal{N}(A)} + \widetilde{Q}^i$ we can write

$$\sum_{i=0}^{k-1} Q^i_{m,1}R = \sum_{i=0}^{k-1} \widetilde{Q}^i R.$$

As $\left\|\widetilde{Q}\right\|_2 < 1$, we have

$$\sum_{i=0}^{\infty} \widetilde{Q}^i R = \left(\sum_{i=0}^{\infty} \widetilde{Q}^i\right) R = \left(I - \widetilde{Q}\right)^{-1} R.$$

Consequently

$$\left(I-\widetilde{Q}\right)^{-1}R=\sum_{i=0}^{\infty}Q_{m,1}^{i}R.$$

We prove that $G=\left(I-\widetilde{Q}\right)^{-1}R$ is a $(1,2,4)$-inverse of A. We have

$$\begin{aligned}GA&=\left(I-\widetilde{Q}\right)^{-1}RA=\left(I-\widetilde{Q}\right)^{-1}(I-Q)\\&=I-\left(I-\widetilde{Q}\right)^{-1}P_{\mathcal{N}(A)}=I-P_{\mathcal{N}(A)}=P_{\mathcal{R}(A^{T})}\end{aligned}$$

and

$$AGA=A\left(I-P_{\mathcal{N}(A)}\right)=A.$$

Since $Re_i\in\mathcal{N}^{\perp}(A)$,

$$\begin{aligned}GAG&=\left(I-\widetilde{Q}\right)^{-1}RA\left(I-\widetilde{Q}\right)^{-1}R\\&=\left(I-\widetilde{Q}\right)^{-1}(I-Q)\left(I-\widetilde{Q}\right)^{-1}R\\&=\left(I-\widetilde{Q}\right)^{-1}\left(I-P_{\mathcal{N}(A)}\right)R\\&=\left(I-\widetilde{Q}\right)^{-1}R=G.\end{aligned}$$

■

REMARK 4.13 *G is not necessarily the Moore-Penrose inverse of A. Consider the inconsistent linear system*

$$Ax=\begin{bmatrix}1&0\\0&1\\1&0\end{bmatrix}x=\begin{bmatrix}1\\0\\0\end{bmatrix}.$$

Then $Q_{3,1}=0$ *and*

$$R=\begin{bmatrix}0&0&1\\0&1&0\end{bmatrix}.$$

Since $\operatorname{rank}(A)=2$, $\widetilde{Q}=Q_{3,1}=0$ *and* $G=R$. *The third Penrose equation is not satisfied because the matrix*

$$AG=\begin{bmatrix}0&0&1\\0&1&0\\0&0&1\end{bmatrix}$$

is not symmetric.

For a similar observation, see Theorem 4 of Whitney and Meany [357]. Finally, we prove the convergence of the Kaczmarz method.

COROLLARY 4.14 *For any matrix $A \in \mathbb{R}^{m\times n}$ with nonzero rows and any m-dimensional column vector b, the algorithm (4.8) generates a convergent sequence of vectors $\{x_{km}\}_{k=0}^{\infty}$ such that*

$$\lim_{k\to\infty} x_{km} = P_{\mathcal{N}(A)}x_0 + Gb, \tag{4.10}$$

where $x_0 \in \mathbb{R}^n$ is an arbitrary vector.

Proof. It follows from Corollary 4.11 and Theorem 4.12 that the first and the second terms of expression (4.9) converge to $P_{\mathcal{N}(A)}x_0$ and Gb, respectively, as k tends to infinity. ■

The linear system has a solution if and only if $AGb = b$. In this case $P_{\mathcal{N}(A)}x_0 + Gb$ is a solution of the system for arbitrary x_0, and Gb is the minimum norm solution. It is clear that the whole sequence $\{x_i\}_{i=0}^{\infty}$ is linearly convergent.

Both convergence theorems of this section used the fact that the subsequence $\{x_{km}\}_{k=0}^{\infty}$ satisfies a stationary iteration of the form

$$x_{km} = Qx_{(k-1)m} + c.$$

Tanabe [338] gave useful characterizations of the class of the linear stationary iterative processes for solving an inconsistent or consistent singular system of linear equations. A somewhat simplified version of the previous convergence analysis for a wider class of projection methods can be found in Maess [241].

Assume again that $A \in \mathbb{R}^{n\times n}$ and A is nonsingular. Vasil'chenko and Svetlanov [351] developed the following block generalization of the Kaczmarz method for the equation $Ax = b$. Let A and b be row partitioned in the form

$$A = \begin{bmatrix} A_1^T \\ A_2^T \\ \vdots \\ A_m^T \end{bmatrix}, \quad b = \begin{bmatrix} b_1 \\ b_2 \\ \vdots \\ b_m \end{bmatrix} \quad \left(A_i^T \in \mathbb{R}^{n_i\times n},\ b_i \in \mathbb{R}^{n_i}\right). \tag{4.11}$$

Consider the subsystem

$$A_j^T x_j = b_j, \tag{4.12}$$

which has maximum row rank. We seek a particular solution of the form $x_j = x_{j-1} + A_j z$. Substituting this into the system we get

$$A_j^T\left(x_{j-1} + A_j z\right) = b_j, \quad z = \left(A_j^T A_j\right)^{-1}\left(b_j - A_j^T x_{j-1}\right)$$

and

$$x_j = x_{j-1} + A_j\left(A_j^T A_j\right)^{-1}\left(b_j - A_j^T x_{j-1}\right) = x_{j-1} - A_j^+\left(A_j^T x_{j-1} - b_j\right).$$

Taking the blocks in cyclic order we obtain the block Kaczmarz method

$$x_{j+1} = x_j - \left(A_i^T\right)^+ \left(A_i^T x_j - b_i\right) \quad (i \equiv j \pmod{m} + 1). \tag{4.13}$$

For $m = n$, the method coincides with the Kaczmarz method.

THEOREM 4.15 *If $A \in \mathbb{R}^{n\times n}$ is nonsingular, then method (4.13) is convergent for any $x_0 \in \mathbb{R}$.*

Proof. For the absolute errors we have the recursion

$$s_{j+1} = \left(I - \left(A_i^T\right)^+ A_i^T\right) s_j = P_{\mathcal{N}(A_i^T)} s_j$$

and for $j = km$,

$$s_{j+m} = P_{\mathcal{N}(A_m^T)} \cdots P_{\mathcal{N}(A_1^T)} s_j.$$

The rest follows from the von Neumann-Halperin theorem of Section 7.3.1. ■

Tewarson [339] investigated the implementation of the Kaczmarz method on sparse linear systems, while McCormick [252] extended it to Hilbert spaces.

2.2 The method of Cimmino

While the Kaczmarz method cyclically projects the current approximation to the next hyperplane, x_j is (orthogonally) reflected to all hyperplanes in the Cimmino method [88]:

$$\delta_i = x_j - 2\frac{e_i^T r(x_j)}{\|A^T e_i\|_2^2} A^T e_i \quad (i = 1, \ldots, n).$$

The new approximation x_{j+1} is taken as the center of the n reflected points:

$$x_{j+1} = x_j - \frac{2}{n}\sum_{i=1}^{n} \frac{e_i^T r(x_j)}{\|A^T e_i\|_2^2} A^T e_i \quad (j \geq 0). \tag{4.14}$$

The Cimmino algorithm can be written in different alternative forms. Let $D = \operatorname{diag}\left(1/\|A^T e_i\|_2^2\right)$. Then

$$\begin{aligned} x_{j+1} &= x_j - \frac{2}{n}\sum_{i=1}^{n} \frac{e_i^T (Ax_j - b)}{\|A^T e_i\|_2^2} A^T e_i \\ &= x_j - \frac{2}{n}\left(\sum_{i=1}^{n} \frac{A^T e_i e_i^T}{\|A^T e_i\|_2^2}\right) Ax_j + \frac{2}{n}\sum_{i=1}^{n} b_i \frac{A^T e_i}{\|A^T e_i\|_2^2} \\ &= \left(I - \frac{2}{n} A^T D A\right) x_j + \frac{2}{n} A^T D b, \end{aligned} \tag{4.15}$$

or

$$x_{j+1} = x_j - \frac{2}{n}(DA)^T(Ax_j - b). \tag{4.16}$$

Thus the Cimmino method is a stationary iteration. The following convergence result is taken from Gastinel [155] (see, also Bodewig [49]).

THEOREM 4.16 *Assume that $A \in \mathbb{R}^{n\times n}$ is nonsingular and $\lambda_i\left(A^TDA\right) < n$ holds for all i. The Cimmino method is convergent linearly for any $x_0 \in \mathbb{R}^n$.*

Proof. The Cimmino algorithm (4.15) has the form (4.2). So we need to prove only

$$\rho\left(I - \frac{2}{n}A^TDA\right) < 1.$$

By definition $\left\|xy^T\right\|_2 = \|x\|_2\|y\|_2$. Hence

$$\left\|A^TDA\right\|_2 \le \sum_{i=1}^{n}\frac{\left\|A^Te_ie_i^TA\right\|_2}{\|A^Te_i\|_2^2} = n$$

implying that any eigenvalue of A^TDA satisfies $0 < \lambda_i\left(A^TDA\right) \le n$. It follows that

$$\lambda_i\left(I - \frac{2}{n}A^TDA\right) = 1 - \frac{2}{n}\lambda_i\left(A^TDA\right).$$

Hence the condition $\lambda_i\left(A^TDA\right) < n$ implies $\left|\lambda_i\left(I - (2/n)A^TDA\right)\right| < 1$ for all i. ■

Some authors (see, e.g., [57]) consider the Cimmino method as a variant of the Kaczmarz algorithm. If we equip the reflected points with positive weights w_i ($\sum_{i=1}^n w_i = 1$) and define x_{j+1} as the gravity center of these mass points, then we obtain the weighted Cimmino algorithm

$$x_{j+1} = x_j - 2\sum_{i=1}^{n} w_i\frac{e_i^T(Ax_j - b)}{\|A^Te_i\|_2^2}A^Te_i \quad (j \ge 0). \tag{4.17}$$

Ansorge [21] proved the convergence of this Cimmino algorithm when each w_i varies with j in a given special way. Otherwise the convergence follows from the convergence theory of the Cimmino method given in Chapter 6.

Block versions of the Cimmino algorithm were used as preconditioner by Bramley and Sameh [53], and Arioli, Duff and Sadkane [111].

2.3 The method of Altman

The less known algorithm of Altman [12] is similar to the Kaczmarz method. It is based on a cyclic sequence of projections and a summation. Let $y_0 = b$, $x_0 = 0$,

$$P_i = \frac{Ae_ie_i^TA^T}{\|Ae_i\|_2^2} \quad (i = 1,\ldots,n), \tag{4.18}$$

and define the sequences

$$y_{j+1} = y_j - P_i y_j \quad (i \equiv j \pmod n + 1) \tag{4.19}$$

and

$$x_{j+1} = x_j + \frac{e_i^T A^T y_j}{\|Ae_i\|_2^2} e_i \quad (i \equiv j \pmod n + 1) \tag{4.20}$$

for $j \geq 0$.

Comparing the Altman algorithm with the Kaczmarz method we can observe that the iterate y_{j+1} lies in $\mathcal{R}^{\perp}(Ae_i)$ and not in the hyperplane $a_i^T x = b_i$. The next proof will reveal that $x_j = A^{-1}(b - y_j)$.

THEOREM 4.17 *If A is nonsingular, the Altman method is linearly convergent for any initial vector x_0 with rate $(\gamma_{Alt}(A))^{1/n}$, where $0 \leq \gamma_{Alt}(A) < 1$ is a constant.*

Proof. We first show that $y_j \to 0$. Relation $y_{j+1} = (I - P_i) y_j$ implies that $\|y_{j+1}\|_2 \leq \|y_j\|_2$. So it suffices to see the subsequence $\{y_{kn}\}_{k=0}^{\infty}$. For $k \geq 1$,

$$y_{kn} = (I - P_n) \cdots (I - P_1) y_{(k-1)n} = Q y_{(k-1)n} = Q^k y_0.$$

Let $z_i = Ae_i / \|Ae_i\|_2$ and apply the Meany inequality (Theorem 2.77 of Chapter 2). It follows that $\|Q\|_2 \leq \sqrt{1 - \det(Z_n^T Z_n)} = \gamma_{Alt}(A) < 1$ and

$$\|y_{kn}\|_2 \leq \|Q\|_2^k \|y_0\|_2 \leq (\gamma_{Alt}(A))^k \|y_0\|_2 .$$

Thus $y_n \to 0$. By definition $P_i y_j = y_j - y_{j+1}$ and

$$\begin{aligned} Ax_{j+1} &= Ax_j + P_i y_j = Ax_j + (y_j - y_{j+1}) \\ &= Ax_0 + \sum_{i=0}^{j} (y_j - y_{j+1}) \\ &= y_0 - y_{j+1}. \end{aligned}$$

Hence $Ax_{j+1} = b - y_{j+1} \to b$ and $x_j \to \omega = A^{-1}b$. As $x_{j+1} - \omega = -A^{-1} y_{j+1}$, the speed of convergence follows from the estimate

$$\|x_{kn} - \omega\|_2 \leq \|A^{-1}\|_2 \|y_{kn}\|_2 \leq \|A^{-1}\|_2 (\gamma_{Alt}(A))^k \|b\|_2 . \tag{4.21}$$

■

Altman [12] proved the convergence without rate. He gave however a new representation of the orthogonal projections and applied the method to the linear least squares problem [11]. His results can be outlined as follows

THEOREM 4.18 *Suppose that $A \in \mathbb{R}^{m \times n}$ $(n \leq m)$ is of full rank and $P_i = Ae_i e_i^T A^T / \|Ae_i\|_2^2$ $(i = 1, \ldots, n)$. Let $Q = (I - P_n) \cdots (I - P_1)$. Then $Q^i \to I - P_{\mathcal{R}(A)}$ linearly with rate $\gamma_{Alt}(A)$, where $0 \leq \gamma_{Alt}(A) < 1$ is a constant.*

Proof. Define $Z = [z_1, \ldots, z_n]$ with $z_i = Ae_i / \|Ae_i\|_2$ and consider the sequence

$$y_{j+1} = y_j - P_i y_j = (I - P_i)\, y_j \quad (i \equiv j \;(\operatorname{mod} n) + 1).$$

For any $y_0 \in \mathbb{R}^m$ the sequence $\{\|y_j\|\}_{j=0}^{\infty}$ is monotone decreasing. If $y_0 \in \mathcal{R}(A)$, then $\{y_j\}_{j=0}^{\infty} \subset \mathcal{R}(A)$,

$$\|y_{jn}\|_2 = \left\|Qy_{(j-1)n}\right\|_2 \le \sqrt{1 - \det\left(Z^T Z\right)}\, \left\|y_{(j-1)n}\right\|_2 = \gamma_{Alt}(A)\, \left\|y_{(j-1)n}\right\|_2$$

and

$$\|y_{jn}\|_2 \le \left(\gamma_{Alt}(A)\right)^j \|y_0\|_2 .$$

Hence $y_j \to 0$ linearly with rate $\left(\gamma_{Alt}(A)\right)^{1/n}$ provided that $y_0 \in \mathcal{R}(A)$. Assume that $z \in \mathbb{R}^m$ is arbitrary and $z = u + v$, where $u \in \mathcal{R}(A)$ and $v \in \mathcal{R}^{\perp}(A)$. Define the sequences $\widehat{y}_0 = u$, $y_0 = z$,

$$\widehat{y}_{j+1} = (I - P_i)\, \widehat{y}_j, \quad y_{j+1} = (I - P_i)\, y_j \quad (i \equiv j \;(\operatorname{mod} n) + 1),$$

respectively. Since $v \in \mathcal{R}^{\perp}(A)$, $y_j = \widehat{y}_j + v$. It follows that $y_j \to v = \left(I - P_{\mathcal{R}(A)}\right) z$ linearly with rate $\left(\gamma_{Alt}(A)\right)^{1/n}$. Thus for any $z \in \mathbb{R}^m$,

$$\begin{aligned} \left\|\left[Q^l - \left(I - P_{\mathcal{R}(A)}\right)\right] z\right\|_2 &= \left\|Q^l z - v\right\|_2 = \|\widehat{y}_{ln}\|_2 \le \left(\gamma_{Alt}(A)\right)^l \|\widehat{y}_0\|_2 \\ &\le \left(\gamma_{Alt}(A)\right)^l \|z\|_2 \end{aligned}$$

implying that

$$\left\|Q^l - \left(I - P_{\mathcal{R}(A)}\right)\right\|_2 \le \left(\gamma_{Alt}(A)\right)^l .$$

■

Consider now the overdetermined linear system

$$Ax = b \quad \left(A \in \mathbb{R}^{m \times n},\, m > n\right)$$

where A has maximum rank. Let $y_0 = b$, $x_0 = 0 \in \mathbb{R}^n$,

$$y_{j+1} = (I - P_i)\, y_j, \quad x_{j+1} = x_j + \frac{e_i^T A y_j}{\|Ae_i\|_2^2} e_i \quad (i \equiv j \;(\operatorname{mod} n) + 1).$$

Since $P_i y_j = y_j - y_{j+1}$, we have

$$Ax_{j+1} = Ax_j + P_i y_j = Ax_j + (y_j - y_{j+1}) = y_0 - y_{j+1} \to P_{\mathcal{R}(A)} y_0 .$$

It follows that $x_{j+1} = \left(A^T A\right)^{-1} A^T (y_0 - y_{j+1}) \to \left(A^T A\right)^{-1} A^T b$, which is the least squares solution of the overdetermined system. Hence we obtained

THEOREM 4.19 *The Altman method converges to the least squares solution linearly if A has maximum column rank.*

REMARK 4.20 *Both convergence results can be collected into one, if the property $Q^l \to \left(I - P_{\mathcal{R}(A)}\right)$ is used.*

Popa [289] suggested an extension of the Kaczmarz method to handle the inconsistency of $Ax = b$ ($A \in \mathbb{R}^{m\times n}$). The method, which is apparently a combination of the Altman and Kaczmarz methods, can be written in the form

$$\begin{aligned} y_{k+1} &= \phi\left(y_k\right), \\ b^{k+1} &= b - y_{k+1}, \\ x_{k+1} &= K\left(x_k, b^{k+1}\right). \end{aligned}$$

Here $y_0 = b$, $x_0 = 0$, and

$$\phi\left(y\right) = \phi_1 \circ \phi_2 \circ \ldots \circ \phi_n\left(y\right) \quad \left(y \in \mathbb{R}^n\right), \quad \phi_j\left(y\right) = y - \frac{e_j^T A^T y}{\|Ae_j\|_2^2} Ae_j.$$

Furthermore $K\left(x_k, b^{k+1}\right)$ is one major iteration step with the Kaczmarz method, using the right-hand side vector b^{k+1}. Popa showed that $x_k \to A^+ b$ provided that A has full rank. Elfving [117] generalized the idea to other iterative procedures. For other extensions, see Popa [290], [291].

2.4 The projection methods of Gastinel

Gastinel's class of projection methods was developed in a sequence of papers [152], [153],[154], [155], [156]. It exploits two ideas, the first of which can be found in Householder [205].

If $v_j \in \mathbb{R}^n$ ($v_j \neq 0$) is any vector, then equation

$$v_j^T A x = v_j^T b$$

is also a hyperplane passing through the point ω. The new iteration x_{j+1} will be the orthogonal projection of x_j onto this hyperplane. Hence we obtain the class of projection methods

$$x_{j+1} = x_j - \frac{v_j^T r\left(x_j\right)}{\|A^T v_j\|_2^2} A^T v_j \quad (j \geq 0) \tag{4.22}$$

due to Householder [205]. Note that $x_{j+1} - x_j$ is orthogonal to $x_{j+1} - \omega$. The Kaczmarz method is a special case of the Householder class, if v_j is taken to be e_i cyclically. For the choice $v_j = r\left(x_j\right)$ we obtain the steepest descent or gradient method (see, e.g., [205], [206], [155], and [57]). Using the technique of Theorems 4.4 and 4.17 we can prove the convergence of algorithm (4.22) for certain parameter choices.

THEOREM 4.21 *Assume that A and $V = [v_1, \ldots, v_n]$ are nonsingular. If the vectors v_j are selected cyclically so that $v_j = v_i$ $(i \equiv j \ (\mathrm{mod}\, n) + 1)$, then algorithm (4.22) is convergent linearly for any x_0 with rate $(\gamma_{Hou}(A, V))^{1/n}$, where $0 \leq \gamma_{Hou}(A, V) < 1$ is a constant.*

Proof. Taking the difference of

$$x_{j+1} = x_j - \frac{v_j^T r(x_j)}{\|A^T v_j\|_2^2} A^T v_j$$

and

$$\omega = \omega - \frac{v_j^T r(\omega)}{\|A^T v_j\|_2^2} A^T v_j$$

we obtain

$$x_{j+1} - \omega = \left(I - \frac{A^T v_j v_j^T A}{\|A^T v_j\|_2^2}\right)(x_j - \omega).$$

Again we have $\|x_{j+1} - \omega\|_2 \leq \|x_j - \omega\|_2$,

$$\begin{aligned} x_{kn} - \omega &= \left(I - \frac{A^T v_n v_n^T A}{\|A^T v_n\|_2^2}\right) \cdots \left(I - \frac{A^T v_1 v_1^T A}{\|A^T v_1\|_2^2}\right)\left(x_{(k-1)n} - \omega\right) \\ &= Q\left(x_{(k-1)n} - \omega\right), \end{aligned}$$

and by the Meany inequality $\|Q\|_2 \leq \gamma_{Hou}(A, V) < 1$. Hence

$$\|x_{kn} - \omega\|_2 \leq \|Q\|_2^k \|x_0 - \omega\|_2 \leq (\gamma_{Hou}(A, V))^k \|x_0 - \omega\|_2.$$

■

We investigate the convergence of two particular cases in the next section. Assume that $v = v(r)$ and

$$\phi(r) = v(r)^T r = \|r\| \tag{4.23}$$

holds for some norm $\|.\|$. Such choice of $v(r)$ was called *norm decomposition* by Gastinel who gave the following particular choices. Let

$$\mathrm{sign}(x) = \begin{cases} 1, & x \geq 0, \\ -1, & x < 0. \end{cases}$$

Then the norm decompositions are given by

$$v(r) = [\mathrm{sign}(r_1), \ldots, \mathrm{sign}(r_n)]^T \tag{4.24}$$

for $\phi(r) = \|r\|_1 = \sum_{i=1}^{n} |r_i|$,

$$v(r) = [0, \ldots, 0, \operatorname{sign}(r_i), 0, \ldots, 0]^T, \quad |r_i| = \max_{1 \le j \le n} |r_j| \tag{4.25}$$

for $\phi(r) = \|r\|_\infty = \max_{1 \le i \le n} |r_i|$, and

$$v(r) = \frac{r}{\|r\|_2} \tag{4.26}$$

for $\phi(r) = \|r\|_2 = \left(\sum_{i=1}^{n} |r_i|^2\right)^{1/2}$. Note that

$$\|v(r)\|_1 \le n, \quad \|v(r)\|_\infty \le 1, \quad \|v(r)\|_2 \le 1 \tag{4.27}$$

hold for the above choices.

Using the concept of norm decomposition Gastinel defined the iterative projection method

$$x_{j+1} = x_j - \frac{\phi(r(x_j))}{\|A^T v(r(x_j))\|_2^2} A^T v(r(x_j)) \quad (j = 0, 1, \ldots) \tag{4.28}$$

for $p = 1, 2, \infty$.

THEOREM 4.22 *(Gastinel). If A is nonsingular, the iteration method (4.28) is convergent linearly for any $x_0 \in \mathbb{R}^n$ with the speed*

$$\|x_j - \omega\|_2 \le \left(\sqrt{1 - \frac{K^2}{cond_2^2(A)}}\right)^j \|x_0 - \omega\|_2, \tag{4.29}$$

where $K > 0$ is a constant depending on ϕ.

Proof. Let $v_j = v(r_j)$. As $x_{j+1} - x_j$ is orthogonal to $x_{j+1} - \omega$ we can write

$$\begin{aligned}\|x_{j+1} - \omega\|_2^2 &= \|x_j - \omega\|_2^2 - \|x_{j+1} - x_j\|_2^2 \\ &= \|x_j - \omega\|_2^2 - \frac{\phi^2(r_j)}{\|A^T v_j\|_2^2} \\ &= \|x_j - \omega\|_2^2 \left(1 - \frac{\|r_j\|_p^2}{\|x_j - \omega\|_2^2} \cdot \frac{1}{\|A^T v_j\|_2^2}\right).\end{aligned}$$

The norm equivalence theorem implies $\|r_j\|_p \ge K \|r_j\|_2$ $(K > 0)$ and

$$\frac{\|r_j\|_p}{\|x_j - \omega\|_2} \ge K \frac{\|A(x_j - \omega)\|_2}{\|x_j - \omega\|_2} \ge \frac{K}{\|A^{-1}\|_2}.$$

Also $\left\|A^T v_j\right\|_2 \leq \|A\|_2$. Hence

$$\frac{\|r_j\|_p^2}{\|x_j-\omega\|_2^2}\cdot\frac{1}{\|A^T v_j\|_2^2}\geq\frac{K^2}{\|A^{-1}\|_2^2}\cdot\frac{1}{\|A\|_2^2}$$

and

$$\|x_{j+1}-\omega\|_2^2\leq\|x_j-\omega\|_2^2\left(1-\frac{K^2}{\mathrm{cond}_2^2\,(A)}\right).$$

■

Gastinel [154] also gave a block generalization of his algorithm, which corresponds to (4.42) due to Householder and Bauer [207]. Gastinel and Joly [156] extended the Gastinel algorithm to Banach spaces.

2.5 The projection methods of Householder and Bauer

Using earlier results of Householder [205] Householder and Bauer [207], [206] developed a general class of projection methods that contained all the algorithms known at the time. They took a variational approach that can be summarized as follows (for more on the variational approach, see Saad [306] and Brezinski [56], [57]).

Let $Y_j \in \mathbb{R}^{n\times n_j}$ $(n_j \leq n)$ be a parameter matrix of maximum rank. The next iteration x_{j+1} is selected such that

$$x_{j+1}=x_j+\delta_j\quad(\delta_j\in\mathcal{R}\,(Y_j))\tag{4.30}$$

and

$$\|x_{j+1}-\omega\|_G\to\min,\tag{4.31}$$

where $\|z\|_G=\left(z^TGz\right)^{1/2}$ is a weighted (elliptic) norm and $G\in\mathbb{R}^{n\times n}$ is a symmetric positive definite matrix. As $\delta_j = Y_j u_j$ for some vector u_j we have to solve the weighted linear squares problem

$$\|x_{j+1}-\omega\|_G^2=\|(x_j-\omega)+Y_ju_j\|_G^2\to\min.\tag{4.32}$$

A vector x minimizes $\|Ax-b\|_W$ if and only if x is a solution of $A^TGAx = A^TGb$. Since $Y_j^TGY_j$ is nonsingular, the minimizer of (4.32) is given by $u_j=-\left(Y_j^TGY_j\right)^{-1}Y_j^TG\,(x_j-\omega)$. Hence

$$x_{j+1}=x_j-Y_j\left(Y_j^TGY_j\right)^{-1}Y_j^TG\,(x_j-\omega).$$

Note that

$$x_{j+1}-\omega=\left(I-Y_j\left(Y_j^TGY_j\right)^{-1}Y_j^TG\right)(x_j-\omega),$$

that is the error vector $x_j - \omega$ is projected onto $\mathcal{R}^{\perp}(GY_j)$ along $\mathcal{R}(Y_j)$. It is now assumed that $\mathcal{R}(GY_j) \subset \mathcal{R}(A^T)$. This holds if and only if $GY_j = A^T V_j$ for some matrix V_j. Substituting this choice we have the algorithm

$$x_{j+1} = x_j - Y_j \left(V_j^T A Y_j\right)^{-1} V_j^T r_j, \tag{4.33}$$

which is the general form of the Householder, Bauer class of projection methods.

It is easy to see that both

$$s_{j+1} = \left(I - Y_j \left(V_j^T A Y_j\right)^{-1} V_j^T A\right) s_j \tag{4.34}$$

and

$$r_{j+1} = \left(I - AY_j \left(V_j^T A Y_j\right) V_j^T\right) r_j \tag{4.35}$$

are oblique projections of s_j and r_j, respectively.

Since $GY_j = A^T V_j$,

$$\begin{aligned} \|s_{j+1}\|_G^2 &= s_j^T \left(I - GY_j \left(Y_j^T G Y_j\right)^{-1} Y_j^T\right) G s_{j+1} \\ &= s_j^T \left(I - GY_j \left(Y_j^T G Y_j\right)^{-1} Y_j^T\right) G s_j \\ &= \|s_j\|_G^2 - s_j^T A^T V_j \left(V_j^T A Y_j\right)^{-1} V_j^T A s_j & (4.36) \\ &= \|s_j\|_G^2 - r_j^T V_j \left(V_j^T A Y_j\right)^{-1} V_j^T r_j. & (4.37) \end{aligned}$$

The Householder, Bauer class may also be obtained in the following way (see, also Gastinel [154]). The exact solution ω satisfies the hyperplanes

$$V_j^T A (x - \omega) = 0.$$

The next iteration x_{j+1} is the common point of these hyperplanes and the manifold

$$y = x_j + Y_j z \quad \left(z \in \mathbb{R}^{n_j}\right)$$

parallel to $\mathcal{R}(Y_j)$ through x_j. The common point x_{j+1} satisfies the equation

$$V_j^T A (x_j + Y_j z) = V_j^T b$$

whose solution is $z = -\left(V_j^T A Y_j\right)^{-1} V_j^T r_j$ provided that $V_j^T A Y_j$ is nonsingular. Hence

$$x_{j+1} = x_j - Y_j \left(V_j^T A Y_j\right)^{-1} V_j^T r_j,$$

which is exactly the algorithm (4.33).

There is no general convergence result for the class (4.33). We have only particular results.

Case A. If $G = A^T A$, then $V_j = AY_j$ and the method (4.33) takes the form

$$x_{j+1} = x_j - Y_j \left(Y_j^T A^T A Y_j\right)^{-1} Y_j^T A^T r_j. \tag{4.38}$$

In this case

$$r_{j+1} = \left(I - AY_j \left(Y_j^T A^T A Y_j\right)^{-1} Y_j^T A^T\right) r_j, \tag{4.39}$$

that is r_{j+1} is the orthogonal projection of r_j. Hence the residual error is reduced at each step in the subclass.

If Y_j is taken to be a single vector, y_j, the method changes to

$$x_{j+1} = x_j - \frac{y_j^T A^T r_j}{\|Ay_j\|_2^2} y_j \quad (j \geq 0), \tag{4.40}$$

which is essentially the projection algorithm of de la Garza (see [206], p. 116). Householder and Bauer suggest the parameter choice $y_j = e_i$. This can be done either in rotation or by selecting the particular e_i that maximizes the quantity

$$\|s_j\|_G^2 - \|s_{j+1}\|_G^2 = \frac{\left(y_j^T A^T r_j\right)^2}{\|Ay_j\|_2^2}. \tag{4.41}$$

THEOREM 4.23 *Assume that A and $Y = [Y_1, \ldots, Y_m]$ are nonsingular. If the matrices Y_j are selected cyclically so that $Y_j = Y_i$ $(i \equiv j \ (mod\, m) + 1)$, then algorithm (4.38) is convergent for any x_0*

Proof. Since $r_{j+1} = P_{\mathcal{R}^\perp(AY_j)} r_j$ and $r_{j+m} = P_{\mathcal{R}^\perp(AY_m)} \cdots P_{\mathcal{R}^\perp(AY_1)} r_j$ for $j = km$, the rest follows from the von Neumann-Halperin theorem of Section 7.3.1. ■

Case B. If $G = I$, then $Y_j = A^T V_j$ and the method (4.33) takes the form

$$x_{j+1} = x_j - A^T V_j \left(V_j^T A A^T V_j\right)^{-1} V_j^T r_j \quad (j \geq 0). \tag{4.42}$$

In this case

$$s_{j+1} = \left(I - A^T V_j \left(V_j^T A A^T V_j\right)^{-1} V_j^T A\right) s_j, \tag{4.43}$$

that is s_{j+1} is the orthogonal projection of s_j. Here the (absolute) error is reduced at each step.

THEOREM 4.24 *Assume that A and $V = [V_1, \ldots, V_m]$ are nonsingular. If the matrices Y_j are selected cyclically so that $V_j = V_i$ $(i \equiv j \ (mod\, m) + 1)$, then algorithm (4.42) is convergent for any x_0*

Proof. Since $s_{j+1} = P_{\mathcal{R}^\perp(A^T V_j)} s_j$ and $s_{j+m} = P_{\mathcal{R}^\perp(A^T V_m)} \cdots P_{\mathcal{R}^\perp(A^T V_1)} s_j$ for $j = km$, the rest follows from the von Neumann-Halperin theorem of Section 7.3.1. ■

We prove the following particular but very interesting convergence results.

THEOREM 4.25 *(Householder, Bauer). Let ϕ_j be defined by*

$$\sin(\phi_j) = \frac{\left\| V_j \left(V_j^T V_j \right)^{-1} V_j^T r_j \right\|_2}{\|r_j\|_2}. \tag{4.44}$$

Then

$$\|s_{j+1}\|_2 \leq \cos(\psi_j) \|s_j\|_2, \tag{4.45}$$

where

$$\cot(\psi_j/2) = cond_2(A) \cot(\phi_j/2) \tag{4.46}$$

and

$$\cos(\psi_j) \geq \frac{cond_2^2(A) - 1}{cond_2^2(A) + 1}. \tag{4.47}$$

Proof. Let

$$H_j = I - A^T V_j \left(V_j^T A A^T V_j \right)^{-1} V_j^T A$$

and apply the generalized Wielandt inequality (see Chapter 2, formula (2.73) or [31]). Since H_j is an orthogonal projector, and hence idempotent, it is possible to write

$$\|s_{j+1}\|_2^2 = \|H_j s_j\|_2^2 = s_j^T H_j s_j = r_j^T M^{-1} w_j,$$

where

$$M = AA^T, \quad w_j = AH_j s_j.$$

The relation $\mathcal{N}(H_j) = \mathcal{R}(A^T V_j)$ implies $w_j^T V_j = s_j^T H_j A^T V_j = 0$. Hence w_j is orthogonal to $\mathcal{R}(V_j)$. It was proved in Chapter 2 (Section 2.6.1) that if the angle between w_j and r_j is at least ϕ_j, then

$$\|H_j s_j\|_2 \leq \cos(\psi_j) \|s_j\|_2,$$

where ψ_j is defined above. But w_j is orthogonal to V_j. Hence if ϕ_j is defined by (4.44), then $\pi/2 - \phi_j$ is the angle between r_j and the space of V_j. Then the angle between r_j and w_j cannot be less than ϕ_j. If $r_j \in \mathcal{R}(V_j)$, then $\phi_j = \pi/2$ and $\cos(\psi_j)$ is minimal and given by

$$\cos(\psi_j) = \left(\text{cond}_2^2(A) - 1\right) / \left(\text{cond}_2^2(A) + 1\right).$$

■

REMARK 4.26 *In order to get a formal convergence result we must assume that $\{V_j\}$ is such that $\cos(\psi_j) \leq q < 1$ holds for all j.*

If each V_j is a single column v_j, then algorithm (4.42) has the form

$$x_{j+1} = x_j - \frac{v_j^T r_j}{\|A^T v_j\|_2^2} A^T v_j, \tag{4.48}$$

which is exactly the method (4.22). In this case

$$\|s_{j+1}\|_2^2 = \|s_j\|_2^2 - \left(r_j^T v_j\right)^2 / v_j^T A A^T v_j. \tag{4.49}$$

We prove two special convergence results not included in Theorem 4.21.

THEOREM 4.27 *(Householder, Bauer). Assume that each step $v_j = e_i$ is selected such that e_i maximizes $\left|r_j^T e_i\right|$. Then algorithm (4.48) is linearly convergent for any x_0 and*

$$\|s_{j+1}\|_2 \leq \left(\sqrt{1 - \frac{1}{n \cdot cond_2^2(A)}}\right) \|s_j\|_2 . \tag{4.50}$$

Proof. We exploit that

$$\|s_{j+1}\|_2^2 / \|s_j\|_2^2 = 1 - \left(r_j^T v_j\right)^2 / \left[s_j^T s_j v_j^T A A^T v_j\right].$$

For the assumed parameter selection the inequality

$$\left|r_j^T v_j\right| = \left|r_j^T e_i\right| = \|r_j\|_\infty \geq n^{-1/2} \|r_j\|_2$$

holds. Also

$$e_i^T A A^T e_i \leq \left\|A^T e_i\right\|_2^2 \leq \|A\|_2^2 .$$

But

$$\|r_j\|_2 \geq \|s_j\|_2 / \left\|A^{-1}\right\|_2 ,$$

and therefore, it follows that

$$\begin{aligned}\left(r_j^T v_j\right)^2 / \left[s_j^T s_j v_j^T A A^T v_j\right] &= \left(r_j^T e_j\right)^2 / \left[s_j^T s_j e_j^T A A^T e_j\right] \\ &\geq n^{-1} \|r_j\|_2^2 / \left[\|s_j\|_2^2 \left\|A^T e_j\right\|_2^2\right] \\ &\geq n^{-1} / \left[\left\|A^{-1}\right\|_2^2 \|A\|_2^2\right]\end{aligned}$$

and

$$\|s_{j+1}\|_2^2 / \|s_j\|_2^2 \leq 1 - n^{-1} \left[\text{cond}_2 (A)\right]^{-2} .$$

■

Since the condition $v_j^T r_j = \|r_j\|_\infty$ is not assumed, the result is a variant of Theorem 4.22. A similar result was obtained by McCormick [252] (see, also [21]), who also investigated a certain "threshold" strategy for choosing e_i.

THEOREM 4.28 *(Householder, Bauer). Assume that each step $v_j = r_j$ is selected, then algorithm (4.48) (the steepest descent method) is linearly convergent for any x_0 and*

$$\|s_{j+1}\|_2 \leq \cos(\psi) \|s_j\|_2, \tag{4.51}$$

where

$$\cos(\psi) = \frac{cond_2^2(A) - 1}{cond_2^2(A) + 1}. \tag{4.52}$$

Proof. The Kantorovich inequality (formula (2.80) in Chapter 2, Section 2.6) implies that

$$\left(r_j^T v_j\right)^2 / \left[s_j^T s_j v_j^T A A^T v_j\right] = \frac{\|r_j\|_2^4}{\left(r_j^T A^{-T} A^{-1} r_j\right)\left(r_j^T A A^T r_j\right)} \geq \sin^2(\psi),$$

where ψ is defined by $\cot(\psi/2) = \mathrm{cond}_2(A)$ or

$$\cos(\psi) = \left(\mathrm{cond}_2^2(A) - 1\right) / \left(\mathrm{cond}_2^2(A) + 1\right).$$

Therefore,

$$\|s_{j+1}\|_2^2 / \|s_j\|_2^2 \leq 1 - \sin^2(\psi).$$

■

Case C. We assume that A is symmetric and positive definite, and $G = A$. Then $Y_j = V_j$ and the method (4.33) takes the form

$$x_{j+1} = x_j - Y_j \left(Y_j^T A Y_j\right)^{-1} Y_j^T r_j \quad (j \geq 0). \tag{4.53}$$

Unlike in Cases A and B, neither s_{j+1} nor r_{j+1} are orthogonal projection of s_j or r_j, respectively.

THEOREM 4.29 *(Householder, Bauer). If each step $Y_j = e_i$, where e_i maximizes $\left|e_i^T r_j\right|$, then the method (4.53) is linearly convergent for any x_0 and*

$$\|s_{j+1}\|_A \leq \left(\sqrt{1 - \frac{1}{n \cdot cond_2(A)}}\right) \|s_j\|_A. \tag{4.54}$$

Proof. We have

$$\|s_{j+1}\|_A^2 = \|s_j\|_A^2 - r_j^T Y_j \left(Y_j^T A Y_j\right)^{-1} Y_j^T r_j, \tag{4.55}$$

$$\left|Y_j^T r_j\right| = \left|e_i^T r_j\right| = \|r_j\|_\infty \geq n^{-1/2} \|r_j\|_2$$

and

$$e_i^T A e_i \leq \|A e_i\|_2 \leq \|A\|_2.$$

Let $A = R^T R$. Then

$$\|s_j\|_A^2 = s_j^T A s_j = s_j^T R^T R s_j = \|R s_j\|_2^2$$

and

$$\|r_j\|_2^2 = \|A s_j\|_2^2 = \left\|R^T R s_j\right\|_2^2 .$$

By simple arguments

$$\|R s_j\|_2 = \left\|R^{-T} R^T R s_j\right\|_2 \leq \left\|R^{-T}\right\|_2 \left\|R^T R s_j\right\|_2 = \left\|R^{-1}\right\|_2 \|r_j\|_2 .$$

Hence

$$\|r_j\|_2^2 \geq \frac{\|R s_j\|_2^2}{\|R^{-1}\|_2^2} = \frac{\|s_j\|_A^2}{\|A^{-1}\|_2}$$

and

$$\begin{aligned} r_j^T Y_j \left(Y_j^T A Y_j\right)^{-1} Y_j^T r_j &= r_j^T e_i \left(e_i^T A e_i\right)^{-1} e_i^T r_j \\ &= \left(e_i^T r_j\right)^2 / e_i^T A e_i \geq n^{-1} \|r_j\|_2^2 / \|A\|_2 \\ &\geq n^{-1} \|s_j\|_A^2 / \mathrm{cond}_2(A) . \end{aligned}$$

This implies that

$$\|s_{j+1}\|_A^2 \leq \left(1 - \frac{1}{n \cdot \mathrm{cond}_2(A)}\right) \|s_j\|_A^2 .$$

■

For other special cases we refer to Householder [206]. Related material and developments can be found in Peters [276], and Wainwright and Keller [353].

The Householder, Bauer class of projection methods also appears in the context of multiparameter descent methods [58], where the efficient choice of Y_j is further analyzed [59].

3. Mixed methods

We consider a method mixed, if the iterative projection method is modified with a relaxation parameter or an acceleration procedure. In both cases the aim is to improve the convergence. We do not study the acceleration techniques here. Instead we refer to Tompkins [342], Liu [235] and Brezinski [57].

3.1 Relaxed Kaczmarz methods

The first of such methods is due to Whitney and Meany [357]. They called the method single pattern LMS (least mean squares) algorithm.

Let $A \in \mathbb{R}^{m\times n}$, $b \in \mathbb{R}^m$ and consider the equation $Ax = b$. The relaxation algorithm of Whitney and Meany has the form

$$\begin{aligned} x_{j+1} &= x_j - \mu e_i^T r(x_j) A^T e_i \\ &= \left(I - \mu A^T e_i e_i^T A\right) x_j + \mu b_i A^T e_i, \end{aligned} \tag{4.56}$$

where I is the $n \times n$ identity matrix, μ is a scalar, $1 \le i \le n$ and $i \equiv j \pmod m + 1$. If the matrix A is normalized so that $\left\|A^T e_i\right\|_2 = 1$ for all i, and $\mu = 1$, we obtain the original Kaczmarz method. Note that $A^T e_k e_k^T A$ is not a projector, if $\left\|A^T e_k\right\|_2 \neq 1$ for some k.

Consider the normalized problem $D^{1/2}Ax = D^{1/2}b$, where

$$D = \operatorname{diag}\left(1/\left\|A^T e_i\right\|_2^2\right).$$

Then the above algorithm takes the form

$$\begin{aligned} x_{j+1} &= x_j - \mu \tfrac{e_i^T r(x_j)}{\|A^T e_i\|_2^2} A^T e_i \\ &= \left(I - \mu P_i\right) x_j + \mu \tfrac{b_i}{\alpha_i} A^T e_i, \end{aligned} \tag{4.57}$$

where

$$P_i = \frac{A^T e_i e_i^T A}{\|A^T e_i\|_2^2}, \quad \alpha_i = \left\|A^T e_i\right\|_2^2. \tag{4.58}$$

Clearly, (4.57) is a relaxed version of the Kaczmarz method applied to the system $Ax = b$. We prove the convergence of (4.57) using the original proof of Whitney and Meany with the necessary modifications.

Let $Q_{k,j}(\mu) = (I - \mu P_k)\cdots(I - \mu P_j)$ $(k \ge j)$ and $Q_{k,j}(\mu) = I$ $(k < j)$. Then recursion (4.57) can be written in the form

$$x_{km} = Q_{m,1}(\mu)\, x_{(k-1)m} + \mu \sum_{i=1}^{m} Q_{m,i+1}(\mu) \frac{b_i}{\alpha_i} A^T e_i \quad (k = 1, 2, \ldots). \tag{4.59}$$

Define the $n \times m$ matrix

$$R(\mu) = \mu \left[Q_{m,2}(\mu) \frac{A^T e_1}{\alpha_1}, \ldots, Q_{m,m+1}(\mu) \frac{A^T e_m}{\alpha_m}\right].$$

Then

$$x_{km} = Q_{m,1}(\mu)\, x_{(k-1)m} + R(\mu)\, b. \tag{4.60}$$

It is easy to see that

$$R(\mu)A = \sum_{i=1}^{m} Re_i e_i^T A = \sum_{i=1}^{m} Q_{m,i+1}(\mu)\,\mu P_i$$
$$= \sum_{i=1}^{m} \left(Q_{m,i+1}(\mu) - Q_{m,i}(\mu)\right) = I - Q_{m,1}(\mu).$$

In place of (4.59) we consider the recurrence relation

$$W_k = Q_{m,1}(\mu)\,W_{k-1} + R(\mu), \tag{4.61}$$

where W_0 is an arbitrary $n \times m$ matrix. If (4.61) is multiplied on the right by b, one obtains (4.59), with $x_{km} = W_k b$ $(k \geq 0)$.

THEOREM 4.30 *(Whitney, Meany). For any matrix $A \in \mathbb{R}^{m\times n}$ and any vector $b \in \mathbb{R}^m$ let the matrices W_k $(k \geq 1)$ be defined by (4.61), where $Q_{m,1}(\mu)$ and $R(\mu)$ are defined as above, W_0 is arbitrary, and $0 < \mu < 2$. Then:*
(i) $W_k \to (I - A^+A)\,A_0 + W(\mu)$.
(ii) $W(\mu) \in A^{\{1,2,4\}}$, but $W(\mu) \notin A^{\{3\}}$, in general.
(iii) If $\operatorname{rank}(A) = n$, then $W(\mu) \to \left(A^T DA\right)^{-1} A^T D$ when $\mu \to 0$.

Proof. The eigenvalues of P_i are 1 and 0, so the eigenvalues ν of $I - \mu P_i$ satisfy $-1 < \nu \leq 1$. Hence

$$\|(I - \mu P_i)\,y\|_2 \leq \|y\|_2\,.$$

Equality occurs if and only if y is orthogonal to $A^T e_i$, in which case

$$(I - \mu P_i)\,y = y.$$

It follows that $\|Q_{m,1}(\mu)\,y\|_2 \leq \|y\|_2$, with equality if and only if y is orthogonal to $A^T e_1, \ldots, A^T e_m$. In this case, $Q_{m,1}(\mu)\,y = y$. We can establish that $Q_{m,1}(\mu)\,y = y$ if and only if $y \in \mathcal{R}^{\perp}\left(A^T\right) = \mathcal{N}(A)$. If $y \in \mathcal{R}\left(A^T\right)$, then for some $\alpha < 1$,

$$\left\|Q_{m,1}^k(\mu)\,y\right\|_2 \leq \alpha^k \|y\|_2 \quad (k \geq 1). \tag{4.62}$$

Now the recurrence relation (4.61) leads to the formula

$$W_k = Q_{m,1}^k(\mu)\,W_0 + \left(\sum_{i=0}^{k-1} Q_{m,1}^i(\mu)\right) R(\mu). \tag{4.63}$$

Since the columns of $R(\mu)$ belong to $\mathcal{R}\left(A^T\right)$, (4.62) shows that the infinite series

$$\sum_{i=0}^{\infty} Q_{m,1}^i(\mu)\,R(\mu)$$

converges to a limit $W(\mu)$. Then

$$\begin{aligned} W(\mu) A &= \lim_{k\to\infty} \left(\sum_{i=0}^{k-1} Q_{m,1}^i (\mu) \right) R(\mu) A \\ &= \lim_{k\to\infty} \left(\sum_{i=0}^{k-1} Q_{m,1}^i (\mu) \right) (I - Q_{m,1}(\mu)) \\ &= \lim_{k\to\infty} \left(I - Q_{m,1}^k (\mu) \right). \end{aligned} \tag{4.64}$$

Since the columns of $Q_{m,1}^k(\mu) R(\mu)$ belong to $\mathcal{R}(A^T)$ for $k = 1, 2, \ldots$, the columns of $W(\mu)$ belong to $\mathcal{R}(A^T)$ (see, also Tanabe's proof in Section 4.2.1). Hence from (4.62), $Q_{m,1}^k(\mu) W(\mu) \to 0$, when $k \to \infty$. Multiplying (4.64) on the right by $W(\mu)$, one has

$$W(\mu) AW(\mu) = \lim_{k\to\infty} \left(W(\mu) - Q_{m,1}^k (\mu) W(\mu) \right) = W(\mu).$$

Since the columns of A^+ belong to $\mathcal{R}(A^T)$, it follows from (4.62) that $Q_{m,1}^k(\mu) A^+ \to 0$ when $k \to \infty$. Multiplying (4.64) on the right by A^+A, one has

$$W(\mu) A = W(\mu) AA^+A = \lim_{k\to\infty} \left(A^+A - Q_{m,1}^k (\mu) A^+A \right) = A^+A. \tag{4.65}$$

This shows that $W(\mu) A$ is symmetric and that $AW(\mu) A = AA^+A = A$. Thus $W(\mu)$ satisfies the first, second and fourth conditions of Penrose. Writing

$$W_0 = (I - A^+A + A^+A) W_0 = \left(P_{\mathcal{N}(A)} + P_{\mathcal{R}(A^T)} \right) W_0$$

we may deduce from (4.63), (4.64), and (4.65) that

$$W_k \to (I - A^+A) W_0 + W(\mu) \tag{4.66}$$

when $k \to \infty$. If $\operatorname{rank}(A) = n$, then $\mathcal{R}(A^T)$ has dimension n. So for some number $\alpha < 1$ and every $y \in \mathbb{R}^n$, $\|Q_{m,1}(\mu) y\|_2 \le \alpha \|y\|_2$. The infinite series $\sum_{k=0}^{\infty} Q_{m,1}^k(\mu)$ therefore converges to $(I - Q_{m,1}(\mu))^{-1}$, and hence, $W(\alpha) = (I - Q_{m,1}(\mu))^{-1} R(\mu)$. Now when $\mu \to 0$,

$$\frac{1}{\mu} (I - Q_{m,1}(\mu)) \to \sum_{i=0}^{m} P_i = A^T DA$$

and

$$\frac{1}{\mu} R(\mu) \to A^T D.$$

It follows that when $\mu \to 0$,

$$W(\mu) = \left[\frac{I - Q_{m,1}(\mu)}{\mu}\right]^{-1} \frac{R(\mu)}{\mu} \to \left(A^T D A\right)^{-1} A^T D.$$

It is easy to check that $\left(A^T D A\right)^{-1} A^T D \in A^{\{1,2,4\}}$. ∎

EXAMPLE 4.31 *Let*

$$A = \begin{bmatrix} 1 & 0 \\ 1 & 0 \\ 0 & 1 \end{bmatrix}$$

and let $\mu = 1$. *Then* $Q_{3,1}(1) = 0$ *and so*

$$W(1) = R(1) = \begin{bmatrix} 0 & 1 & 0 \\ 0 & 0 & 1 \end{bmatrix}.$$

Hence

$$AW(1) = \begin{bmatrix} 0 & 1 & 0 \\ 0 & 1 & 0 \\ 0 & 0 & 1 \end{bmatrix},$$

which is not symmetric. Thus $W(\mu)$ *need not satisfy the third condition of Penrose.*

THEOREM 4.32 *(Whitney, Meany). For any matrix* $A \in \mathbb{R}^{m\times n}$ *and any vector* $b \in \mathbb{R}^m$, *let the vectors* x_k $(k \geq 1)$ *be defined by (4.57) with* $0 < \mu < 2$ *and* x_0 *arbitrary. Then*

$$\lim_{k\to\infty} x_k = x(\mu) = P_{\mathcal{N}(A)} x_0 + W(\mu) b$$

and the speed of convergence is linear. If $Ax = b$ *has a solution, then* $x(\mu)$ *is a solution for any* x_0, *and* $W(\mu) b$ *is the minimum norm solution. If* $rank(A) = n$, *then* $x(\mu) \to P_{\mathcal{N}(A)} x_0 + \left(A^T D A\right)^{-1} A^T D b$.

Proof. If $b \neq 0$, let the matrix W_0 be determined so that $W_0 b = x_0$, and let the matrices W_k be defined as in the previous theorem. Then

$$x_{km} = W_k b \to \left(I - A^+ A\right) W_0 b + W(\mu) b = \left(I - A^+ A\right) x_0 + W(\mu) b = x(\mu).$$

∎

Nicolaides [267] showed first that the standard *SOR* (*S*uccessive *O*ver*Re*laxation) method can be written in the form

$$x_{j+1} = P_n x_j + C, \tag{4.67}$$

where

$$P_m = \left(I - \mu e_m e_m^T D^{-1} A\right) \cdots \left(I - \mu e_2 e_2^T D^{-1} A\right)\left(I - \mu e_1 e_1^T D^{-1} A\right). \tag{4.68}$$

Since a similar expression occurs in the relaxed Kaczmarz algorithm, the theory of SOR methods can be used in the convergence analysis. Dax [96] proved the following result.

THEOREM 4.33 *(Dax). Let the sequence generated by the scheme*

$$x_{j+1} = x_j - \mu \frac{e_i^T r(x_j)}{\|A^T e_i\|_2^2} A^T e_i \quad (0 < \mu < 2,\ i \equiv j \ (mod\, m) + 1).$$

Then the following claims hold:
(a) The sequence $\{x_{km}\}_{k=0}^{\infty}$ *always converges.*
(b) If $Ax = b$ *is solvable then the limit point* $\widehat{x}$ *solves this system.*
(c) If $Ax = b$ *is solvable and* $x_0 \in \mathcal{R}(A^T)$, *then* $\widehat{x}$ *is the minimal norm solution of* $Ax = b$.
(d) If $Ax = b$ *is solvable, then the sequence* $\{\|x_{km} - \widehat{x}\|\}_{k=0}^{\infty}$ *is monotone decreasing.*
(e) If $Ax = b$ *is solvable, then there exist positive constants* α *and* β *such that*

$$\|x_{km} - \widehat{x}\| \leq \alpha \|Ax_{km} - b\|, \tag{4.69}$$

$$\|x_{km} - \widehat{x}\| \leq \beta \left\|x_{(k+1)m} - x_{km}\right\|. \tag{4.70}$$

A comparison with the Whitney, Meany result shows that claims (d) and (e) are the only improvements. Dax notes that for consistent systems all subsequences $\{x_{km+i}\}_{k=0}^{\infty}$ $(1 \leq i \leq m)$ have a common limit. On the other hand, when $Ax = b$ is inconsistent each subsequence may have a different limit. Dax proves the result using his convergence theory developed for linear stationary iterative processes [96].

Further results can be found in Whitney and Meany [357], Herman, Lent and Lutz [195], and Trummer [346], [347], [348].

Consider the system

$$Ax = b,$$

where $A \in \mathbb{R}^{lm \times n}$ and $b \in \mathbb{R}^{lm}$. Partition A and b as

$$A = \begin{bmatrix} A_1^T \\ A_2^T \\ \vdots \\ A_m^T \end{bmatrix}, \quad b = \begin{bmatrix} b_1 \\ b_2 \\ \vdots \\ b_m \end{bmatrix}, \tag{4.71}$$

with $A_i^T \in \mathbb{R}^{l \times n}$, $b_i \in \mathbb{R}^l$. The block Kaczmarz method of Eggermont, Herman and Lent is defined by the recursion

$$x_{j+1} = x_j - \mu A_i \left(A_i^T x_j - b_i\right) \quad (i \equiv j \ (\mathrm{mod}\, m) + 1,\ j \geq 0). \tag{4.72}$$

In the special case $l = 1$, and if the rows of A are scaled to have length one, this is the Kaczmarz algorithm with constant relaxation. The following result holds [80], [78].

THEOREM 4.34 *(Censor, Eggermont and Gordon). For all μ small enough,*

$$x^*(\mu) = \lim_{k\to\infty} x_{km} \tag{4.73}$$

exists, and

$$\lim_{\mu\to 0} x^*(\mu) = A^+b + \left(I - A^+A\right)x^0. \tag{4.74}$$

The proof of theorem is similar to those of Whitney and Meany [357], and Tanabe [337].

Assume that the rows of A are scaled to have length 1. So we have the equivalent system

$$D^{1/2}Ax = D^{1/2}b.$$

The above theorem says that $z^* = \lim_{\mu\to 0} x^*(\mu)$ equals

$$z^* = \left(D^{1/2}A\right)^+ D^{1/2}b + \left(I - \left(D^{1/2}A\right)^+\left(D^{1/2}A\right)\right)x^0.$$

It follows that

$$A^TDAz^* = A^TDb,$$

i.e., x^* is a weighted least squares solution of the system $Ax = b$.

Kamath and Sameh [216] investigated a relaxed block Kaczmarz method of *SSOR* type . The *SSOR*=symmetrized *SOR* methods are essentially such that any two consecutive major iterations are executed so that the second iteration takes the equations in reversed order (see, e.g., [366], [176]). The first iteration is called the forward sweep, while the second one is referred to as the backward sweep.

The method is related to algorithm (4.13) of Vasil'chenko and Svetlanov [351]. Let n be even, and let A and b be partitioned as

$$A = \begin{bmatrix} A_1^T \\ A_2^T \end{bmatrix}, \quad b = \begin{bmatrix} b_1 \\ b_2 \end{bmatrix} \quad \left(A_i^T \in \mathbb{R}^{(n/2)\times n},\ b_i \in \mathbb{R}^{n/2},\ i = 1,2\right). \tag{4.75}$$

Let

$$V_1^T = \left[I_{n/2}, 0\right], \quad V_2^T = \left[0, I_{n/2}\right]. \tag{4.76}$$

Then one (major) iteration of the Kamath, Sameh projection method, which is a block version of the row *SSOR* method of Björck and Elfving [47] (see also Section 4.3.3), has the form

$$x_{j+1} = x_j - \mu A_i\left(A_i^TA_i\right)^{-1}\left(A_i^Tx_j - b_i\right), \tag{4.77}$$

where $j = 1, 2, 3, 4$ and $i = \min(j, 5 - j)$. We can write one iteration of the block row *SSOR* method in the form

$$x_{k+1} = Qx_k + Rb, \tag{4.78}$$

where $Q = Q_1Q_2Q_2Q_1$, in which $Q_i = I - \mu P_i$ $(i = 1, 2)$ with $P_i = A_iA_i^+$, $A_i^+ = \left(A_i^TA_i\right)^{-1}A_i^T$, and

$$R = \mu\left[\left(I + Q_1Q_2Q_2\right)\left(A_1^+\right)^T, Q_1\left(Q_2 + I\right)\left(A_2^+\right)^T\right].$$

THEOREM 4.35 *(Kamath). Assume that A is nonsingular. Let $b \in \mathcal{R}(A)$ and $x_0 \in \mathcal{R}\left(A^T\right)$. Then the iteration (4.78) converges towards the solution of $Ax = b$ for $0 < \mu < 2$.*

Kamath and Sameh [216] showed that the optimal μ is 1. Note that the case $\mu = 1$ corresponds to the symmetrized block Kaczmarz method (4.13) of Vasil'chenko and Svetlanov [351]. Kamath and Sameh applied the projection method with conjugate gradient acceleration. Bramley and Sameh [53] also investigated the above procedure with m row blocks. They pointed out that the choice $\mu = 1$ may not be optimal if $m \geq 3$.

For further types of the relaxed block Kaczmarz method, see Elfving [116] and Popa [290].

3.2 Relaxed Cimmino methods

Relaxed Cimmino type methods were investigated by Ansorge [21], Dax [96], Sloboda [320] and Elfving [116].

We need the following result on the solutions of normal equations. Let $A \in \mathbb{R}^{m\times n}$, $b \in \mathbb{R}^m$ and assume that $A^Te_i \neq 0$ for all i.

THEOREM 4.36 *(Dax). Let $W \in \mathbb{R}^{p\times m}$, $V \in \mathbb{R}^{n\times n}$ ($\det(V) \neq 0$) and $\widehat{A} = WAV$. Consider the iteration*

$$x_{j+1} = x_j - \mu VV^TA^TW^TW\left(Ax_j - b\right) \quad (j \geq 0) \tag{4.79}$$

with

$$0 < \mu < 2/\rho\left(\widehat{A}^T\widehat{A}\right). \tag{4.80}$$

Then the following claims hold.
(a) The sequence $\{x_j\}_{j=0}^{\infty}$ converges, and the limit point $\widehat{x}$ solves the least-squares problem

$$\|W(Ax - b)\|_2^2 \to \min. \tag{4.81}$$

(b) The sequences $\{\|W(Ax_j - b)\|\}_{j=0}^{\infty}$ and $\left\{\left\|V^{-1}(x_j - \widehat{x})\right\|\right\}_{j=0}^{\infty}$ are monotone decreasing.

(c) The distance $\|x_j - \widehat{x}\|$ *satisfies the bound*

$$\|x - \widehat{x}\| \leq \left\|VV^T A^T W^T W\left(Ax_j - b\right)\right\| / \sigma = \left\|x_{j+1} - x_j\right\| / (\mu\sigma), \qquad (4.82)$$

where σ *denotes the smallest eigenvalue of* $\widehat{A}^T \widehat{A}$ *that differs from zero.*
(d) If $x_0 \in \mathcal{R}\left(VV^T A^T W^T W A\right)$, *then* $\widehat{x}$ *solves the problem*

$$\min_{y \in \{x | A^T W^T W A y = A^T W^T W b\}} \left\|V^{-1} y\right\|. \qquad (4.83)$$

The result extends Theorem 16 of Tanabe [338]. It implies the following convergence result for a Cimmino type method.

COROLLARY 4.37 *(Dax). Assume that* $W = diag\left(w_1, \ldots, w_m\right)$ *and consider the Cimmino type iteration*

$$x_{j+1} = x_j - \mu A^T W^2\left(Ax_j - b\right) = x_j - \mu \sum_{i=1}^{m} a_i w_i^2\left(a_i^T x_j - b_i\right), \qquad (4.84)$$

where

$$0 < \mu < 2 / \left(\sum_{i=1}^{m} w_i^2 \left\|a_i\right\|_2^2\right). \qquad (4.85)$$

Then the following claims hold:
(a) The sequence $\left\{x_j\right\}_{j=0}^{\infty}$ *converges and the limit point* $\widehat{x}$ *solves (4.81).*
(b) The sequences $\left\{\left\|W\left(Ax_j - b\right)\right\|\right\}_{j=0}^{\infty}$ *and* $\left\{\left\|x_j - \widehat{x}\right\|\right\}_{j=0}^{\infty}$ *are monotone decreasing.*
(c) The distance $\|x_j - \widehat{x}\|$ *satisfies the bound*

$$\|x - \widehat{x}\| \leq \left\|A^T W^2\left(Ax_j - b\right)\right\| / \sigma = \left\|x_{j+1} - x_j\right\| / (\mu\sigma), \qquad (4.86)$$

where here σ *denotes the smallest eigenvalue of* $A^T W^2 A$.
(d) If $x_0 \in \mathcal{R}\left(A^T W\right)$, *then* $\widehat{x}$ *solves the problem*

$$\min_{y \in \{x | A^T W^2 A y = A^T W^2 b\}} \|y\|. \qquad (4.87)$$

Moreover, if $w_i \neq 0$ *for* $i = 1, \ldots, m$, *then the last property holds whenever* $x_0 \in \mathcal{R}\left(A^T\right)$.

Proof. In this case $V = I$ and $\widehat{A} = WA$. We need to prove only that $\rho\left(\widehat{A}^T \widehat{A}\right) \leq \sum_{i=1}^{m} w_i^2 \left\|a_i\right\|_2^2$. Let x be a unit vector, i.e., $\|x\|_2 = 1$. Then the Cauchy-Schwarz inequality implies

$$x^T \widehat{A}^T \widehat{A} x = x^T A^T W^2 A x = \sum_{i=1}^{m} w_i^2 \left(a_i^T x\right)^2 \leq \sum_{i=1}^{m} w_i^2 \left\|a_i\right\|_2^2. \qquad (4.88)$$

The rest follows from the Rayleigh-Ritz theorem. ■

Note that equality is possible in (4.88) if and only if $a_i = x$ for $i = 1, \ldots, m$, that is, when all vectors are parallel to one another. Hence if $\operatorname{rank}(A) > 1$, $\rho\left(\widehat{A}^T \widehat{A}\right) < \sum_{i=1}^m w_i^2 \|a_i\|_2^2$. Thus the relaxation parameter

$$\mu = 2/\left(\sum_{i=1}^m w_i^2 \|a_i\|_2^2\right)$$

is also allowed in the corollary above.

We show that method (4.84) is indeed related to the Cimmino method. Introduce the notation $\widetilde{w}_i = w_i \|a_i\|_2 / \left(\sum_{i=1}^m w_i^2 \|a_i\|_2^2\right)^{1/2}$, $\widetilde{a}_i = a_i / \|a_i\|_2$, and $\widetilde{b}_i = b_i / \|a_i\|_2$ $(i = 1, \ldots, m)$. The method (4.84) can be written as

$$x_{j+1} = x_j - \theta \sum_{i=1}^m \widetilde{w}_i^2 \widetilde{a}_i \left(\widetilde{a}_i^T x_j - \widetilde{b}_i\right), \tag{4.89}$$

where

$$\sum_{i=1}^m \widetilde{w}_i^2 = 1, \quad 0 < \theta \le 2. \tag{4.90}$$

Sloboda [320] investigated the underdetermined linear system

$$Ax = b \quad \left(A \in \mathbb{R}^{m \times n},\, b \in \mathcal{R}(A),\, m \le n,\, \operatorname{rank}(A) = m\right) \tag{4.91}$$

and defined the relaxed Cimmino method

$$x_{j+1} = x_j - \mu (DA)^T (Ax_j - b), \tag{4.92}$$

where $D = \operatorname{diag}\left(1/\left\|A^T e_i\right\|\right) \in \mathbb{R}^{m \times m}$.

THEOREM 4.38 *(Sloboda). Let $x_0 \in \mathbb{R}^n$ be arbitrary and let $\left\|I - \mu AA^T\right\| < 1$. Then algorithm (4.92) converges to a solution of (4.91). If $x_0 = 0$, then this solution is the minimum norm solution of (4.91).*

Proof. Assume that the rows of A are normalized such that $D = I$. Then algorithm (4.92) can be written in the form

$$x_{j+1} = \left(I - \mu A^T A\right) x_j + \mu A^T b.$$

We can prove by induction that

$$x_j = \left(I - \mu A^T Q_j A\right) x_0 + \mu A^T Q_j b,$$

where $Q_j = \sum_{i=0}^{j-1} \left(I - \mu AA^T\right)^i$. The assumption $\left\|I - \mu AA^T\right\| < 1$ implies that $\{Q_j\}$ is convergent and

$$\lim_{j \to \infty} Q_j = \left(I - \left(I - \mu AA^T\right)\right)^{-1} = \frac{1}{\mu}\left(AA^T\right)^{-1}.$$

Hence

$$\lim_j x_j = \left(I - A^T \left(AA^T\right)^{-1} A\right) x_0 + A^T \left(AA^T\right)^{-1} b = \left(I - A^+ A\right) x_0 + A^+ b,$$

which is the solution of $Ax = b$. The choice $x_0 = 0$ gives the minimum norm solution. ∎

Sloboda [320] also investigated the possible values of μ for different classes of matrices. We quote here only the following results.

DEFINITION 4.39 *The matrix $A \in \mathbb{R}^{m\times n}$ $(m \le n)$ is k-diagonally dominant $(k \ge 1)$ if*

$$k \sum_{j=1, j\neq i}^{n} |a_{ij}| < |a_{ii}| \quad (i = 1, \ldots, m). \tag{4.93}$$

THEOREM 4.40 *(Sloboda). Let AA^T be a k-diagonally dominant matrix with $k \ge 1$ and let $x_0 = 0$. Then algorithm (4.92) converges to the minimum norm solution of $Ax = b$ for $0 < \mu \le 2k/(k+1)$.*

Assume again that $D = I$. Let $\lambda_{\min}$ and $\lambda_{\max}$ be the minimal and the maximal eigenvalue of AA^T, respectively. Since

$$\lambda_{\max} \le \max_i \left[a_i^T a_i + \sum_{j=1, j\neq i} \left|a_i^T a_j\right| \right] < 1 + \frac{1}{k} = \frac{k+1}{k}$$

we obtain that for $\mu \le 2k/(k+1)$,

$$\left\|I - \mu AA^T\right\| = \max\left(\left|1 - \mu\lambda_{\min}\right|, \left|1 - \mu\lambda_{\max}\right|\right) < 1.$$

Sloboda [320] proved that $\mu = 1$ is the optimal relaxation parameter, if AA^T is k-diagonally dominant and $D = I$. Applications of these results in ODE and BVP discretizations are given in Sloboda and Sgallari [321].

Elfving [116] developed a relaxed block Cimmino method. Assume that equation $Ax = b$ $(A \in \mathbb{R}^{m\times n})$ is consistent, that is $b \in \mathcal{R}(A)$. It will be shown in the next section that the minimum norm solution $x = A^+ b$ can be computed by solving the system

$$AA^T y = b, \quad x = A^T y. \tag{4.94}$$

Further it follows from the Urquhart theorem that if $b \in \mathcal{R}(A)$, then $x = A^+ b = A^{(1,4)} b$. If $\operatorname{rank}(A) = m$, then $A^+ = A^{(1,4)} = A^T \left(AA^T\right)^{-1}$. The block Jacobi method [366] applied to (4.94) with the partition

$$A = \begin{bmatrix} A_1^T \\ A_2^T \\ \vdots \\ A_p^T \end{bmatrix}, \quad b = \begin{bmatrix} b_1 \\ b_2 \\ \vdots \\ b_p \end{bmatrix} \quad \left(A_i^T \in \mathbb{R}^{m_i \times n},\ b_i \in \mathbb{R}^{m_i}\right) \tag{4.95}$$

then becomes

$$x_{j+1} = x_j - \sum_{i=1}^{p} A_i \left(A_i^T A_i\right)^{-1} \left(A_i^T x_j - b_i\right).$$

Dropping the assumption that A has full rank, Elfving defined the Cimmino (block row Jacobi) type method

$$x_{j+1} = x_j - \mu \sum_{i=1}^{p} \left(A_i^T\right)^{(1,4)} \left(A_i^T x_j - b_i\right). \tag{4.96}$$

Elfving [116] proved that if $b \in \mathcal{R}(A)$ and $x_0 \in \mathcal{R}\left(A^T\right)$, then method (4.96) converges to the minimum norm solution for any $\mu > 0$ small enough. A *SOR* version of the algorithm contains the Kaczmarz method.

3.3 Relaxed Householder, Bauer methods

Björck and Elfving [47] extended subclasses (4.40) and (4.48) of the Householder, Bauer methods to rectangular in connection with efficient implementation of *SOR* methods. Let $A \in \mathbb{R}^{m \times n}$, $b \in \mathbb{R}^m$ and $x = A^+ b$ the Moore-Penrose or pseudoinverse solution of $Ax = b$.

THEOREM 4.41 *(Björck, Elfving). Assume that the system $Ax = b$ is consistent, i.e., $b \in \mathcal{R}(A)$. Then the solution x of*

$$x = A^T y, \quad AA^T y = b \tag{4.97}$$

is the pseudoinverse solution $x = A^+ b$.

Proof. By construction we have $x \in \mathcal{R}\left(A^T\right)$. Since $\mathcal{R}\left(AA^T\right) = \mathcal{R}(A)$ the system $AA^T y = b$ is also consistent and it follows that $0 = AA^T y - b = Ax - b \perp \mathcal{R}(A)$. Thus, by Lemma 1.80, x is the pseudoinverse solution. ∎

THEOREM 4.42 *(Björck, Elfving). Let x be a solution of*

$$A^T A x = A^T b, \tag{4.98}$$

and assume that $x \in \mathcal{R}\left(A^T\right)$. Then x is the pseudoinverse solution $A^+ b$.

Proof. It follows directly that $A^T (Ax - b) = 0$, that is $Ax - b \perp \mathcal{R}(A)$. Thus by Lemma 1.80 x is the pseudoinverse solution. ∎

First let $v_j \in \mathbb{R}^m$ $(0 \neq v_j \notin \mathcal{N}\left(A^T\right))$ and compute

$$x_{j+1} = x_j - \alpha_j A^T v_j \quad \left(\alpha_j = \frac{v_j^T r_j}{\|A^T v_j\|_2^2},\ j \geq 1\right), \tag{4.99}$$

which is the formal generalization of (4.48).

If the system $Ax = b$ is consistent, then $AA^+b = b$ and it follows that $s_{j+1}^T A^T v_j = 0$, where $s_j = A^+b - x_j$. Therefore

$$\|s_{j+1}\|_2^2 = \|s_j\|_2^2 - \frac{\left(v_j^T r_j\right)^2}{\|A^T v_j\|_2^2} \le \|s_j\|_2^2 ,$$

in agreement with the regular case ($m = n$). Thus for consistent systems the class of methods (4.99) is error reducing.

For the second class, which is a generalization of (4.40), let $y_j \in \mathbb{R}^n$ ($0 \ne y_j \notin \mathcal{N}(A)$) and compute

$$x_{j+1} = x_j - \alpha_j y_j \qquad \left(\alpha_j = \frac{y_j^T A^T r_j}{\|Ay_j\|_2^2},\ j \ge 1\right). \tag{4.100}$$

It follows immediately that $r_{j+1}^T A y_j = 0$. Therefore

$$\|r_{j+1}\|_2^2 = \|r_j\|_2^2 - \frac{\left(y_j^T A^T r_j\right)^2}{\|Ay_j\|_2^2} \le \|r_j\|_2^2 ,$$

that is the class (4.100) is residual reducing.

Assume now that A has nonzero rows and let $v_j = e_j$ in (4.99). Then algorithm (4.99) becomes the Kaczmarz method

$$x_{j+1} = x_j - \frac{e_j^T r_j}{\|A^T e_j\|_2^2} A^T e_j \quad (j = 1, \ldots, m). \tag{4.101}$$

Now consider the Gauss-Seidel method for the system (4.97), where in the jth minor iteration step, y_j is updated by

$$\Delta y_j = e_j \left(e_j^T A A^T y_j - b_j\right) / \|A^T e_j\|_2^2 .$$

Then $x_j = A^T y_j$ and $x_{j+1} = x_j + A^T \Delta y_j$ lead to the Kaczmarz method above. Thus for the choice $x_0 = A^T y_0$, the Kaczmarz method is equivalent to the Gauss-Seidel method for (4.97). It follows that one step of the row *SOR* method applied to (4.97) can be written as

$$x_{j+1} = x_j - \mu \frac{e_j^T r_j}{\|A^T e_j\|_2^2} A^T e_j \quad (j = 1, \ldots, m), \tag{4.102}$$

which is a relaxed version of the Kaczmarz method considered in Section 4.3.1.

Assume that A has nonzero columns and let $y_j = e_j$ in cyclic order. Then (4.100) becomes

$$x_{j+1} = x_j - \frac{e_j^T A^T r_j}{\|Ae_j\|_2^2} e_j \quad (j = 1, \ldots, n). \tag{4.103}$$

Note that in the jth step only the jth component of x_j is changed. It is easy to show that (4.103) is equivalent to the Gauss-Seidel method applied to (4.98). Introducing an acceleration or relaxation parameter μ we get the column *SOR* method for (4.98)

$$x_{j+1} = x_j - \mu \frac{e_j^T A^T r_j}{\|Ae_j\|_2^2} e_j \quad (j = 1, \ldots, n). \tag{4.104}$$

Björck and Elfving stress that (4.102) and (4.104) are very efficient ways to implement the successive overrelaxation (*SOR*) methods for (4.97) and (4.98) when A is a sparse matrix. For details we refer to [47].

Björck and Elfving symmetrized the above projection/*SOR* methods to improve the efficiency. If a sweep forward in (4.102) is followed by a sweep backward, i.e.,

$$x_{j+1} = x_j - \mu \frac{e_i^T r_j}{\|A^T e_i\|_2^2} A^T e_i \quad (j = 1, \ldots, 2m,\ i = \min(j, 2m + 1 - j)), \tag{4.105}$$

then we obtain the row *SSOR* method for (4.97). In a complete double sweep the unknowns are transformed to the form

$$x^{(k+1)} = Q_{SSOR} x^{(k)} + R_{SSOR} b, \tag{4.106}$$

where

$$Q_{SSOR} = Q_1 Q_2 \cdots Q_m Q_m \cdots Q_2 Q_1, \quad Q_i = I - \mu A^T e_i e_i^T A / \left\|A^T e_i\right\|_2^2. \tag{4.107}$$

The matrices Q_i and Q_{SSOR} are symmetric.

Similarly, the column *SSOR* method for (4.98) can be written as

$$x_{j+1} = x_j - \mu \frac{e_i^T A^T r_j}{\|Ae_i\|_2^2} e_i \quad (j = 1, \ldots, 2n,\ i = \min(j, 2n + 1 - j)). \tag{4.108}$$

In a double sweep $x^{(k)}$ and the corresponding residual vector $r^{(k)}$ are transformed into

$$x^{(k+1)} = \overline{Q}_{SSOR} x^{(k)} + \overline{R}_{SSOR} b, \quad r^{(k+1)} = \overline{P}_{SSOR} r^{(k)}, \tag{4.109}$$

where

$$\overline{P}_{SSOR} = P_1 P_2 \cdots P_n P_n \cdots P_2 P_1, \quad P_i = I - \mu A e_i e_i^T A^T / \|Ae_i\|_2^2. \tag{4.110}$$

The matrix $\overline{P}_{SSOR}$ is symmetric.

Björck and Elfving [47] apply the general *SSOR* theory [366], [176] to accelerate the above two processes. The idea of the double sweep also appears elsewhere (see, Section 4.3.1 or Kamath and Sameh [216]).

We stress the similarity of the above expressions with those in the alternating projection method .

García-Palomares [151] also suggested a relaxed projection method of the form

$$x_{j+1} = x_j - \mu_j \frac{v_j^T r_j}{v_j^T AMA^T v_j} MA^T v_j \quad (j \geq 0)\,, \tag{4.111}$$

where M is symmetric and positive definite. For $M = I$ the method is a relaxed version of (4.99). Otherwise it can be derived from (4.33). Let $G = M^{-1}$. Then $Y_j = MA^T V_j$ and (4.33) becomes

$$x_{j+1} = x_j - MA^T V_j \left(V_j^T AMA^T V_j\right)^{-1} V_j^T r_j.$$

If $V_j = v_j$ we obtain (4.111) without relaxation. For more on algorithm (4.111), see [317].

Chapter 5

PROJECTION METHODS FOR NONLINEAR ALGEBRAIC EQUATIONS

We investigate two types of projection methods for solving nonlinear algebraic equations of the form

$$F(x) = 0 \quad (F : \mathbb{R}^m \to \mathbb{R}^m), \tag{5.1}$$

where

$$F(x) = [f_1(x), \dots, f_m(x)]^T. \tag{5.2}$$

These methods are generalizations of those of Chapters 3 and 4 and coincide with them on linear systems of the form $F(x) = Ax - b$. It is also a common feature that the methods are of row action type (see, e.g., [77]).

For the Jacobian matrix we use the notation

$$F'(x) = [\partial f_i(x) / \partial x_j]_{i,j=1}^{m} = A(x) \tag{5.3}$$

to invoke the similarities with the linear case. Vector $\omega \in \mathbb{R}^m$ denotes any solution of the equation. We use the following assumptions:

$$\exists A(x)^{-1} \quad (x \in S(\omega, \delta_0)), \tag{5.4}$$

$$\|F(x) - F(y)\| \leq K_0 \|x - y\| \quad (x, y \in S(\omega, \delta_0)) \tag{5.5}$$

and

$$\|A(x) - A(y)\| \leq K_1 \|x - y\|^{\alpha} \quad (x, y \in S(\omega, \delta_0)), \tag{5.6}$$

where $K_0, K_1 \geq 0$, $0 < \alpha \leq 1$ and $\delta_0 > 0$. Condition (5.4) implies that ω is an isolated solution of $F(x) = 0$. We need the following simple result.

LEMMA 5.1 *Assume that $F : \mathbb{R}^m \to \mathbb{R}^m$ satisfies condition (5.6) and let $Z \in \mathbb{R}^{m \times m}$ be arbitrary. Then for any $x, y \in S(\omega, \delta_0)$ there is a matrix $\widehat{A} \in \mathbb{R}^{m \times m}$ such that*

$$F(y) = F(x) + Z(y - x) + \left(\widehat{A} - Z\right)(y - x) \tag{5.7}$$

and $\left\|\widehat{A}-A(x)\right\| \leq \sqrt{m}K_1\|x-y\|^{\alpha}$. *If* $\|Z-A(x)\| \leq \Gamma$, *then*

$$\left\|\widehat{A}-Z\right\| \leq \Gamma+\sqrt{m}K_1\|x-y\|^{\alpha}. \tag{5.8}$$

Proof. Since for any i,

$$f_i(y) = f_i(x) + \nabla f_i(x+\vartheta_i(y-x))^T(y-x) \quad (0<\vartheta_i<1),$$

we have $F(y) = F(x) + \widehat{A}(y-x)$ with

$$\widehat{A} = \left[\nabla f_1(x+\vartheta_i(y-x)), \ldots, \nabla f_m(x+\vartheta_i(y-x))\right]^T.$$

This clearly implies the equality. The rest follows from condition (5.6) and the triangle inequality. ■

Let us consider the following iteration method for solving $F(x)=0$:

$$x_{k+1} = x_k - P_k\left(V_k^T A_k P_k\right)^{-1} V_k^T F(x_k) \quad (k=0,1,\ldots), \tag{5.9}$$

where $A_k \in \mathbb{R}^{m\times m}$, $P_k, V_k \in \mathbb{R}^{m\times m_k}$ and $m_k \leq m$.

When $F(x) = Ax-b$ and $A_k = A$ the iteration changes to

$$x_{k+1} = x_k - P_k\left(V_k^T A P_k\right)^{-1} V_k^T r_k \quad (k \geq 0),$$

which has the form of the Householder, Bauer class of projection methods. If $m_k \equiv m$, and the matrices A_k, V_k and P_k are invertible, then iteration (5.9) becomes the modified Newton method

$$x_{k+1} = x_k - A_k^{-1}F(x_k) \quad (k \geq 0). \tag{5.10}$$

For the choice $A_k = A(x_k)$ we obtain the Newton method.

Assume that $m_k < m$ and substitute

$$F(x_k) = F(\omega) + A_k(x_k-\omega) + \left(\widehat{A}_k - A_k\right)(x_k-\omega)$$

into (5.9). Then we obtain the recursion

$$x_{k+1}-\omega = B_k(x_k-\omega) + C_k(x_k-\omega), \tag{5.11}$$

where

$$B_k = I - P_k\left(V_k^T A_k P_k\right)^{-1} V_k^T A_k \tag{5.12}$$

is a projection and

$$C_k = P_k^T\left(V_k^T A_k P_k\right)^{-1} V_k^T\left(\widehat{A}_k - A_k\right). \tag{5.13}$$

Assuming that

$$\left\| P_k^T \left(V_k^T A_k P_k \right)^{-1} V_k^T \right\| \leq K_2 \quad (k \geq 0) \tag{5.14}$$

and

$$\left\| A_k - A(\omega) \right\| \leq \gamma \left\| x_k - \omega \right\|^{\alpha} \quad (k \geq 0) \tag{5.15}$$

we easily derive the bounds

$$\left\| C_k \right\| \leq K_2 \left(\gamma + \sqrt{m} K_1 \right) \left\| x_k - \omega \right\|^{\alpha} = K_3 \left\| x_k - \omega \right\|^{\alpha} \quad (k \geq 0) \tag{5.16}$$

and

$$\left\| x_{k+1} - \omega \right\| \leq \left\| B_k \right\| \left\| x_k - \omega \right\| + K_3 \left\| x_k - \omega \right\|^{1+\alpha} \quad (k \geq 0).$$

Since $\|B_k\| \geq 1$, we cannot prove convergence in the standard way. We need however special results from Chapters 3 and 4.

We investigate two subclasses of algorithm (5.9). In the first section we deal with the nonlinear Kaczmarz and Cimmino algorithms. These methods have local convergence with linear speed. If the f_i's are convex, then we can rewrite the nonlinear system $F(x) = 0$ as a convex feasibility problem. The convex feasibility problem, which is investigated in Chapter 6, can also be solved by other variants of the Kaczmarz and Cimmino methods. In the second section we develop a nonlinear generalization of the conjugate direction methods of Chapter 3. The nonlinear conjugate direction methods have local convergence of order $1 + \alpha$. In special cases these methods possess monotone and essentially global convergence in the natural partial ordering of $\mathbb{R}^n$. We also point out the reason for the different convergence behavior of the two subclasses.

Although we do not investigate the matter we have to mention that certain Krylov subspace projection methods for linear systems are used in conjunction with the inexact Newton method [65].

1. Extensions of classical iterative projection methods

We investigate nonlinear versions of the Kaczmarz and Cimmino methods studied in Chapter 4.

Tompkins [342] suggested first a nonlinear version of the Kaczmarz method for solving $F(x) = 0$. The Tompkins, Kaczmarz algorithm has the form

$$x_{k+1} = x_k - \frac{f_i(x_k)}{\left\| \nabla f_i(x_k) \right\|^2} \nabla f_i(x_k) \quad (i \equiv k \pmod{m} + 1). \tag{5.17}$$

Tompkins did not prove the convergence of the method. It was McCormick [251], [252] who first showed the local linear convergence of the modified

Kaczmarz method

$$x_{k+1} = x_k - \frac{f_{i(k)}(x_k)}{\left\|\nabla f_{i(k)}(x_k)\right\|^2} \nabla f_{i(k)}(x_k) \quad (k \geq 0) \tag{5.18}$$

for various selection strategies $i = i(k)$ which include the cyclic selection

$$i \equiv k \pmod{m} + 1 \tag{5.19}$$

and the optimal (maximum error) selection

$$\left|f_{i(k)}(x_k)\right| = \max_{1 \leq i \leq m} \left|f_i(x_k)\right|. \tag{5.20}$$

For other strategies we refer to [252] and [77].

Meyn [256] proved the local convergence of the relaxed nonlinear Kaczmarz method

$$x_{k+1} = x_k - \mu \frac{f_i(x_k)}{\left\|\nabla f_i(x_k)\right\|^2} \nabla f_i(x_k) \quad (0 < \mu < 2,\ i \equiv k \pmod{m} + 1). \tag{5.21}$$

Martinez [247], [248], [249], [106] investigated several versions of the nonlinear block Kaczmarz method with relaxation. Assume that $F(x)$, $F'(x)$ and I_m are partitioned in the following way:

$$F(x) = \begin{bmatrix} F_1(x) \\ \vdots \\ F_r(x) \end{bmatrix} \quad \left(F_i : \mathbb{R}^m \to \mathbb{R}^{m_i},\ i = 1, \dots, r\right), \tag{5.22}$$

$$A(x) = \begin{bmatrix} A_1(x) \\ \vdots \\ A_r(x) \end{bmatrix} \quad \left(A_i \in \mathbb{R}^{m_i \times m},\ i = 1, \dots, r\right) \tag{5.23}$$

and

$$I_m = [E_1, \dots, E_r] \quad \left(E_i \in \mathbb{R}^{m \times m_i},\ i = 1, \dots, r\right). \tag{5.24}$$

Hence $F_i(x) = E_i^T F(x)$ and $A_i(x) = E_i^T A(x)$. Then we can define the following relaxed block Kaczmarz method

$$x_{k+1} = x_k - \mu_k A_{i(k)}^T(x_k) \left[A_{i(k)}(x_k) A_{i(k)}^T(x_k)\right]^{-1} F_{i(k)}(x_k) \quad (k \geq 0), \tag{5.25}$$

where $0 < \varepsilon_1 \leq \mu_k \leq 2 - \varepsilon_1$. The algorithm can also be written in the form

$$x_{k+1} = x_k - \mu_k A^T(x_k) E_{i(k)} \left[E_{i(k)}^T A(x_k) A^T(x_k) E_{i(k)}\right]^{-1} E_{i(k)}^T F(x_k), \tag{5.26}$$

which corresponds to (5.9). Martinez proved the local convergence for the cyclic selection $i(k)$ and for the selection $i = i(k)$ defined by

$$\left\|F_{i(k)}(x_k)\right\| \geq \theta \max_{1 \leq i \leq r} \|F_i(x_k)\|. \tag{5.27}$$

For the Cimmino method we know fewer results. Hart and Motzkin [188] developed the Cimmino type method

$$x_{k+1} = x_k - \mu_k \sum_{i=1}^{m} \lambda_i \frac{f_i(x_k)}{\|\nabla f_i(y_k)\|^2} \nabla f_i(y_k) \quad (k \geq 0), \tag{5.28}$$

where $\mu_k > 0$, $\lambda_i > 0$ $(i = 1, \ldots, m)$ and $y_k \in \mathbb{R}^m$ is a parameter vector. Setting $y_k = x_k$ we obtain

$$x_{k+1} = x_k - \mu_k \sum_{i=1}^{m} \lambda_i \frac{f_i(x_k)}{\|\nabla f_i(x_k)\|^2} \nabla f_i(x_k) \quad (k \geq 0), \tag{5.29}$$

which is indeed a relaxed Cimmino method.

THEOREM 5.2 *(Hart, Motzkin). Let $\xi = \sum_{i=1}^{m} \lambda_i$, $0 < \mu_k < 2m/\xi$ and assume that F is continuously differentiable on $S(\omega, \delta_0)$ and satisfies condition (5.4). Then there exist numbers $0 < q < 1$ and $\delta^* > 0$ $(\delta^* \leq \delta_0)$ such that for any $x_0 \in S(\omega, \delta^*)$ and $\{y_k\} \subset S(\omega, \delta^*)$ the sequence $\{x_k\}$ converges to ω and $\|x_k - \omega\| \leq q^k \|x_0 - \omega\|$ $(k \geq 0)$.*

Further references on the nonlinear Cimmino method are [271], [256] and [235]. Ortega and Rheinbolt [271] pointed out that a special case of (5.29) is connected to the Gauss-Newton iteration. Meyn [256] proved the local convergence of algorithm (5.29). Liu [235] developed acceleration schemes for the Kaczmarz and Cimmino algorithms.

Next we prove the convergence of the relaxed nonlinear block Kaczmarz method in the case of constant relaxation and cyclic selection. Thus we investigate the method

$$x_{k+1} = x_k - \mu A^T(x_k) E_i \left[E_i^T A(x_k) A^T(x_k) E_i\right]^{-1} E_i^T F(x_k), \tag{5.30}$$

where $k \geq 0$,

$$0 < \mu < 2, \quad i \equiv k \pmod{m} + 1. \tag{5.31}$$

The presented new proof, which is different from those of Martinez, enables us to compare the nonlinear Kaczmarz type methods with the nonlinear conjugate direction methods of next section.

In the case of algorithm (5.30) iteration (5.11) changes to

$$s_{k+1} = B_k s_k + C_k s_k, \tag{5.32}$$

where $s_k = x_k - \omega$, $B_k = I - \mu R^{(k)}$,

$$R^{(k)} = R_i(x_k) = A^T(x_k) E_i \left[E_i^T A(x_k) A^T(x_k) E_i\right]^{-1} E_i^T A(x_k) \quad (5.33)$$

is an orthogonal projection on $\mathcal{R}\left(A^T(x_k) E_i\right)$ and

$$C_k = A^T(x_k) E_i \left[E_i^T A(x_k) A^T(x_k) E_i\right]^{-1} E_i^T \left(\widehat{A}_k - A(x_k)\right). \quad (5.34)$$

For $0 < \mu < 2$, $\left\|I - \mu R^{(k)}\right\|_2 \leq 1$. If $\left\|A^{-1}(x)\right\| \leq K_2$ for all $x \in S(\omega, \delta_0)$ and $x_k \in S(\omega, \delta_0)$, then

$$\left\|\widehat{A}_k - A(x_k)\right\| \leq \left(\sqrt{m} + 1\right) K_1 \left\|x_k - \omega\right\|^\alpha,$$

$$\left\|A^T(x_k) E_i \left[E_i^T A(x_k) A^T(x_k) E_i\right]^{-1} E_i^T\right\| \leq \left\|A^{-1}(x_k)\right\| \leq K_2$$

and

$$\left\|C_k\right\| \leq \left(\sqrt{m} + 1\right) K_1 K_2 \left\|x_k - \omega\right\|^\alpha = K_3 \left\|x_k - \omega\right\|^\alpha. \quad (5.35)$$

We need the following technical result.

LEMMA 5.3 *Consider the sequence $s_{k+1} = B_k s_k + C_k s_k$ $(k = 1, 2, \ldots)$, where $B_k, C_k \in \mathbb{R}^{m \times m}$, $\|B_k\| \leq K_B$, $\|C_k\| \leq K_C$ $(k = 1, 2, \ldots)$. Then for $k \geq 1$,*

$$s_{k+1} = B_k B_{k-1} \cdots B_1 s_1 + D_k s_1, \quad (5.36)$$

where $\|D_k\| \leq k K_C (K_B + K_C)^{k-1}$.

Proof. For $k = 1$, $s_2 = B_1 s_1 + C_1 s_1$. For $k := k + 1$,

$$\begin{aligned} s_{k+1} &= (B_k + C_k)(B_{k-1} \cdots B_2 B_1 s_1 + D_{k-1} s_1) \\ &= B_k B_{k-1} \cdots B_1 s_1 + C_k B_{k-1} \cdots B_2 B_1 s_1 + (B_k + C_k) D_{k-1} s_1 \\ &= B_k B_{k-1} \cdots B_1 s_1 + D_k s_1. \end{aligned}$$

Using elementary operations we have

$$\|D_k\| \leq K_C K_B^{k-1} + (k-1) K_C (K_B + K_C)^{k-1} \leq k K_C (K_B + K_C)^{k-1}.$$

■

In our case $\|B_i\| \leq 1$ for any i and $0 < \mu < 2$. If $x_i \in S(\omega, \delta_o)$, then $\|C_i\| \leq K_3 \|s_i\|^\alpha \leq K_3 \delta_0^\alpha$. Consider the subsequence $\{x_{nr}\}_{n=0}^\infty$ and set $k = nr$. The solution of recursion (5.32) is given by

$$s_{k+r} = B_{k+r-1} \cdots B_{k+1} B_k s_k + D_{k+r,k} s_k, \quad (5.37)$$

where $\|D_{k+r,k}\| \leq r K_C (1 + K_C)^{r-1}$. The indices of $D_{k+r,k}$ denote the fact that the summation started from s_k.

Substituting $B_{k+j} = I - \mu R_{j+1}(x_{k+j})$ $(j = 0, \ldots, r-1)$ we have

$$s_{k+r} = (I - \mu R_r(x_{k+r-1})) \cdots (I - \mu R_1(x_k)) s_k + D_{k+r,k} s_k, \tag{5.38}$$

where $R_i(x)$ is the orthogonal projection on $\mathcal{R}\left(A^T(x) E_i\right)$. We show that $\|B_{k+r-1} \cdots B_{k+1} B_k\| \leq q_1 < 1$ holds under certain conditions.

LEMMA 5.4 *Let $R_i \in \mathbb{R}^{m \times m}$ be the orthogonal projection on $\mathcal{R}\left(A_i^T\right)$ for $i = 1, \ldots, r$ and assume that $A^T = \left[A_1^T, \ldots, A_r^T\right] \in \mathbb{R}^{m \times m}$ is nonsingular. Then for any fixed $0 < \mu < 2$,*

$$\|(I - \mu R_r) \cdots (I - \mu R_1)\|_2 < 1. \tag{5.39}$$

Proof. We exploit the proof of Theorem 4.30 of Whitney and Meany. For $0 < \mu < 2$, $\|I - \mu R_i\|_2 \leq 1$ for all i. Hence $\|(I - \mu R_i) y\|_2 \leq \|y\|_2$. The equality holds if and only if $R_i y = 0$, that is if $y \in \mathcal{R}^{\perp}\left(A_i^T\right)$. Using induction we prove that $\|(I - \mu R_r) \cdots (I - \mu R_1) y\|_2 = \|y\|_2$ holds if and only if $y \in \mathcal{R}^{\perp}\left(A_i^T\right)$ $(i = 1, \ldots, r)$. For $r = 1$, the claim is true. Assume that $\|(I - \mu R_j) \cdots (I - \mu R_1) y\|_2 = \|y\|_2$ holds if and only if $y \in \mathcal{R}^{\perp}\left(A_i^T\right)$ $(i = 1, \ldots, j)$. Since

$$\begin{aligned} \|y\|_2 &= \|(I - \mu R_{j+1})\left[(I - \mu R_j) \cdots (I - \mu R_1) y\right]\|_2 \\ &\leq \|(I - \mu R_j) \cdots (I - \mu R_1) y\|_2 \leq \|y\|_2, \end{aligned}$$

the assumption implies $y \in \mathcal{R}^{\perp}\left(A_i^T\right)$ for $i = 1, \ldots, j$. But for such y,

$$(I - \mu R_j) \cdots (I - \mu R_1) y = y$$

and we obtain the relation $\|(I - \mu R_{j+1}) y\|_2 = \|y\|_2$ which implies $y \in \mathcal{R}^{\perp}\left(A_{j+1}^T\right)$. Thus we showed that $\|(I - \mu R_r) \cdots (I - \mu R_1) y\|_2 \leq \|y\|_2$ holds with equality if and only if $y \in \mathcal{R}^{\perp}\left(A^T\right) = \{0\}$. Hence for any $0 \neq y \in \mathbb{R}^m$,

$$\|(I - \mu R_r) \cdots (I - \mu R_1) y\|_2 < \|y\|_2$$

and

$$\|(I - \mu R_r) \cdots (I - \mu R_1)\|_2 = \max_{\|y\|_2 \leq 1} \|(I - \mu R_r) \cdots (I - \mu R_1) y\|_2 < 1.$$

■

Taking $A = A(\omega)$ we obtain that

$$\|(I - \mu R_r(\omega)) \cdots (I - \mu R_1(\omega))\|_2 < 1.$$

The continuity of the Jacobian matrix $A(x)$ at $x = \omega$ implies the existence of numbers $0 < \delta_1 < \delta_0$ and $0 < q_1 < 1$ such that

$$\left[A^T(x_{k+r-1}) E_r, \ldots, A^T(x_k) E_1\right]$$

is nonsingular and

$$\|(I - \mu R_r(x_{k+r-1})) \cdots (I - \mu R_1(x_k))\|_2 \le q_1 \tag{5.40}$$

holds for all $x_k, \ldots, x_{k+r-1} \in \overline{S(\omega, \delta_1)}$.

LEMMA 5.5 *If $x_k \in S(\omega, \delta_2)$ and $0 < \delta_2 \le \frac{\delta_1}{2}(1 + 2K_0K_2)^{-r}$, then $x_{k+j} \in S\left(\omega, \frac{\delta_1}{2}\right)$ and $\|s_{k+j}\| \le (1 + 2K_0K_2)^j \|s_k\|$ for $j = 1, \ldots, r$.*

Proof. Since

$$x_{k+1} = x_k - \mu A^T(x_k) E_i \left[E_i^T A(x_k) A^T(x_k) E_i\right]^{-1} E_i^T F(x_k)$$

and

$$\omega = \omega - \mu A^T(x_k) E_i \left[E_i^T A(x_k) A^T(x_k) E_i\right]^{-1} E_i^T F(\omega)$$

we obtain the recursion

$$s_{k+1} = s_k - \mu A^T(x_k) E_i \left[E_i^T A(x_k) A^T(x_k) E_i\right]^{-1} E_i^T \left[F(x_k) - F(\omega)\right].$$

This implies the inequality

$$\|s_{k+1}\| \le \|s_k\| + \mu K_2 \|F(x_k) - F(\omega)\| \le (1 + 2K_0K_2)\|s_k\|.$$

Hence $\|s_{k+j}\| \le (1 + 2K_0K_2)^j \|s_k\| \le \delta_1/2$ for $j = 1, \ldots, r$. ∎

We now give two estimates for $K_C \ge \max_{k \le j \le k+r} \|C_j\|$. If $x_k, \ldots, x_{k+r} \in S(\omega, \delta_0)$, then for $0 \le j \le r$,

$$\|C_{k+j}\| \le K_3 \|s_{k+j}\|^\alpha \le K_3 (1 + 2K_0K_2)^{r\alpha} \|s_k\|^\alpha = K_4 \|s_k\|^\alpha \le K_4 \delta_0^\alpha = K_5.$$

If $x_k \in S(\omega, \delta_2)$, then $x_{k+j} \in S\left(\omega, \frac{\delta_1}{2}\right)$ and

$$\|s_{k+r}\| \le q_1 \|s_k\| + \|D_{r+k,k}\| \|s_k\|$$

where

$$\|D_{r+k,k}\| \le r K_C (1 + K_C)^{r-1} \le r K_4 (1 + K_5)^{r-1} \|s_k\|^\alpha = K_6 \|s_k\|^\alpha.$$

Hence

$$\|s_{k+r}\| \le q_1 \|s_k\| + K_6 \|s_k\|^{1+\alpha}.$$

For any $q_1 < q < 1$ there is a number $\delta_3 > 0$ such that $q_1 + K_6 \delta_3^\alpha = q$. Then $\|s_k\| \le \delta_3$ implies

$$\|s_{k+r}\| \le (q_1 + K_6 \|s_k\|^\alpha) \|s_k\| \le q \|s_k\| < \delta_3.$$

Let $\delta^* = \min\{\delta_2, \delta_3\}$. Then $x_0 \in S(\omega, \delta^*)$ implies that $x_r \in S(\omega, \delta^*)$ and $\|s_r\| \le q \|s_0\| < \delta^*$. Consequently, $x_{nr} \in S(\omega, \delta^*)$ and $\|s_{nr}\| \le q^n \|s_0\|$ hold for $n \ge 0$. Hence $x_{nr} \to \omega$ linearly as $n \to \infty$.

THEOREM 5.6 *Suppose that conditions (5.4)-(5.6) are satisfied and $0 < \mu < 2$. Then there exists a number $0 < \delta^* < \delta_0$ such that for any $x_0 \in S(\omega, \delta^*)$ the relaxed Kaczmarz method (5.30) converges to ω linearly.*

Proof. For $x_0 \in S(\omega, \delta^*)$, $\|x_{nr} - \omega\| \le q^n \|x_0 - \omega\|$ holds ($n \ge 0$). For any $i = nr + j$, $1 \le j < r$, we have

$$\|x_{nr+j} - \omega\| \le (1 + 2K_0K_2)^{r-1} \|x_{nr} - \omega\| = \beta q^n$$

$$\le \beta \frac{\left(q^{\frac{1}{r}}\right)^{nr+j}}{\left(q^{\frac{1}{r}}\right)^{j}} \le \left(\frac{\beta}{\left(q^{\frac{1}{r}}\right)^{r-1}} \right) \left(q^{\frac{1}{r}}\right)^{nr+j}.$$

∎

If we restrict μ such that $0 < \varepsilon_1 \le \mu \le 2 - \varepsilon_1$, then bound (5.40) remains true with another $q_1 < 1$ for any $\mu \in [\varepsilon_1, 2 - \varepsilon_1]$. Hence the method

$$x_{k+1} = x_k - \mu_k A^T(x_k) E_i \left[E_i^T A(x_k) A^T(x_k) E_i\right]^{-1} E_i^T F(x_k) \tag{5.41}$$

with

$$0 < \varepsilon_1 \le \mu_k \le 2 - \varepsilon_1, \quad i \equiv k \pmod m + 1 \tag{5.42}$$

is also linearly convergent under the assumptions (5.4)-(5.6).

The convergence of the Kaczmarz method depends on the product matrix $(I - \mu R_r) \cdots (I - \mu R_1)$, which is not 0, in general. Hence the Kaczmarz method has only linear convergence with rate $\approx \|(I - \mu R_r) \cdots (I - \mu R_1)\|$. Estimates of this quantity are given in Chapter 7.

2. Nonlinear conjugate direction methods

We generalize the finitely terminating Galerkin-Petrov or conjugate direction methods of Chapter 3 as follows. Assume that $1 \le r(i) \le m$, $\sum_{k=1}^{r(i)} m_k^{(i)} = m$,

$$P_k^{(i)}, V_k^{(i)} \in \mathbb{R}^{m \times m_k^{(i)}} \quad (k = 1, \dots, r(i)), \tag{5.43}$$

$$V_j^{(i)T} A_j^{(i)} P_k^{(i)} = 0 \quad (j < k), \tag{5.44}$$

and

$$\det\left(V_k^{(i)T} A_k^{(i)} P_k^{(i)}\right) \ne 0 \quad (k = 1, \dots, r(i)). \tag{5.45}$$

The last two conditions imply that

$$\left[V_j^{(i)T} A_j^{(i)} P_k^{(i)}\right]_{j,k=1}^{r(i)} \tag{5.46}$$

is a nonsingular block lower triangular matrix. Hence we impose a kind of A-conjugacy.

Algorithm 1 (Nonlinear Conjugate Direction Method)

$x_1 \in \mathbb{R}^m$
for $i = 1, 2, \ldots$
  $y_1 = x_i$
  for $k = 1, \ldots, r(i)$

$$y_{k+1} = y_k - P_k^{(i)} \left(V_k^{(i)T} A_k^{(i)} P_k^{(i)} \right)^{-1} V_k^{(i)T} F(y_k)$$

  end
  $x_{i+1} = y_{r(i)+1}$
end

If $F(x) = Ax - b$ and $A_j^{(1)} = A$ ($j = 1, \ldots, r(1)$) the algorithm coincides with Algorithm CDM and finitely terminates. If $V_k^{(i)}$'s are partial permutation matrices, then we have a row-action method (see, e.g., [77]).

Note that $r(i)$, the matrices $P_k^{(i)}$, $V_k^{(i)}$ and the partition $\left\{ m_1^{(i)}, \ldots, m_{r(i)}^{(i)} \right\}$ can change with the major iteration i. For simplicity we drop the upper index i whenever possible and use only $r(i)$ as a reminder.

THEOREM 5.7 *Assume that conditions (5.4)-(5.6) hold and let the matrices* V_k, P_k *be such that*

$$\left\| P_k \left(V_k^T A_k P_k \right)^{-1} V_k^T \right\| \le K_2 \quad (k = 1, \ldots, r(i);\ i \ge 1) \tag{5.47}$$

holds. There exist $\Gamma^*, \delta^* > 0$ *such that if*

$$\|A_k - A(\omega)\| \le \Gamma^* \quad (k = 1, \ldots, r(i);\ i \ge 1) \tag{5.48}$$

and $x_1 \in S(\omega, \delta^*)$, *then Algorithm 1 converges to* ω *with a linear speed. If* $\|A_k - A(\omega)\| \le K_6 \|x_i - \omega\|^\alpha$ $(k = 1, \ldots, r(i), \quad i \ge 1)$, *the order of convergence is at least* $1 + \alpha$.

Proof. Let $K_3 = (1 + K_0 K_2)^m$ and $\delta_1 \le \delta_0 / (2K_3)$. Consider the identity

$$s_{k+1} = s_k - P_k \left(V_k^T A_k P_k \right)^{-1} V_k^T \left[F(y_k) - F(\omega) \right], \tag{5.49}$$

where $s_k = y_k - \omega$ $(k = 1, \ldots, r(i))$. If $y_1 \in S(\omega, \delta_1)$ then conditions (5.5) and (5.47) imply

$$\|s_{k+1}\| \le (1 + K_0 K_2) \|s_k\| \le (1 + K_0 K_2)^k \|s_1\| \le K_3 \|s_1\| \le \frac{\delta_0}{2}$$

and $y_{k+1} \in S(\omega, \delta_0/2)$ for $k = 1, \ldots, r(i)$. Let $\gamma_k = \|A_k - A(\omega)\|$ and consider the expansion

$$F(y_k) = F(\omega) + A_k(y_k - \omega) + (\hat{A}_k - A_k)(y_k - \omega), \tag{5.50}$$

where

$$\left\|A_k - \hat{A}_k\right\| \leq \gamma_k + \sqrt{m}K_1 \left\|s_k\right\|^{\alpha}.$$

By substituting expansion (5.50) into (5.49) we have the nonlinear recursion

$$s_{k+1} = B_k s_k + C_k s_k \tag{5.51}$$

with $B_k = I_m - R_k$, $R_k = P_k(V_k^T A_k P_k)^{-1} V_k^T A_k$ and

$$C_k = P_k(V_k^T A_k P_k)^{-1} V_k^T (A_k - \hat{A}_k).$$

Observe that R_k is an oblique projection of rank m_k, $\mathcal{N}(R_k) = \mathcal{R}^{\perp}(A_k^T V_k)$ and $\mathcal{R}(R_k) = \mathcal{R}(P_k)$. We can establish the following two bounds:

$$\begin{aligned} \|B_k\| &\leq \|I_m\| + \|R_k\| \leq \sqrt{m} + K_2 \|A_k\| \\ &\leq \sqrt{m} + K_2 \left(\|A(\omega)\| + \max_k \gamma_k\right) = K_B \end{aligned}$$

and

$$\begin{aligned} \|C_k\| &\leq K_2 \left\|A_k - \hat{A}_k\right\| \\ &\leq K_2 \left(\gamma_k + \sqrt{m}K_1 \|s_k\|^{\alpha}\right) \\ &\leq K_2 \left((\max_k \gamma_k) + \sqrt{m}K_1 \left(\tfrac{\delta_0}{2}\right)^{\alpha}\right) = K_C. \end{aligned}$$

The solution of recursion (5.51) is given by Lemma 5.3:

$$s_{k+1} = B_k \cdots B_1 s_1 + D_k s_1, \tag{5.52}$$

where $\|D_k\| \leq k K_C \left(K_B + K_C\right)^{k-1}$. Thus we have the bound

$$\|s_{k+1}\| \leq \|B_k \cdots B_1 s_1\| + k K_C \left(K_B + K_C\right)^{k-1} \|s_1\|. \tag{5.53}$$

The first part of the bound is of $O\left(\|s_1\|\right)$ provided that $B_k \cdots B_1 \neq 0$. If K_C is small enough, say $O\left(\|s_1\|^{\alpha}\right)$, then the second part of the bound is a quantity of $O\left(\|s_1\|^{1+\alpha}\right)$. Consequently, the error of the last minor iterate depends on the properties of the matrix $Q_k = B_k \cdots B_1$. As $R_j R_k = 0$ for $j < k$, Theorem 3.8 implies that Q_k is also a projection with

$$\mathcal{R}(Q_k) = \cap_{j=1}^{k} \mathcal{N}(R_j) \quad \mathcal{N}(Q_k) = \cup_{j=1}^{k} \mathcal{R}(R_j).$$

Properties $\mathcal{R}(R_j) = \mathcal{R}(P_j)$ and $\operatorname{rank}(P) = m$ imply

$$\mathcal{N}(Q_k) = \mathcal{R}([P_1, \ldots, P_k]). \tag{5.54}$$

The relation

$$\mathcal{R}(Q_k) = [\cup_{j=1}^{k} \mathcal{R}(A_j^T V_j)]^{\perp} = \mathcal{R}([P_{k+1}, \ldots, P_{r(i)}]) \tag{5.55}$$

follows from $\mathcal{N}(R_j) = \mathcal{R}^{\perp}\left(A_j^T V_j\right)$ $(j = 1, \ldots, r(i))$ and the orthogonality relation $A_j^T V_j \perp P_t$ $(j < t)$ (see conditions (5.44), (5.45)). For $k = r(i)$ the matrix Q_k is the zero projection with $\mathcal{N}(Q_k) = \mathcal{R}(P)$ and $\mathcal{R}(Q_k) = \{0\}$. Hence $Q_{r(i)} = 0$ and we have the estimate

$$\left\|s_{r(i)+1}\right\| \leq r(i)\, K_C \left(K_B + K_C\right)^{r(i)-1} \|s_1\| .$$

Let $\gamma_k \leq \Gamma$ for all i and k. As

$$\begin{aligned} K_B + K_C &\leq \sqrt{m} + K_2\left(\|A(\omega)\| + 2\left(\max_k \gamma_k\right) + \sqrt{m}K_1\left(\frac{\delta_0}{2}\right)^{\alpha}\right) \\ &\leq \sqrt{m} + K_2\left(\|A(\omega)\| + 2\Gamma + \sqrt{m}K_1\left(\frac{\delta_0}{2}\right)^{\alpha}\right) = K_4 \end{aligned}$$

we can write

$$r(i)\, K_C \left(K_B + K_C\right)^{r(i)-1} \leq K_5\left(\max_k \gamma_k + \|s_1\|^{\alpha}\right),$$

where $K_5 = mK_2 \max\left\{1, \sqrt{m}K_1K_3^{\alpha}\right\} K_4^{m-1}$. Hence

$$\|x_{i+1} - \omega\| \leq K_5\left(\max_k \gamma_k + \|x_i - x^*\|^{\alpha}\right)\|x_i - \omega\| . \tag{5.56}$$

Let $0 < \Gamma^* \leq \Gamma$ and $\delta^* \leq \delta_1$ be so small that $K_5\left(\Gamma^* + (\delta^*)^{\alpha}\right) \leq 1/2$. Then

$$\|x_{i+1} - \omega\| \leq \frac{1}{2}\|x_i - \omega\| .$$

If $x_1 \in S(\omega, \delta^*)$ and $\gamma_k \leq \Gamma^*$ for all i and k, then

$$\|x_{i+1} - \omega\| \leq \frac{1}{2}\|x_i - \omega\| \leq \cdots \leq \left(\frac{1}{2}\right)^{i}\|x_1 - \omega\| .$$

Thus we proved the linear convergence. If $\max_k \gamma_k \leq K_6 \|s_1\|^{\alpha}$, then

$$K_5\left(\max_k \gamma_k + \|s_1\|^{\alpha}\right) \leq K_7 \|s_1\|^{\alpha}$$

with $K_7 = K_5 \max\{1, K_6\}$ and

$$\left\|s_{r(i)+1}\right\| \leq K_7 \|s_1\|^{1+\alpha} .$$

Assume now that $0 < \delta^* \leq \delta_1$ satisfies $K_7(\delta^*)^{\alpha} \leq 1/2$. Then for $y_1 = x_i \in S(\omega, \delta^*)$ the inequality $\|x_{i+1} - \omega\| \leq (1/2)\|x_i - \omega\|$ holds. If $x_1 \in S(\omega, \delta^*)$

then we have $x_i \to \omega$ $(i \to +\infty)$. The order of convergence is given by the much sharper estimation

$$\|x_{i+1} - \omega\| \leq K_7^{-1/\alpha} \left(K_7^{1/\alpha} \|x_i - \omega\| \right)^{1+\alpha} \leq K_7^{-1/\alpha} \left(K_7^{1/\alpha} \delta^* \right)^{(1+\alpha)^i}$$

where $K_7^{1/\alpha}\delta^* \leq (1/2)^{1/\alpha} < 1$. ■

REMARK 5.8 *If* $\max_k \gamma_k \leq K_6 \|s_1\|^\beta$ *with* $0 < \beta \leq \alpha$, *then the order of convergence is* $1 + \beta$.

Assume now that

$$\max_k \gamma_k \leq K_6 \|s_1\|^\beta, \quad (i \geq 1,\, 0 < \beta \leq \alpha). \tag{5.57}$$

The proof of Theorem 5.7 reveals that in the minor iteration steps the error vector s_{k+1} has a nonvanishing first order component $B_k s_k$ that dominates the $(1+\beta)$-order component $C_k s_k$. The quantity $\|B_k\|$ is minimal if and only if $\mathcal{R}(P_k) = \mathcal{R}(A_k^T V_k)$.

Consider the estimate

$$\|s_{k+1}\| \leq \|B_k \cdots B_1 s_1\| + k K_C \left(K_B + K_C\right)^{k-1} \|s_1\|. \tag{5.58}$$

The second part of the bound is of $O\left(\|s_1\|^{1+\beta}\right)$. Hence the dominating part of the error s_{k+1} is given by $B_k \cdots B_1 s_1$ in the term of s_1. The projection $Q_k = B_k \cdots B_1$ maps the error vector $s_1 = x_i - \omega$ onto $\mathcal{R}\left(\left[P_{k+1}, \dots, P_{r(i)}\right]\right)$ along $\mathcal{R}\left([P_1, \dots, P_k]\right)$. Therefore $\|B_k \cdots B_1 s_1\|$ depends on s_1 and the choice of $P_1, \dots, P_k$.

We point out that the property $B_{r(i)} \cdots B_1 = 0$ makes the difference between the convergence behavior of the nonlinear conjugate direction methods and the Kaczmarz method, where the corresponding matrix

$$(I - \mu R_r) \cdots (I - \mu R_1)$$

is nonzero.

The upper estimation $\|B_k \cdots B_1 s_1\| \leq \|B_k \cdots B_1\| \, \|s_1\|$ has a minimal error constant if and only if the projection $B_k \cdots B_1$ is symmetric. The matrix Q_k is symmetric if and only if $\mathcal{R}(Q_k) = \mathcal{N}^\perp(Q_k)$. From (5.54) and (5.55) it follows that Q_k is symmetric for all $k = 1, \dots, r(i)$ if and only if $P^T P$ is a diagonal matrix. It is noted that $\|B_k \cdots B_1\| \leq \mathrm{cond}(P)$ holds in any submultiplicative unitarily invariant matrix norm.

Finally we note that condition (5.47) is equivalent with the condition

$$\|R_k\| = \left\| P_k \left(V_k^T A_k P_k\right)^{-1} V_k^T A_k \right\| \leq \Gamma_2 \quad (k = 1, \dots, r(i);\ i \geq 1) \tag{5.59}$$

if $\|A_k - A(\omega)\| \leq \Gamma^*$ and Γ^* is small enough. Therefore condition (5.47) which is the key of the convergence is satisfied if and only if the oblique projections R_k are uniformly bounded.

2.1 Particular methods

In practice $A_k^{(i)}$ is an approximation to $A(\omega)$. Before discussing any particular method we have to remark that the rank reduction procedure and the other conjugation procedures mentioned in Chapter 3 can be used to produce directions V_k and P_k that satisfy (5.44), (5.45).

2.1.1 Methods with fixed direction matrices

Assume that $A_k^{(i)} = A$, $V_k^{(i)} = V_k$, $P_k^{(i)} = P_k$, $r(i) = r$ and

$$\left\{m_1^{(i)}, \ldots, m_{r(i)}^{(i)}\right\} = \{m_1, \ldots, m_r\}$$

for all $i \geq 1$. Then Algorithm 1 has the form

Algorithm 2

$x_1 \in \mathbb{R}^m$
for $i = 1, 2, \ldots$
  $y_1 = x_i$
  for $k = 1, \ldots, r$
    $y_{k+1} = y_k - P_k \left(V_k^T A P_k\right)^{-1} V_k^T F(y_k)$
  end
  $x_{i+1} = y_{r+1}$
end

By setting

$$K_2 = \max_{1 \leq k \leq r} \left\| P_k \left(V_k^T A_k P_k\right)^{-1} V_k^T \right\| \tag{5.60}$$

we satisfy condition (5.47). Hence Algorithm 2 has linear convergence rate provided that approximations $x_1 \approx \omega$ and $A \approx A(\omega)$ are good enough.

The multistep version of Algorithm 2 uses V_k, P_k and A_k for a fixed number of iterations and then recalculates them. The form of the algorithm is the following

Algorithm 3 (Multistep version of Algorithm 2)

$x_1 \in \mathbb{R}^m$
for $i = 1, 2, \ldots$
  $z_1 = x_i$
  for $j = 1, \ldots, t+1$
    $y_1^{(j)} = z_j$
    for $k = 1, \ldots, r$
      $y_{k+1}^{(j)} = y_k^{(j)} - P_k^{(i)} \left(V_k^{(i)T} A_k^{(i)} P_k^{(i)}\right)^{-1} V_k^{(i)T} F\left(y_k^{(j)}\right)$
    end
    $z_{j+1} = y_{r+1}^{(j)}$
  end
  $x_{i+1} = z_{t+2}$
end

THEOREM 5.9 *Assume that conditions (5.4)-(5.6) hold with* $\alpha = 1$ *and the matrices* V, P *are such that*

$$\left\| P_k \left(V_k^T A_k P_k \right)^{-1} V_k^T \right\| \leq K_2 \quad (k = 1, \ldots, r;\ i \geq 1) \tag{5.61}$$

holds. There exist $\Gamma^*, \delta^* > 0$ *such that if*

$$\| A_k - A(\omega) \| \leq \Gamma^* \quad (k = 1, \ldots, r;\ i \geq 1) \tag{5.62}$$

and $x_1 \in S(\omega, \delta^*)$, *then Algorithm 3 converges to* ω *with a linear speed. If* $\| A_k - A(\omega) \| \leq K_6 \| x_i - \omega \|$ $(k = 1, \ldots, r,\ \ i \geq 1)$, *the order of convergence is at least* $t + 2$.

Proof. By repeating the arguments of the proof of Theorem 5.7 we obtain the inequality

$$\| z_{j+1} - \omega \| \leq K_5 \left(\max_k \gamma_k + \| z_j - \omega \| \right) \| z_j - \omega \| \quad (j = 1, \ldots, t+1).$$

If $0 < \Gamma^* \leq \Gamma$ and $\delta^* \leq \delta_1$ are such that $K_5 (\Gamma^* + (\delta^*)^\alpha) \leq 1/2$, then $\| z_{j+1} - \omega \| \leq (1/2) \| z_j - \omega \|$ and

$$\| x_{i+1} - \omega \| \leq \frac{1}{2^{t+1}} \| x_i - \omega \| .$$

Thus we have the linear convergence speed. If $\| A_k - A(\omega) \| \leq K_6 \| x_i - \omega \|$, then by induction we can prove that $\| z_j - \omega \| \leq K^{(j)} \| z_1 - \omega \|^j$ $(j \geq 2)$. Hence we have

$$\| x_{i+1} - \omega \| \leq K^{(t+2)} \| x_i - \omega \|^{t+2}$$

which is one order better than before. Thus the convergence order is not less than $t+2$. It is noted that Γ^* and δ^* must be significantly smaller than those of Theorem 5.7. ∎

The multistep version improves the speed of Algorithm 2. Similar results on multistep versions of classical methods can be found in Ortega and Rheinboldt [271].

2.1.2 Methods with Quasi-Newton update matrices

Let $r(i) = m$, $A_k^{(i)} = A^{(i)}$ $(k = 1, \ldots, m)$ and calculate $A^{(i+1)}$ via the Broyden update :

$$\begin{aligned} x_{i+1} &= y_{m+1}, \\ s_i &= x_{i+1} - x_i, \\ Y_i &= F(x_{i+1}) - F(x_i), \\ A^{(i+1)} &= A^{(i)} + \left(Y_i - A^{(i)} s_i \right) s_i^T / \left(s_i^T s_i \right). \end{aligned}$$

Let A^T have the QR factorization $A^T = QR$. The pair (I, Q) is then A-conjugate because $IAQ = R^T$ is lower triangular. Hence we can define the following special case of Algorithm 1.

Algorithm 4

$x_1 \in \mathbb{R}^m$, $A^{(1)} \in \mathbb{R}^{m\times m}$
for $i = 1, 2, \ldots$
  $y_1 = x_i$
  Calculate $QR = \left[A^{(i)}\right]^T$ and set $P = Q$
  for $k = 1, \ldots, m$
    $y_{k+1} = y_k - Pe_k\left(e_k^T A^{(i)} Pe_k\right)^{-1} e_k^T F(y_k)$
  end
  $x_{i+1} = y_{m+1}$
  $Y_i = F(x_{i+1}) - F(x_i)$
  $s_i = x_{i+1} - x_i$
  $A^{(i+1)} = A^{(i)} + \left(Y_i - A^{(i)} s_i\right) s_i^T / \left(s_i^T s_i\right)$
end

Note that $e_k^T A^{(i)} Pe_k = r_{kk}$, where $R = [r_{ij}]_{i,j=1}^m$. We now use the theory of quasi-Newton methods [101] to prove that $A^{(i)}$ remains a sufficiently good approximation to $A(\omega)$. We assume that $\alpha = 1$. The lemma of bounded deterioration [101] (Lemma 8.2.1, pp. 175, 176) says, that if $x_{i+1}, x_i \in S(\omega, \delta_0)$ and $x_i \neq \omega$, then

$$\left\|A^{(i+1)} - A(\omega)\right\| \leq \left\|A^{(i)} - A(\omega)\right\| + \frac{K_1}{2}\left(\|x_{i+1} - \omega\| + \|x_i - \omega\|\right) \quad (5.63)$$

holds for the Broyden-update. Suppose that $\Gamma^* = 2\delta^*$ and $K_5(\Gamma^* + \delta^*) = 3K_5\delta^* \leq 1/2$. Let $\varepsilon = \min\left\{\delta^*, \frac{2\delta^*}{3K_1}\right\}$, $\|x_1 - \omega\| \leq \varepsilon$, $\left\|A^{(1)} - A(\omega)\right\| \leq \delta^*$. We prove by induction that

$$\left\|A^{(i)} - A(\omega)\right\| \leq \Gamma^* \quad (i \geq 1).$$

By the lemma of bounded deterioration we have

$$\left\|A^{(2)} - A(\omega)\right\| \leq \left\|A^{(1)} - A(\omega)\right\| + \frac{3K_1}{4}\varepsilon \leq \delta^* + \frac{\delta^*}{2} < \Gamma^*.$$

Assume that $\left\|A^{(i)} - A(\omega)\right\| \leq \delta^* + \left(\sum_{j=1}^{i-1} 2^{-j}\right)\delta^*$. Again by the lemma we get

$$\left\|A^{(i+1)} - A(\omega)\right\| \leq \left\|A^{(i)} - A(\omega)\right\| + \frac{3K_1}{2^{i+1}}\varepsilon \leq \delta^* + \left(\sum_{j=1}^{i} 2^{-j}\right)\delta^* < \Gamma^*.$$

Hence we proved that the quasi-Newton update matrices $A^{(i)}$ are bounded and are sufficiently close to $A(\omega)$.

THEOREM 5.10 *Assume that conditions (5.4)-(5.6) hold with* $\alpha = 1$. *There exist* $\varepsilon, \delta_1 > 0$ *such that if* $\left\| A^{(1)} - A(\omega) \right\| \leq \delta_1$ *and* $\|x_1 - \omega\| \leq \varepsilon$, *then Algorithm 4 converges to* ω *linearly.*

Proof. Since condition (5.47) becomes

$$\left\| Pe_k \left(e_k^T A^{(i)} Pe_k \right)^{-1} e_k^T \right\| \leq 1/\left| r_{kk} \right|,$$

the convergence follows from Theorem 5.7 and the continuity of the QR factorization. ∎

The advantage of Algorithm 4 is that the QR factorization of $A^{(i+1)}$ can be calculated from the QR factorization of $A^{(i)}$ in $O\left(m^2\right)$ arithmetic operations (see, e.g., [101], [165]). Derivation and variants of Algorithm 4 and related results can be found in papers [149], [140], [141]. It is worth noting that O'Leary [268] showed that the Broyden method itself when applied to linear systems is a kind of projection method.

2.1.3 The nonlinear ABS methods

The generation of direction matrices V_k, P_k is done through the rank reduction algorithm in the following way.

Algorithm 5 (The block nonlinear ABS method)
$x_1 \in \mathbb{R}^m$
for $i = 1, 2, \ldots$
  $y_1 = x_i$, $H_1 = I$
  for $k = 1, \ldots, r(i)$
    $u_k = \sum_{j=1}^{k} \tau_{jk} y_j \quad \left(\tau_{jk} \geq 0,\ \sum_{j=1}^{k} \tau_{jk} = 1 \right)$
    $P_k = H_k^T Z_k$
    $y_{k+1} = y_k - P_k \left(V_k^T A(u_k) P_k \right)^{-1} V_k^T F(y_k)$
    $H_{k+1} = H_k - H_k A^T(u_k) V_k \left(W_k^T H_k A^T(u_k) V_k \right)^{-1} W_k^T H_k$
  end
  $x_{i+1} = y_{r(i)+1}$
end

The parameters $Z_k, W_k \in \mathbb{R}^{m \times m_k}$ are subject to the conditions

$$\det\left(P_k^T A^T(u_k) V_k \right) \neq 0 \quad \text{and} \quad \det\left(W_k^T H_k A^T(u_k) V_k \right) \neq 0. \tag{5.64}$$

Using the properties of the rank reduction algorithm one can show that matrices $A_k = A(u_k)$, P_k and V_k satisfy conditions (5.44) and (5.45). It is also easy to see that

$$\begin{aligned} \|A_k - A(\omega)\| = \|A(u_k) - A(\omega)\| &\leq \sqrt{m} K_1 \max_{1 \leq j \leq k} \|s_j\|^\alpha \\ &\leq \sqrt{m} K_1 K_3^\alpha \|x_i - \omega\|^\alpha . \end{aligned}$$

Hence by Theorem 5.7, Algorithm 5 has local convergence of order $1+\alpha$.

We mention the following special cases when $r(i) \equiv m$.

(i) Implicit LU or Brown method :

$$V_k = W_k = Z_k = e_k \quad (k = 1, \ldots, m); \tag{5.65}$$

(ii) Huang method :

$$Z_k = W_k = A(u_k) V_k \quad (k = 1, \ldots, m); \tag{5.66}$$

(iii) Symmetrized conjugate direction methods :

$$W = Z, \quad V_k = C(u_k) P_k \quad (k = 1, \ldots, m), \tag{5.67}$$

where $C(x)$ is continuous and $G(x) = C^T(x) A(x)$ is symmetric and positive definite in $S(\omega, \delta_0)$.

If conditions (5.4)-(5.6) are satisfied, then it can be shown [137] that method (i) is locally convergent if $A(\omega)$ is strongly nonsingular, method (ii) is locally convergent if $[V_1, \ldots, V_m]$ is nonsingular, and method (iii) is locally convergent if Z is nonsingular.

The nonlinear ABS methods were developed in [3] and [2] (see, also [4]). Theorem 5.7 is a generalization of the convergence results [134], [136], [137] developed for the nonlinear ABS methods. A very efficient special case of the block implicit LU method is given in [145].

2.2 Monotone convergence

In certain special cases the nonlinear conjugate direction methods have monotone and essentially global convergence in the natural partial ordering. We also show that a particular conjugate direction method is faster than Newton's method also in the partial ordering.

We require the following concepts and results. The function $F : \mathbb{R}^m \to \mathbb{R}^m$ is said to be convex on a convex set $D \subseteq \mathbb{R}^m$ if

$$F(\alpha x + (1 - \alpha y)) \le \alpha F(x) + (1 - \alpha) F(y) \tag{5.68}$$

holds for all $x, y \in D$ and $\alpha \in [0,1]$. We recall the following basic result [271], [270]. Assume that $F : \mathbb{R}^m \to \mathbb{R}^m$ is differentiable on the convex set D. Then F is convex on D if and only if

$$F(y) - F(x) \ge F'(x)(y - x) \tag{5.69}$$

holds for all $x, y \in D$. The function $F : \mathbb{R}^m \to \mathbb{R}^k$ is said to be isotone, if $x \le y$ implies $F(x) \le F(y)$.

The monotone convergence will be proven under the following conditions:
(a) $F : \mathbb{R}^m \to \mathbb{R}^m$ *is continuously differentiable and convex on* $\mathbb{R}^m$*;*
(b) $A(x) = F'(x)$ *is a nonsingular M-matrix for all* $x \in \mathbb{R}^m$ *and* $A : \mathbb{R}^m \to \mathbb{R}^{m\times m}$ *is isotone.*

If there is a solution ω of the equation $F(x) = 0$, then under the condition (a) $F(z) \geq 0$ implies $z \geq \omega$, and $F(z) \leq 0$ implies $z \leq \omega$. Furthermore the solution ω is unique if it exists (see, e.g., [270]). Here we assume that ω exists. Characterizations of functions satisfying (a) and (b) can be found in [271], [270], [314], [131], [132] and the references therein.

Consider the following special case of Algorithm 1:

$$\left.\begin{array}{rcl} y_1 & = & x_i \\ y_{k+1} & = & y_k - p_k\left(e_k^T A_i p_k\right)^{-1} e_k^T F(y_k) \quad (k = 1,\dots,m) \\ x_{i+1} & = & y_{m+1} \end{array}\right\} (i \geq 1), \tag{5.70}$$

where

$$A_i = A(x_i), \quad P = [p_1,\dots,p_m] = Q_i U_{A_iQ_i}^{-1}. \tag{5.71}$$

Such directions p_k can be produced by the rank reduction or the other conjugation procedures studied in Chapter 3. The nonlinear ABS algorithm (Algorithm 5) produces the p_k's within the minor iteration loop.

LEMMA 5.11 *If* $Q_i \geq 0$ *and* A_iQ_i *is a nonsingular M-matrix, then*
(i) $e_k^T A_i p_k > 0$ $(k = 1,\dots,m)$*;*
(ii) $P \geq 0$.

Proof. We recall the fact that a nonsingular matrix A is an M-matrix if and only if there exist lower and upper triangular M-matrices R and S, respectively, such that $A = RS$ ([121], [43], [314], [361]). Since R and S are also nonsingular, it follows that the factors of the LDU factorization $A = L_A D_A U_A$ are all nonsingular M-matrices. Hence if A_iQ_i is a nonsingular M-matrix, then $U_{A_iQ_i}^{-1} \geq 0$ and $P = Q_i U_{A_iQ_i}^{-1} \geq 0$. Also, we have

$$e_k^T A_i p_k = e_k^T A_i Q_i U_{A_iQ_i}^{-1} e_k = e_k^T L_{A_iQ_i} D_{A_iQ_i} e_k = e_k^T D_{A_iQ_i} e_k > 0.$$

■

The following result establishes a monotonicity property of M-matrices [139], [138].

THEOREM 5.12 *Let* A *and* B *be two M-matrices such that* $A \leq B$ *and assume that* A *is nonsingular or irreducible singular and* B *is nonsingular. Let* $A = L_A V_A$ *and* $B = L_B V_B$ *be the LU factorizations of* A *and* B *such that both* L_A *and* L_B *are unit lower triangular. Then*

$$L_A \leq L_B, \quad V_A \leq V_B. \tag{5.72}$$

In addition, if $V_A = D_A U_A$ and $V_B = D_B U_B$, where U_A and U_B are unit upper triangular, D_A and D_B are diagonal matrices, then

$$D_A \leq D_B, \quad U_A \leq U_B. \tag{5.73}$$

Proof. We prove the result by induction. Assume that $k \times k$ matrices A and B are nonsingular such that $L_A \leq L_B$ and $V_A \leq V_B$ hold. Let

$$A' = \begin{bmatrix} A & c \\ r^T & a \end{bmatrix}, \quad B' = \begin{bmatrix} B & p \\ q^T & b \end{bmatrix} \quad \left(c, r, p, q \in \mathbb{R}^k\right).$$

Assume that A' and B' are also nonsingular M-matrices. They have the LU-factorizations

$$A' = \begin{bmatrix} L_A & 0 \\ r^T V_A^{-1} & 1 \end{bmatrix} \begin{bmatrix} V_A & L_A^{-1} c \\ 0 & a - r^T A^{-1} c \end{bmatrix},$$

$$B' = \begin{bmatrix} L_B & 0 \\ q^T V_B^{-1} & 1 \end{bmatrix} \begin{bmatrix} V_B & L_B^{-1} p \\ 0 & b - q^T B^{-1} p \end{bmatrix}.$$

By assumption $L_A \leq L_B$, $V_A \leq V_B$, $c \leq p \leq 0$, $r \leq q \leq 0$ and $0 < a \leq b$. Relation $L_A^{-1} \geq L_B^{-1} \geq 0$ implies $L_A^{-1} c \leq L_B^{-1} p \leq 0$. Inequality $V_A^{-1} \geq V_B^{-1} \geq 0$ implies $r^T V_A^{-1} \leq q^T V_B^{-1} \leq 0$. Notice that $a - r^T A^{-1} c = \det(A') / \det(A) > 0$ and $b - q^T B^{-1} p = \det(B') / \det(B) > 0$. Finally $0 < a - r^T A^{-1} c \leq b - q^T B^{-1} p$ follows from $A^{-1} \geq B^{-1} \geq 0$. Thus we proved the first part of the theorem for the nonsingular case. If A' is an irreducible singular matrix of order m, then by Theorem 4.16 of Berman and Plemmons [43] (p.156) A' has rank $m-1$, each principal submatrix of A' other than A' itself is a nonsingular M-matrix, $a = r^T A^{-1} c$ and A' has the LU-factorization

$$A' = \begin{bmatrix} L_A & 0 \\ r^T V_A^{-1} & 1 \end{bmatrix} \begin{bmatrix} V_A & L_A^{-1} c \\ 0 & 0 \end{bmatrix}$$

with singular upper triangular matrix $V_{A'}$. As $0 \leq b - q^T B^{-1} p$ the theorem also holds in this case. Let $V_A = D_A U_A$ and $V_B = D_B U_B$. As A^T and B^T are also M-matrices that satisfy $A^T \leq B^T$, we have the LU-factorizations $A^T = U_A^T (D_A L_A^T)$ and $B^T = U_B^T (D_B L_B^T)$ with $U_A^T \leq U_B^T$ and $D_A L_A^T \leq D_B L_B^T$. This implies $U_A \leq U_B$ and $0 \leq D_A \leq D_B$. The same reasoning applies to A' and B' if they are nonsingular. If A' is irreducible singular, then

$$D_{A'} U_{A'} = \begin{bmatrix} D_A & 0 \\ 0 & 0 \end{bmatrix} \begin{bmatrix} U_A & D_A^{-1} L_A^{-1} c \\ 0 & 0 \end{bmatrix},$$

$$D_{B'}U_{B'} = \begin{bmatrix} D_B & 0 \\ 0 & b - q^T B^{-1} p \end{bmatrix} \begin{bmatrix} U_B & D_B^{-1} L_B^{-1} p \\ 0 & 1 \end{bmatrix},$$

from which $0 \le D_{A'} \le D_{B'}$ immediately follows. As $D_A^{-1} L_A^{-1} c \le D_B^{-1} L_B^{-1} p \le 0$ the rest of the theorem also follows. ∎

From Theorem 4.16 of Berman and Plemmons [43] (p. 156) it follows that Theorem 5.12 does not hold if B is irreducible singular and $A \neq B$. Theorem 5.12 is not true for general matrices. The matrices

$$A = \begin{bmatrix} 1 & -4 & 9 & -12 \\ -4 & 17 & -40 & 57 \\ 9 & -40 & 98 & -148 \\ -12 & 57 & -148 & 242 \end{bmatrix}, \quad B = \begin{bmatrix} 2 & -4 & 9 & -12 \\ -4 & 17 & -40 & 57 \\ 9 & -40 & 98 & -148 \\ -12 & 57 & -148 & 242 \end{bmatrix}$$

are monotone, $A \le B$ and

$$L_A = \begin{bmatrix} 1 & 0 & 0 & 0 \\ -4 & 1 & 0 & 0 \\ 9 & -4 & 1 & 0 \\ -12 & 9 & -4 & 1 \end{bmatrix}, \quad L_B = \begin{bmatrix} 1 & 0 & 0 & 0 \\ -2 & 1 & 0 & 0 \\ 9/2 & -22/9 & 1 & 0 \\ -6 & 11/3 & -240/67 & 1 \end{bmatrix}.$$

L_A and L_B are not comparable, implying that Theorem 5.12 does not hold for monotone matrices. The reverse of Theorem 5.12 is also not true. Let

$$L_A = \begin{bmatrix} 1 & 0 & 0 & 0 \\ -4 & 1 & 0 & 0 \\ -9 & -4 & 1 & 0 \\ -12 & -9 & -4 & 1 \end{bmatrix}, \quad L_B = \begin{bmatrix} 1 & 0 & 0 & 0 \\ -4 & 1 & 0 & 0 \\ 0 & -4 & 1 & 0 \\ 0 & 0 & -4 & 1 \end{bmatrix}.$$

Matrices L_A and L_B are M-matrices and $L_A \le L_B$. The product matrix $L_A L_A^T$ is not an M-matrix. Furthermore $L_A L_A^T$ and $L_B L_B^T$ are not comparable. We remark however that if L and R are lower triangular M-matrices such that $LL^T \le RR^T$ holds, then $L \le R$ also holds. This result is due to Schmidt and Patzke [312].

THEOREM 5.13 *Assume that*
(i) F satisfies conditions (a) and (b);
(ii) Algorithm (5.70) satisfies condition (5.71);
(iii) The matrices Q_i are nonnegative such that $Q_i^{-1} \ge D_i A(x_i)$ holds for some nonsingular diagonal $D_i \ge 0$ $(i \ge 1)$;
(iv) There exist two nonsingular matrices Q_∞ and D_∞ such that $Q_i \ge Q_\infty \ge 0$ and $D_i \ge D_\infty \ge 0$ $(i \ge 1)$;
If $x_1 \in \mathbb{R}^m$ is an arbitrary point such that $F(x_1) \ge 0$, then algorithm (5.70) satisfies

$$\omega \le x_{i+1} = y_{m+1} \le y_m \le \cdots \le y_1 = x_i \quad (i \ge 1) \tag{5.74}$$

and $x_i \to \omega$ *as* $i \to +\infty$.

Proof. We will use the simple observation that if $A^{-1} \geq 0$, $B \geq 0$ and $B^{-1} \geq DA$ holds with a nonsingular diagonal matrix $D \geq 0$, then AB is an M-matrix. If, in addition, $C^{-1} \geq 0$ and $C \leq A$, then CB is also an M-matrix that satisfies $CB \leq AB$. We prove the monotone convergence in several steps.
1. Assume that $F(x_i) \geq 0$. Then $x_i \geq \omega$. Let $k = 1$ and $y_1 = x_i$. By assumption $p_1 \geq 0$ and $e_1^T A_i p_1 > 0$. Consequently $\psi_1 = \psi_1(x_i) = p_1 e_1^T / (e_1^T A_i p_1) \geq 0$ and $y_2 = y_1 - \psi_1 F(y_1) \leq y_1$. We show that $y_2 \geq \omega$. Relation (5.69) yields

$$F(y_2) - F(y_1) \geq A(y_1)(y_2 - y_1) = -A(y_1)\psi_1 F(y_1).$$

As $A(y_1)Q_i = A(x_i)Q_i$ is an M-matrix we have

$$A(y_1)\psi_1 = A(y_1)p_1 e_1^T / (e_1^T A(y_1) p_1) = \frac{L_{A(y_1)Q_i} D_{A(y_1)Q_i} e_1 e_1^T}{e_1^T L_{A(y_1)Q_i} D_{A(y_1)Q_i} e_1} \leq e_1 e_1^T$$

from which $A(y_1)\psi_1 F(y_1) \leq e_1 e_1^T F(y_1)$ and $F(y_2) \geq (I - e_1 e_1^T) F(y_1) \geq 0$ follow. The latter condition implies $y_2 \geq \omega$.
2. Assume now that $y_k \leq y_{k-1} \leq \cdots \leq y_1$ and $F(y_k) \geq 0$. Since $\psi_k = \psi_k(x_i) = p_k e_k^T / (e_k^T A_i p_k) \geq 0$, we obtain that $y_{k+1} = y_k - \psi_k(u_k) F(y_k) \leq y_k$. Relation (5.69) implies again that

$$F(y_{k+1}) - F(y_k) \geq A(y_k)(y_{k+1} - y_k) = -A(y_k)\psi_k F(y_k).$$

Since $A(x)$ is isotone,

$$\frac{A(y_k) p_k}{e_k^T A_i p_k} \leq \frac{A(y_1) p_k}{e_k^T A(y_1) p_k} = \frac{L_{A(y_1)Q_i} D_{A(y_1)Q_i} e_k}{e_k^T L_{A(y_1)Q_i} D_{A(y_1)Q_i} e_k} \leq e_k$$

and $A(y_k)\psi_k \leq e_k e_k^T$. Hence $F(y_{k+1}) \geq \left(I - e_k e_k^T\right) F(y_k)$ and $y_{k+1} \geq \omega$.
3. So far we proved that

$$\omega \leq x_{i+1} = y_{m+1} \leq y_m \leq \cdots \leq y_1 = x_i \quad (i \geq 1).$$

Hence there exists $\hat{x} \in \mathbb{R}^m$ such that $x_i \to \hat{x}$ as $i \to +\infty$. We need to prove that $\hat{x} = \omega$. Introduce the notation $y_j^{(i)} = y_j$ ($j = 1, \ldots, m+1$; $i \geq 1$). Then $\lim_{i \to +\infty} y_j^{(i)} = \hat{x}$ holds for $j = 1, \ldots, m+1$. Let

$$\gamma_i = \sum_{j=1}^{m} \psi_j [F(y_1^{(i)}) - F(y_j^{(i)})] \quad (i \geq 1).$$

As $0 \le \psi_j \le \sum_{j=1}^{m} \psi_j$ we take a monotone norm for which $0 \le X \le Y$ implies $\|X\| \le \|Y\|$. Then

$$\|\gamma_i\| \le \left\|\sum_{j=1}^{m} \psi_j\right\| \sum_{j=1}^{m} \left\|F\left(y_1^{(i)}\right) - F\left(y_j^{(i)}\right)\right\| \quad (i \ge 1).$$

The continuity assumption implies that

$$\lim_{i\to+\infty} \sum_{j=1}^{m} \left\|F\left(y_1^{(i)}\right) - F\left(y_j^{(i)}\right)\right\| = 0.$$

By definition

$$\sum_{j=1}^{m} \psi_j = \sum_{j=1}^{k} \frac{p_k e_k^T}{e_k^T A_i p_k} = PD_{A(y_1)Q_i}^{-1} = Q_i U_{A(y_1)Q_i}^{-1} D_{A(y_1)Q_i}^{-1}.$$

If B is a nonsingular M-matrix and $B = L_B D_B U_B$ is its LDU-factorization, then $0 \le D_B^{-1} \le U_B^{-1} D_B^{-1} \le B^{-1}$. Substituting B with $A(y_1) Q_i$ we obtain that

$$0 \le Q_i D_{A(y_1)Q_i}^{-1} \le \sum_{j=1}^{m} \psi_j = Q_i U_{A(y_1)Q_i}^{-1} D_{A(y_1)Q_i}^{-1} \le A^{-1}(y_1).$$

Assumption (iii) and Theorem 5.12 imply $A_i Q_i \le D_i^{-1}$ and $D_{A_i Q_i} \le D_i^{-1}$, respectively. Since $A_i = A(x_i) \ge A(\omega)$ we obtain

$$0 \le Q_\infty D_\infty \le Q_i D_i \le \sum_{j=1}^{m} \psi_j \le A^{-1}(\omega).$$

Consequently $\gamma_i \to 0$ as $i \to +\infty$. Assume that $F(\hat{x}) \ge 0$ and $F(\hat{x}) \ne 0$. Then $\left(\sum_{j=1}^{m} \psi_j\right) F\left(y_1^{(i)}\right) \ge Q_\infty D_\infty F(\hat{x}) \ge 0$ and $Q_\infty D_\infty F(\hat{x}) \ne 0$. On the other hand $\left(\sum_{j=1}^{m} \psi_j\right) F\left(y_1^{(i)}\right) = y_1^{(i)} - y_{m+1}^{(i)} + \gamma_i \to 0$ $(i \to +\infty)$, which is a contradiction. Thus $\hat{x} = \omega$, which was to be proved. ∎

Observe that the decrease of F is characterized by the inequality

$$F(y_{k+1}) \ge (I - e_k e_k^T) F(y_k) \ge 0. \tag{5.75}$$

If $x_0 \in \mathbb{R}^m$ is arbitrary, then $x_1 = x_0 - [F'(x_0)]^{-1} F(x_0)$ satisfies the condition $F(x_1) \ge 0$ (see [270]). Hence Theorem 5.13 is essentially a global convergence theorem.

The parameter matrices Q_i can be chosen so that Q_i^{-1} are M-matrices satisfying $Q_i^{-1} \ge D_i^{-1} A(x_i)$. For the choice $Q_1 \ge 0$, $Q_1^{-1} \ge A(x_1) D^{-1}$ the

matrix $Q_i = Q_1$ satisfies condition (iii) provided that $x_i \le x_1$. For the updated Q_i we mention the following possible choices (see, also [361]):
$Q_i^{-1} = A_i + uv^T$ $(u, v \in \mathbb{R}^m)$,
Q_i^{-1} is a bidiagonal M-matrix obtained from A_i by setting the other elements of A_i to 0,
Q_i^{-1} is a tridiagonal M-matrix obtained from A_i by setting the other elements of A_i to 0.

Assume that a point $z_1 \in \mathbb{R}^m$ exists such that $F(z_1) \le 0$. Then we can define the algorithm

$$\left.\begin{array}{rcl} w_1 &=& z_i \\ w_{k+1} &=& w_k - p_k \left(e_k^T A_i p_k\right)^{-1} e_k^T F(w_k) \quad (k = 1, \ldots, m) \\ z_{i+1} &=& w_{m+1} \end{array}\right\} (i \ge 1), \tag{5.76}$$

as parallel subordinate to algorithm (5.70), (5.71). Note that

$$\psi_k = p_k \left(e_k^T A_i p_k\right)^{-1} e_k^T$$

is defined by algorithm (5.70), (5.71).

The following result can be proved similarly to Theorem 5.13.

THEOREM 5.14 *Assume that the conditions of Theorem 5.13 hold. In addition, assume the existence of a point z_1 such that $F(z_1) \le 0$. Then for the subordinate algorithm (5.76) we have*

$$\omega \ge z_{i+1} = w_{m+1} \ge w_m \ge \cdots \ge w_1 = z_i \quad (i \ge 1) \tag{5.77}$$

and $z_i \to \omega$ as $i \to +\infty$.

The increase of F is characterized by the inequality

$$F(w_{k+1}) \le (I - e_k e_k^T) F(w_k) \le 0. \tag{5.78}$$

Theorems 5.13 and 5.14 together imply that algorithms (5.70), (5.71) and (5.76) produce two-sided approximations to the solution ω in the form of inclusion intervals

$$z_i \le z_{i+1} \le \omega \le x_{i+1} \le x_i \quad (i = 1, \ldots) \tag{5.79}$$

with $x_i - z_i \to 0$ $(i \to +\infty)$.

Next we give an estimation for the inclusion interval $[z_{i+1}, x_{i+1}]$.

THEOREM 5.15 *Under the conditions of Theorems 5.13 and 5.14 the interval*

$$\left[z_i - Q_i U_{A_i Q_i}^{-1} D_{A_i Q_i}^{-1} F(z_i), x_i - Q_i U_{A_i Q_i}^{-1} D_{A_i Q_i}^{-1} F(x_i)\right] \tag{5.80}$$

covers the interval $[z_{i+1}, x_{i+1}]$.

Proof. The inequalities

$$F(y_{k+1}) \geq (I - e_k e_k^T) F(y_k)$$

and

$$F(w_{k+1}) \leq (I - e_k e_k^T) F(w_k)$$

imply

$$F(y_{k+1}) \geq (I - \sum_{j=1}^{k} e_j e_j^T) F(y_1)$$

and

$$F(w_{k+1}) \leq (I - \sum_{j=1}^{k} e_j e_j^T) F(w_1),$$

respectively. Since $\psi_k e_j = 0$ $(k \neq j)$, we have $\psi_k F(y_k) \geq \psi_k F(y_1)$ and $\psi_k F(w_k) \leq \psi_k F(w_1)$. From the inequalities

$$\left(\sum_{j=1}^{m} \psi_j\right) F(y_j) \geq \left(\sum_{j=1}^{m} \psi_j\right) F(y_1) = Q_i U_{A_i Q_i}^{-1} D_{A_i Q_i}^{-1} F(y_1)$$

and

$$\left(\sum_{j=1}^{m} \psi_j\right) F(w_j) \leq \left(\sum_{j=1}^{m} \psi_j\right) F(w_1) = Q_i U_{A_i Q_i}^{-1} D_{A_i Q_i}^{-1} F(w_1)$$

the statement follows. ∎

Finally, we show that a particular method of Algorithm 1 is faster than Newton's method in the partial ordering.

THEOREM 5.16 *Assume that conditions of Theorem 5.13 are satisfied and let* $Q_i = A^{-1}(x_i) D_1$, *where* $D_1 \geq 0$ *is diagonal. Then the corresponding algorithm (5.70), (5.71) is faster than Newton's method in the partial ordering provided that they start from the same initial point* x_1.

Proof. Assume for a moment that x_i is the starting point for both methods. Then by definition $Q_i U_{A_i Q_i}^{-1} D_{A_i Q_i}^{-1} = A_i^{-1} = A^{-1}(x_i)$. Since

$$x_{i+1} \leq x_i - Q_i U_{A_i Q_i}^{-1} D_{A_i Q_i}^{-1} F(x_i) = x_i - A^{-1}(y_1) F(x_i)$$

and $x_{Newton} = x_i - A^{-1}(x_i) F(x_i)$, we proved the statement for one iteration. We need the observation that if $x \leq y$, $F(x) \geq 0$, then $x - A^{-1}(x) F(x) \leq$

$y - A^{-1}(y)F(y)$. If we multiply the inequality $F(x) - F(y) \geq F'(y)(x - y)$ by $A^{-1}(y)$ on the left, then we obtain $y - A^{-1}(y)F(y) \geq x - A^{-1}(y)F(x)$. As $A(y)$ is isotone and an M-matrix, we have $A^{-1}(x) \geq A^{-1}(y)$ which implies $A^{-1}(x)F(x) \geq A^{-1}(y)F(x)$ and $x - A^{-1}(x)F(x) \leq x - A^{-1}(y)F(x)$, from which the observation follows. We now complete the proof as follows. We start from x_1. Algorithm (5.70), (5.71) generates the sequence x_i^{ncd}, while the Newton method generates x_i^{Newton}. Assume that $x_i^{ncd} \leq x_i^{Newton}$. Then

$$\begin{aligned} x_{i+1}^{ncd} &\leq x_{Newton} = x_i^{ncd} - A^{-1}(x_i^{ncd})F(x_i^{ncd}) \\ &\leq x_{i+1}^{Newton} = x_i^{Newton} - A^{-1}(x_i^{Newton})F(x_i^{Newton}) \end{aligned}$$

implies the theorem. ■

Theorems 5.13, 5.14 and 5.15 are special cases of more general results developed for the nonlinear ABS methods, where $e_k^T A_k p_k$ is allowed instead of $e_k^T A_i p_k$ [138]. The monotone convergence of the Newton method was proved by Baluev (see, e.g., [271], [270]). The special nonlinear conjugate direction method of Theorem 5.16 is essentially a modified Newton method belonging to Algorithm 1. Frommer [130], [129] proved a similar result for the Brown method, which belongs to Algorithm 5.

Chapter 6

PROJECTION METHODS IN OPTIMIZATION

Projections appear in many areas of optimization. In linear programming we mention the interior gradient projection method of Pyle and Cline [299], [298] and the interior point method of Karmarkar. Papers [349], [328], [166], [211] on projections are related to the Karmarkar algorithm. In nonlinear programming the gradient projection methods and the convex feasibility problem are perhaps the most important applications of projections. Further typical applications can be found in papers [185], [238], [192], [77], [78], [313], [82], [99].

Here we investigate the gradient projection methods for convex optimization problems and the convex feasibility problem. In the whole chapter we use only the Euclidean norm $\|\cdot\|_2$. Hence we omit the index.

1. Introduction

We summarize the elements of convex analysis. More on convex analysis can be found in [302], [332], [38], [294], [341], [219], [200] and [201], say.

DEFINITION 6.1 *A set $C \subset \mathbb{R}^n$ is called convex if $x, y \in C$ implies*

$$z = \lambda x + (1 - \lambda) y \in C \tag{6.1}$$

for all $0 \leq \lambda \leq 1$.

DEFINITION 6.2 *For $x, y \in \mathbb{R}^n$, let*

$$[x, y] = \{z \mid z = \alpha x + (1 - \alpha) y,\ 0 \leq \alpha \leq 1\} \tag{6.2}$$

and

$$(x, y) = \{z \mid z = \alpha x + (1 - \alpha) y,\ 0 < \alpha < 1\}. \tag{6.3}$$

The line segments $[x, y] \subset \mathbb{R}^n$ and $(x, y) \subset \mathbb{R}^n$ are called closed and open intervals, respectively.

The set $C \subset \mathbb{R}^n$ is convex if and only if $x, y \in C$ implies $[x, y] \subset C$.

DEFINITION 6.3 *The set $K \subset \mathbb{R}^n$ is called a cone if $\lambda x \in K$ whenever $x \in K$ and $\lambda \geq 0$.*

It is easy to verify that $K \subset \mathbb{R}^n$ is a convex cone if and only if $x + y \in K$, $\lambda x \in K$ whenever $x, y \in K$ and $\lambda \geq 0$.

DEFINITION 6.4 *Let $X \subset \mathbb{R}^n$ and $v \in \mathbb{R}^n$. If $p \in X$ is such that*

$$\|p - v\| = \inf_{x \in X} \|x - v\| = d(v, X), \tag{6.4}$$

then p is the projection of v on X.

This projection is also referred to as the metric, minimum norm or the closest point projection .

LEMMA 6.5 *For any closed set X and any point $v \in \mathbb{R}^n$ there is a point $p \in X$ which is a projection of v onto X. Moreover, if X is convex, then the point p is unique.*

Proof. If $v \in X$, then $p = v$. We can assume that $v \notin X$. There is a sequence $\{x_k\} \in X$ such that $\lim_{k \to \infty} \|x_k - v\| = d(v, X)$. The sequence $\{x_k\}$ is bounded and there exists a convergent subsequence $\{x_{k_i}\}$ with $\lim_{i \to \infty} x_{k_i} = p$. Since X is closed, $p \in X$ and $\|p - v\| = d(v, X)$. Finally, assume that X is convex and we have two different projections p and p' such that $\|p - v\| = \|p' - v\| = d(v, X)$. Then $z = (p + p')/2 \in X$. The identity

$$\|x + y\|^2 + \|x - y\|^2 = 2\|x\|^2 + 2\|y\|^2$$

implies

$$\begin{aligned} \|z - v\|^2 &= \left\| \frac{p-v}{2} + \frac{p'-v}{2} \right\|^2 \\ &= \frac{1}{2}\|p - v\|^2 + \frac{1}{2}\|p' - v\|^2 - \frac{1}{4}\|p - p'\|^2 < d^2(v, X), \end{aligned}$$

which contradicts the definition of $d(v, X)$. ■

DEFINITION 6.6 *Let $C \subset \mathbb{R}^n$ be a closed convex set. For each v in $\mathbb{R}^n$ define $R_C(v)$ as the projection of v onto C.*

The projection mapping $R_C(v)$ is not necessarily linear. We show later that $R_C(v)$ is Lipschitz continuous.

THEOREM 6.7 *Let $C \subset \mathbb{R}^n$ be a closed convex set. Then $p = R_C(v)$ if and only if*

$$(x - p)^T (v - p) \leq 0 \tag{6.5}$$

holds for all $x \in C$.

Proof. Let p be the projection of the point v on C. Let $x \in C$ be arbitrary and consider $z = \alpha x + (1-\alpha)\,p$. The convexity of C implies that $z \in C$ for any $\alpha \in [0,1]$. Since

$$\|z - v\|^2 = \alpha^2 \|x-p\|^2 + 2\alpha\,(x-p)^T(p-v) + \|p-v\|^2 \geq \|p-v\|^2,$$

we have

$$\phi(\alpha) = \alpha^2 \|x-p\|^2 + 2\alpha\,(x-p)^T(p-v) \geq 0 \quad (\alpha \in [0,1]).$$

Hence

$$\phi'(0) = 2\,(x-p)^T(p-v) \geq 0,$$

which implies (6.5). Assume now that (6.5) holds. Then for any $x \in C$ we have

$$\begin{aligned} \|x-v\|^2 &= \|(x-p)+(p-v)\|^2 \\ &= \|x-p\|^2 + 2\,(x-p)^T(p-v) + \|p-v\|^2 \geq \|p-v\|^2. \end{aligned}$$

Hence p is the projection of v on C. ■

COROLLARY 6.8 *Let $C \subset \mathbb{R}^n$ be a convex closed set and $p = R_C(v)$. Then for any $x \in C$,*

$$(x-v)^T(x-p) = \|x-p\|^2 + (p-v)^T(x-p) \geq \|x-p\|^2, \tag{6.6}$$

$$\|x-p\| \leq \|x-v\| \tag{6.7}$$

and

$$\|p-v\|^2 \leq \|v-x\|^2 - \|p-x\|^2. \tag{6.8}$$

Proof. Since $(x-p)^T(v-p) \leq 0$ we have

$$\begin{aligned} (x-v)^T(x-p) &= \|x-p\|^2 + [(x-v)-(x-p)]^T(x-p) \\ &= \|x-p\|^2 + (p-v)^T(x-p) \geq \|x-p\|^2 \end{aligned}$$

and

$$\|x-v\|^2 = \|x-p\|^2 + 2\,(x-p)^T(p-v) + \|p-v\|^2 \geq \|x-p\|^2.$$

This proves the first two claims. A simple arrangement gives

$$\begin{aligned} \|p-v\|^2 &= \|v-x\|^2 - \|p-x\|^2 - 2\,(x-p)^T(p-v) \\ &\leq \|v-x\|^2 - \|p-x\|^2. \end{aligned}$$

■

THEOREM 6.9 *If $C \subset \mathbb{R}^n$ is a closed convex set, then*

$$\|R_C(u) - R_C(v)\| \leq \|u - v\| \tag{6.9}$$

for all $u, v \in \mathbb{R}^n$.

Proof. If $R_C(u) = R_C(v)$, the statement becomes the obvious inequality $0 \leq \|u - v\|$. We can assume that $R_C(u) \neq R_C(v)$. Substituting x by $R_C(v)$ in Theorem 6.7 we obtain

$$(R_C(v) - R_C(u))^T (u - R_C(u)) \leq 0.$$

If we replace u and v, then

$$(R_C(u) - R_C(v))^T (v - R_C(v)) \leq 0.$$

Adding the two inequalities one obtains

$$(R_C(u) - R_C(v))^T (R_C(u) - R_C(v) - (u - v)) \leq 0.$$

A simple arrangement and the Cauchy-Schwartz inequality lead us to

$$\begin{aligned} \|R_C(u) - R_C(v)\|^2 &\leq (R_C(u) - R_C(v))^T (u - v) \\ &\leq \|R_C(u) - R_C(v)\| \, \|u - v\| \end{aligned}$$

and

$$\|R_C(u) - R_C(v)\| \leq \|u - v\|.$$

■

Hence we proved the Lipschitz continuity of the projection mapping $R_C : \mathbb{R}^n \to C$. Asplund [23] proved that the metric projection R_C on any closed subset C of a finite-dimensional Euclidean space is almost everywhere differentiable. He used a theorem of A.D. Alexandrov on the differentiability of convex functions (see, e.g., [332], p. 153 or [200], p. 192). The result also follows from a theorem of Rademacher, which states that the locally Lipschitz continuous functions $F : \mathbb{R}^n \to \mathbb{R}^m$ are almost everywhere differentiable (see, e.g., [120], [303]). Related differentiability results can be found in Abatzouglu [5] and Hiriart-Urruty [199].

DEFINITION 6.10 *Let $D \subset \mathbb{R}^n$ be a closed convex set. The mapping $V : D \to D$ is said to be firmly contractive if the following inequality is satisfied for all $x, y \in D$:*

$$(V(x) - V(y))^T (x - y) \geq \|V(x) - V(y)\|^2. \tag{6.10}$$

The proof of the previous theorem indicates that projection R_C is firmly contractive on $\mathbb{R}^n$ provided that C is closed and convex. The firmly contractive mappings are also called firmly nonexpansive (see e.g., [32]).

DEFINITION 6.11 *A function $f : \mathbb{R}^n \to \mathbb{R}$ is said to be convex if for any $x, y \in \mathbb{R}^n$ and $\lambda \in [0,1]$,*

$$f(\lambda x + (1-\lambda) y) \leq \lambda f(x) + (1-\lambda) f(y). \tag{6.11}$$

REMARK 6.12 *If $C \subset \mathbb{R}^n$ is a closed and convex set, then the distance function $d(v, C) = \|R_C(v) - v\|$ is convex on $\mathbb{R}^n$.*

Proof. By definition, for any $z \in C$ and for any $v, w \in \mathbb{R}^n$,

$$\begin{aligned} d(\lambda v + (1-\lambda) w, C) &= \|R_C(\lambda v + (1-\lambda) w) - [\lambda v + (1-\lambda) w]\| \\ &\leq \|z - [\lambda v + (1-\lambda) w]\|. \end{aligned}$$

The choice $z = \lambda R_C(v) + (1-\lambda) R_C(w) \in C$ gives

$$\begin{aligned} d(\lambda v + (1-\lambda) w, C) &\leq \|[\lambda R_C(v) + (1-\lambda) R_C(w)] - [\lambda v + (1-\lambda) w]\| \\ &\leq \lambda \|R_C(v) - v\| + (1-\lambda) \|R_C(w) - w\|. \end{aligned}$$

■

REMARK 6.13 *If $f_1, \ldots, f_k$ are convex functions and $\lambda_1, \ldots, \lambda_k \geq 0$, then function $f = \sum_{i=1}^{k} \lambda_i f_i$ is also convex.*

REMARK 6.14 *If $f : \mathbb{R}^n \to \mathbb{R}^+$ is convex, then $f^2 : \mathbb{R}^n \to \mathbb{R}^+$ is also convex.*

DEFINITION 6.15 *A function $f : \mathbb{R}^n \to \mathbb{R}$ defined for all $x \in \mathbb{R}^n$ is said to be strictly convex if for any $x, y \in \mathbb{R}^n$ and $\lambda \in (0,1)$,*

$$f(\lambda x + (1-\lambda) y) < \lambda f(x) + (1-\lambda) f(y). \tag{6.12}$$

LEMMA 6.16 *Assume that $f : \mathbb{R}^n \to \mathbb{R}$ is a continuously differentiable function. The following statements are equivalent:*
(i) $f(x)$ is a convex function.
(ii) $f(x_2) - f(x_1) \geq \nabla f(x_1)^T (x_2 - x_1)$ for any $x_1, x_2 \in \mathbb{R}^n$.
(iii) $\nabla f(x + \lambda p)^T p$ is a nondecreasing function of λ.
If f is twice continuously differentiable function, then
(iv) $\nabla^2_{xx} f(x)$, the matrix of second derivatives, is positive definite.

LEMMA 6.17 *Assume that $f : \mathbb{R}^n \to \mathbb{R}$ is a continuously differentiable function. The following statements are equivalent:*
(i) $f(x)$ is a strictly convex function.
(ii) $f(x_2) - f(x_1) > \nabla f(x_1)^T (x_2 - x_1)$ for any $x_1, x_2 \in \mathbb{R}^n$, $x_1 \neq x_2$.
(iii) $\nabla f(x + \lambda p)^T p$ is a strictly increasing function of λ.

LEMMA 6.18 *Let f be a continuously differentiable convex function and C be a closed convex set. A point $x^* \in C$ is the minimum point of f in C if and only if the inequality*

$$\nabla f(x^*)^T (x - x^*) \geq 0 \quad (x \in C) \tag{6.13}$$

holds.

Proof. For a continuously differentiable convex function we have

$$f(x) - f(x^*) \geq \nabla f(x^*)^T (x - x^*).$$

Hence

$$f(x) - f(x^*) \geq 0 \quad (x \in C),$$

which shows that x^* is a minimum point of f in C. We now prove the necessity. Let x^* be a minimum point. Then for any $x \in C$ and $0 < \lambda \leq 1$, we have

$$f((1-\lambda)x^* + \lambda x) = f(x^* + \lambda(x - x^*)) \geq f(x^*)$$

or

$$\frac{f(x^* + \lambda(x - x^*)) - f(x^*)}{\lambda} \geq 0.$$

If $\lambda \to 0$ we obtain

$$\nabla f(x^*)^T (x - x^*) \geq 0 \quad (x \in C).$$

■

If f is not convex, then condition (6.13) is only necessary.

LEMMA 6.19 *Let f be a continuously differentiable convex function and C be a closed convex set. A point $x^* \in C$ is the minimum point of f in C if and only if*

$$x^* = R_C(x^* - \lambda \nabla f(x^*)) \quad (\lambda \geq 0). \tag{6.14}$$

Proof. Let x^* be a minimum point and $v = x^* - \lambda \nabla f(x^*)$. Condition (6.13) implies

$$-\lambda \nabla f(x^*)^T (x - x^*) \leq 0 \quad (x \in C).$$

Since $-\lambda \nabla f(x^*) = v - x^*$ we obtain $(v - x^*)^T (x - x^*) \leq 0$. It follows from Theorem 6.7 that x^* is the projection of v on C. Conversely, assume that $x^* = R_C(v)$. Theorem 6.7 implies

$$(x - x^*)^T (v - x^*) \leq 0 \quad (x \in C).$$

Since $v = x^* - \lambda \nabla f(x^*)$ and $\lambda > 0$, we have $\nabla f(x^*)^T (x - x^*) \geq 0$. Hence x^* is a minimum point. ■

DEFINITION 6.20 *A function $f : C \to \mathbb{R}$ ($C \subset \mathbb{R}^n$) is said to be strongly convex if there is a constant $\rho > 0$ such that for any $x, y \in C$ and any $\alpha \in [0,1]$ the inequality*

$$f(\alpha x + (1-\alpha) y) \le \alpha f(x) + (1-\alpha) f(y) - \alpha(1-\alpha)\rho \|x-y\|^2 \quad (6.15)$$

holds.

The function $f(x) = x^T Bx + p^T x$ ($B \in \mathbb{R}^{n\times n}$) is strongly convex if B is positive definite. If a function f is strongly convex and continuous on a closed convex set C, then for any point $y \in C$, the set

$$C_0 = \{x \in C \mid f(x) \le f(y)\}$$

is bounded and there is a unique point x^* such that

$$f(x^*) = \min\{f(x) \mid x \in C\}.$$

THEOREM 6.21 *If $f(x)$ is strongly convex on a convex and closed set C and x^* is the minimum point of f in C, then*
(i) for all $x \in C$ the inequality

$$\|x - x^*\|^2 \le \frac{2}{\rho}(f(x) - f(x^*)) \quad (6{:}16)$$

holds, and if $f \in C^1(C)$, then
(ii) for all $x, y \in C$ there hold

$$(\nabla f(x) - \nabla f(y))^T (x-y) \ge \rho\|x-y\|^2, \quad (6.17)$$

(iii)

$$\|x - x^*\| \le \frac{1}{\rho}\|\nabla f(x)\|, \quad (6.18)$$

(iv)

$$0 \le f(x) - f(x^*) \le \frac{1}{\rho}\|\nabla f(x)\|^2. \quad (6.19)$$

Proof. (i) The definition of strong convexity implies

$$f\left(\frac{1}{2}x + \frac{1}{2}x^*\right) \le \frac{1}{2}f(x) + \frac{1}{2}f(x^*) - \frac{1}{4}\rho\|x-x^*\|^2.$$

Since $f(x^*) \le f\left(\frac{1}{2}x + \frac{1}{2}x^*\right)$, the inequality of claim (i) holds.
(ii) The convexity of f implies that

$$f(x) - f(y) \le \nabla f(x)^T (x-y).$$

By definition of strong convexity we obtain

$$\begin{aligned}\frac{1}{4}\rho\|x-y\|^2 &\leq \frac{1}{2}\left[f(x)-f\left(\frac{1}{2}x+\frac{1}{2}y\right)\right]+\frac{1}{2}\left[f(y)-f\left(\frac{1}{2}x+\frac{1}{2}y\right)\right]\\ &\leq \frac{1}{4}\nabla f(x)^T(x-y)+\frac{1}{4}\nabla f(y)^T(y-x)\\ &= \frac{1}{4}(\nabla f(x)-\nabla f(y))^T(x-y).\end{aligned}$$

(iii) In the minimum points $x^* \in C$ we have (6.13). Hence (ii) implies that

$$\begin{aligned}\rho\|x-x^*\|^2 &\leq (\nabla f(x)-\nabla f(x^*))^T(x-x^*)\\ &\leq \nabla f(x)^T(x-x^*) \leq \|\nabla f(x)\|\,\|x-x^*\|.\end{aligned}$$

It follows now that $\|x-x^*\| \leq \|\nabla f(x)\|/\rho$.
(iv) It follows from the convexity of f and (iii) that

$$0 \leq f(x)-f(x^*) \leq \nabla f(x)^T(x-x^*) \leq \|\nabla f(x)\|\,\|x-x^*\| \leq \frac{1}{\rho}\|\nabla f(x)\|^2.$$

■

For other types of convexity we refer the reader to [302], [332], [200] or [288].

2. Methods for constrained optimization problems

Consider the optimization problem

$$f(x) \to \min \quad (x \in C), \tag{6.20}$$

where $f : \mathbb{R}^n \to \mathbb{R}$ is a convex function and $C \subset \mathcal{D}(f) \subset \mathbb{R}^n$ is a convex closed set. The points of C are said to be feasible. The convexity of f and C are assumed in the whole section unless otherwise stated.

The methods of feasible directions, similarly to certain unconstrained minimization techniques [309], [33], are iterative methods of the form

$$x_{k+1} = x_k + \lambda_k d_k \quad (\lambda_k > 0), \tag{6.21}$$

where
(i) x_k is feasible,
(ii) direction d_k is such that $x_k + \lambda d_k$ is feasible for small $\lambda > 0$, and
(iii)

$$f(x_k + \lambda_k d_k) \leq f(x_k). \tag{6.22}$$

DEFINITION 6.22 *Consider the optimization problem $g(x) \to \min$ ($x \in X$), where $g : \mathbb{R}^n \to \mathbb{R}$ and $X \subset \mathbb{R}^n$ is nonempty. A nonzero vector d is called* *feasible direction at* $x \in X$ *if there exists a* $\delta > 0$ *such that* $x + \lambda d \in X$ *for all* $\lambda \in (0, \delta)$. *Furthermore, d is called an improving feasible direction at* $x \in X$ *if there exists a* $\delta > 0$ *such that* $g(x + \lambda d) < g(x)$ *and* $x + \lambda d \in X$ *for all* $\lambda \in (0, \delta)$.

The methods of feasible directions differ from each other in the selection of λ_k and the direction d_k. There are several types of such methods (see e.g., [369], [343], [283], [308], [34], [294], [35], [309], [219], [33], [222]). Here we investigate the gradient projection method of Goldstein, Levitin and Polyak.

The idea of obtaining a feasible direction by means of a projection is due to Rosen [304], [305]. Rosen considers a sequence of the form

$$x_{k+1} = x_k + \lambda_k P_k \left(-\nabla f\left(x_k\right)\right), \tag{6.23}$$

where P_k is a projection. However P_k is not the closest point projection. Rosen's method is sometimes called the tangent manifold projection process (see, e.g., [112]). For details on the Rosen method we refer to [304], [305], [296], [299], [283], [308], [208], [34], [294], [35], [309], [33], [109] and [110].

The gradient projection method of Goldstein [162], Levitin and Polyak [232] has the form

$$x_{k+1} = R_C\left(x_k - \lambda_k \nabla f\left(x_k\right)\right), \tag{6.24}$$

where R_C denotes the metric projection on C. Thus x_{k+1} is the nearest feasible point to $x_k - \lambda_k \nabla f\left(x_k\right)$. The method is certainly motivated by the result of Lemma 6.19.

The following convergence result [286] is only of theoretical importance.

THEOREM 6.23 *(Polyak). Assume that* $f : C \to \mathbb{R}$ *is convex and continuously differentiable,* $C \subset \mathbb{R}^n$ *is convex and closed. Also assume that there is a minimum point* $x^* \in C$, $f(x^*) = f^*$ *and* $\|\nabla f(x)\| \le c$ *for all* $x \in C_0 = \{z \in C \mid \|z - x_0\| \le \|x^* - x_0\|\}$. *Then the method*

$$x_{k+1} = R_C\left(x_k - \lambda_k \nabla f\left(x_k\right)\right) \tag{6.25}$$

with

$$\lambda_k = \mu_k \frac{f(x_k) - f^*}{\left\|\nabla f(x_k)\right\|^2} \quad (0 < \varepsilon_1 \le \mu_k \le 2 - \varepsilon_2,\ \varepsilon_2 > 0) \tag{6.26}$$

generates the sequence $\{x_k\}$ *such that*

(i) $f(x_k) \to f^*$;

(ii) The convergent subsequences of $\{x_k\}$ *tend to some minimum point* $\widetilde{x} \in C$.

Proof. We exploit the inequalities $\|R_C(u) - R_C(v)\| \leq \|u - v\|$ and $f(x^*) \geq f(x_k) + \nabla f(x_k)^T (x^* - x_k)$. It follows that

$$\begin{aligned}
\|x_{k+1} - x^*\|^2 &= \|R_C(x_k - \lambda_k \nabla f(x_k)) - x^*\|^2 \\
&\leq \|x_k - \lambda_k \nabla f(x_k) - x^*\|^2 \\
&= \|x_k - x^*\|^2 + \lambda_k^2 \|\nabla f(x_k)\|^2 - 2\lambda_k \nabla f(x_k)^T (x_k - x^*) \\
&\leq \|x_k - x^*\|^2 + \mu_k^2 \frac{(f(x_k) - f^*)^2}{\|\nabla f(x_k)\|^2} \\
&\quad - 2\mu_k \frac{f(x_k) - f^*}{\|\nabla f(x_k)\|^2} (f(x_k) - f(x^*)) \\
&\leq \|x_k - x^*\|^2 - \varepsilon_1 \varepsilon_2 \frac{(f(x_k) - f^*)^2}{\|\nabla f(x_k)\|^2}.
\end{aligned}$$

Consequently $\|x_k - x^*\|$ is monotone decreasing and $\|x_k - x^*\| \leq \|x_0 - x^*\|$. Therefore $(f(x_k) - f^*)^2 / \|\nabla f(x_k)\|^2 \to 0$. Since $x_k \in C_0$ and $\|\nabla f(x_k)\| \leq c$, we obtain that $f(x_k) \to f^*$. The sequence $\{x_k\}$ is in the compact set C_0 and we can select a convergent subsequence $\{x_{k_j}\}$ such that $x_{k_j} \to \widetilde{x}$. Since $\widetilde{x} \in C_0$ and $f(x_{k_j}) \to f(\widetilde{x}) = f^*$, $\widetilde{x}$ is a minimum point of f. We prove that all convergent subsequences of $\{x_k\}$ tend to $\widetilde{x}$. Assume that $\{x_{m_j}\}$ is such that $x_{m_j} \to \widehat{x}$ and $\widetilde{x} \neq \widehat{x}$. Since $\|x_k - \widetilde{x}\| \to \rho \geq 0$ as $k \to \infty$, we have $\|x_{m_j} - \widetilde{x}\| \to \|\widehat{x} - \widetilde{x}\| = \rho$ and $\|x_{k_j} - \widetilde{x}\| \to 0 = \rho$. This implies $\widetilde{x} = \widehat{x}$, which is a contradiction. ■

As f^* is generally unknown, the above version of the gradient projection method is not implementable. The theorem, however, shows a typical convergence behavior of such methods (for more details see, e.g., [33]). The following result [287] has more practical value.

THEOREM 6.24 *(Polyak). Assume that $C \subset \mathbb{R}^n$ is convex and closed, $f : C \to \mathbb{R}$ is differentiable on C, $x_0 \in C$, $C_0 = \{x \in C \mid f(x) \leq f(x_0)\}$ and*

$$\|\nabla f(x) - \nabla f(y)\| \leq M \|x - y\| \quad (x, y \in C_0). \tag{6.27}$$

Consider the algorithm

$$x_{k+1} = R_C(x_k - \lambda_k d_k), \tag{6.28}$$

where

$$\|d_k - \nabla f(x_k)\| \le c\,\|x_{k+1} - x_k\|. \tag{6.29}$$

The following statements hold:
(i) If $0 \le \lambda_k \le 2/(M+c)$, *then* $\{f(x_k)\}$ *is monotone decreasing.*
(ii) If C_0 *is bounded,* f *is convex,* $0 < \varepsilon_1 \le \lambda_k \le 2/(M+2c) - \varepsilon_2$ *and* $\varepsilon_1, \varepsilon_2 > 0$, *then*

$$\lim_k f(x_k) = f^* = \inf_x f(x), \quad f(x_k) - f^* = O\left(\frac{1}{k}\right). \tag{6.30}$$

(iii) If f *is twice differentiable and*

$$m\|y\|^2 \le y^T \nabla^2_{xx} f(x)\, y \le M\|y\|^2 \quad (m, M > 0), \tag{6.31}$$

$$\left\|f''(x) - f''(y)\right\| \le L\|x - y\|, \tag{6.32}$$

$$\gamma = \max\{|1 - \lambda m|, |1 - \lambda M|\}, \quad p = \frac{\gamma + c\lambda}{1 - c\lambda} < 1,$$

then for $\lambda_k \equiv \lambda$ *the sequence* $\{x_k\}$ *tends to the minimum point* x^* *with the rate*

$$\|x_k - x^*\| = O\left(p^k\right). \tag{6.33}$$

Both theorems of Polyak are valid in Hilbert spaces (see [286], [287]). Polyak [285] also studies a modification of algorithm (6.24). McCormick and Tapia [250] investigated the convergence of (6.24) when λ_k is any solution to

$$f(R_C(x_k - \lambda \nabla f(x_k))) \to \min \quad (\lambda \ge 0) \tag{6.34}$$

and C is a cone. Dunn [112] gives a thorough convergence analysis that includes the convexity assumptions and the selection of λ_k's as well. Further convergence theorems are given by Phelps [279], [280], Calamai and Moré [73].

We can write $x_{k+1} = R_C(x_k - \lambda_k \nabla f(x_k))$ in the form

$$x_{k+1} = x_k - (x_k - R_C(x_k - \lambda_k \nabla f(x_k))).$$

Introducing the direction $d_k = x_k - R_C(x_k - \lambda_k \nabla f(x_k))$ we obtain the relaxed gradient projection methods of the form

$$\begin{aligned} x_{k+1} &= x_k - \beta_k d_k \quad (0 < \beta \le \beta_k \le 1), \\ d_k &= x_k - R_C(x_k - \lambda_k \nabla f(x_k)). \end{aligned} \tag{6.35}$$

The choice $\beta_k \equiv 1$ gives algorithm (6.24). The gradient projection methods (6.24) and (6.35) reduce to the steepest descent method for $C = \mathbb{R}^n$.

If C is convex, then

$$x_k - \beta d_k = x_k + \beta \left(R_C \left(x_k - \lambda_k \nabla f \left(x_k\right)\right) - x_k\right)$$

$$= \left(1 - \beta\right) x_k + \beta R_C \left(x_k - \lambda_k \nabla f \left(x_k\right)\right) \in C$$

for $0 \leq \beta \leq 1$. Therefore $-d_k$ is a feasible direction at x_k. We exploit that $R_C \left(x_k - \lambda_k \nabla f \left(x_k\right)\right)$ is the projection of $x_k - \lambda_k \nabla f \left(x_k\right)$ on C if and only if

$$\left(x - R_C \left(x_k - \lambda_k \nabla f \left(x_k\right)\right)\right)^T \left[\left(x_k - \lambda_k \nabla f \left(x_k\right)\right) - R_C \left(x_k - \lambda_k \nabla f \left(x_k\right)\right)\right] \leq 0$$

holds for all $x \in C$. For $x = x_k$ it follows that

$$\left\|x_k - R_C \left(x_k - \lambda_k \nabla f \left(x_k\right)\right)\right\|^2 - \lambda_k \left(x_k - R_C \left(x_k - \lambda_k \nabla f \left(x_k\right)\right)\right)^T \nabla f \left(x_k\right) \leq 0$$

and

$$0 \leq \left\|x_k - R_C \left(x_k - \lambda_k \nabla f \left(x_k\right)\right)\right\|^2 \leq \lambda_k \nabla f \left(x_k\right)^T d_k.$$

If $x_k = R_C \left(x_k - \lambda_k \nabla f \left(x_k\right)\right)$, then (by Lemma 6.19) x_k is the minimum point of f. If $x_k \neq R_C \left(x_k - \lambda_k \nabla f \left(x_k\right)\right)$, then $\nabla f \left(x_k\right)^T d_k > 0$. Since

$$f \left(x_k - \beta d_k\right) = f \left(x_k\right) - \beta \nabla f \left(x_k\right)^T d_k + o \left(\beta d_k\right),$$

we obtain $f \left(x_k - \beta d_k\right) < f \left(x_k\right)$ for sufficiently small $\beta > 0$. Hence d_k is an improving feasible direction. Note that Lemma 6.16 implies that

$$f \left(x_k - \beta d_k\right) \geq f \left(x_k\right) - \beta \nabla f \left(x_k\right)^T d_k.$$

Hence the decrease of the objective function f is bounded by $\beta \nabla f \left(x_k\right)^T d_k$ at the point x_k .

We now prove the convergence of the gradient projection algorithm (6.35). The next results are due to Karmanov [218] (see also [219]).

Consider sequences $\{x_k\}$ of $\mathbb{R}^n$ with the following properties:

(P1) $\{x_k\} \subset C$.

(P2) $f \left(x_{k+1}\right) \leq f \left(x_k\right)$ $(k = 0, 1, \ldots)$.

Furthermore assume that:

(P3) $C_0 = \{x \in C \mid f \left(x\right) \leq f \left(x_0\right)\}$ is bounded, $\|x - y\| \leq \delta$ $(x, y \in C_0)$.

(P4) $\|\nabla f \left(x\right)\| \leq \gamma < +\infty$ for all $x \in C_0$.

LEMMA 6.25 *If the sequence $\{\mu_k\}$ is such that*

$$\mu_k > 0, \quad \mu_k - \mu_{k+1} \geq \tau_k \mu_k^2, \quad \tau_k \geq 0, \quad (k = 0, 1, \ldots)$$

and $\tau_j > 0$ for some j, then

$$\mu_m \leq \frac{1}{\sum_{k=0}^{m-1} \tau_k} \quad (m - 1 \geq j). \tag{6.36}$$

Proof. Since

$$\frac{1}{\mu_{k+1}} - \frac{1}{\mu_k} = \frac{\mu_k - \mu_{k+1}}{\mu_k \mu_{k+1}} \geq \tau_k \frac{\mu_k}{\mu_{k+1}} \geq \tau_k$$

and

$$\sum_{k=0}^{m-1} \left(\frac{1}{\mu_{k+1}} - \frac{1}{\mu_k} \right) = \frac{1}{\mu_m} - \frac{1}{\mu_0} \geq \sum_{k=0}^{m-1} \tau_k > 0 \quad (m-1 \geq j)$$

we have

$$\mu_m \leq \left[\sum_{k=0}^{m-1} \tau_k + \frac{1}{\mu_0} \right]^{-1} < \left[\sum_{k=0}^{m-1} \tau_k \right]^{-1}.$$

■

Let $v_k = x_k - \lambda_k \nabla f(x_k)$, where $0 < \lambda_k \leq \lambda'' < +\infty$.

LEMMA 6.26 *For any $x_k \in C$ and $\lambda_k > 0$ we have*

$$\lambda_k \nabla f(x_k)^T (x_k - R_C(v_k)) \geq \| x_k - R_C(v_k) \|^2 \tag{6.37}$$

and

$$\| x_k - R_C(v_k) \| \leq \lambda_k \| \nabla f(x_k) \|. \tag{6.38}$$

Proof. The Lemma easily follows from Corollary 6.8. ■

If $x_k = R_C(v_k)$, then (by Lemma 6.19) $f(x_k) = f^*$. Therefore we can assume that $\| x_k - R_C(v_k) \| > 0$ for any k.

THEOREM 6.27 *(Karmanov). If the conditions (P1)-(P4) hold, then*

$$f(x_m) - f(x^*) \leq \left[c \sum_{k=0}^{m-1} \lambda_k^2 \frac{f(x_k) - f(x_{k+1})}{\| x_k - R_C(v_k) \|^2} \right]^{-1}, \tag{6.39}$$

where $c > 0$ is a constant.

Proof. Let $\mu_k = f(x_k) - f(x^*)$. Lemma 6.16 implies

$$\begin{aligned}
\mu_k &\leq \nabla f(x_k)^T (x_k - x^*) \\
&= \nabla f(x_k)^T (x_k - R_C(v_k)) + \nabla f(x_k)^T (R_C(v_k) - x^*) \\
&= \nabla f(x_k)^T (x_k - R_C(v_k)) + \tfrac{1}{\lambda_k} (v_k - R_C(v_k))^T (x^* - R_C(v_k)) \\
&\quad - \tfrac{1}{\lambda_k} (x_k - R_C(v_k))^T (x^* - R_C(v_k)).
\end{aligned}$$

Since $(v_k - R_C(v_k))^T (x - R_C(v_k)) \le 0$ for all $x \in C$, we obtain that

$$\begin{aligned} \mu_k &\le \nabla f(x_k)^T (x_k - R_C(v_k)) - \tfrac{1}{\lambda_k} (x_k - R_C(v_k))^T (x^* - R_C(v_k)) \\ &\le \left(\|\nabla f(x_k)\| + \tfrac{1}{\lambda_k} \|x^* - R_C(v_k)\| \right) \|x_k - R_C(v_k)\| \\ &\le \left(\|\nabla f(x_k)\| + \tfrac{1}{\lambda_k} \|x^* - x_k\| + \tfrac{1}{\lambda_k} \|x_k - R_C(v_k)\| \right) \|x_k - R_C(v_k)\| . \end{aligned}$$

It follows from condition (P4) and the second inequality of the last lemma that

$$\mu_k \le \left(2\gamma + \frac{\delta}{\lambda_k} \right) \|x_k - R_C(v_k)\| . \tag{6.40}$$

Furthermore

$$\mu_k - \mu_{k+1} = f(x_k) - f(x_{k+1}) = \frac{f(x_k) - f(x_{k+1})}{\|x_k - R_C(v_k)\|^2} \|x_k - R_C(v_k)\|^2 ,$$

which implies

$$\mu_k - \mu_{k+1} \ge \frac{f(x_k) - f(x_{k+1})}{\|x_k - R_C(v_k)\|^2} \mu_k^2 \left(2\gamma + \frac{\delta}{\lambda_k} \right)^{-2} .$$

Using Lemma 6.25 one gets

$$\mu_m = f(x_m) - f(x^*) \le \left[\sum_{k=0}^{m-1} \frac{\lambda_k^2}{(2\gamma\lambda_k + \delta)^2} \frac{f(x_k) - f(x_{k+1})}{\|x_k - R_C(v_k)\|^2} \right]^{-1} .$$

Since $\lambda_k \le \lambda''$, $2\gamma\lambda_k + \delta \le 2\gamma\lambda'' + \delta$ and

$$\sum_{k=0}^{m-1} \frac{\lambda_k^2}{(2\gamma\lambda_k + \delta)^2} \frac{f(x_k) - f(x_{k+1})}{\|x_k - R_C(v_k)\|^2} \ge \frac{1}{(2\gamma\lambda'' + \delta)^2} \sum_{k=0}^{m-1} \lambda_k^2 \frac{f(x_k) - f(x_{k+1})}{\|x_k - R_C(v_k)\|^2}$$

we easily obtain (6.39) with $c = (2\gamma\lambda'' + \delta)^{-2}$. ∎

REMARK 6.28 *If f is strongly convex on C, then Theorem 6.21 implies the estimate*

$$\|x_m - x^*\|^2 \le \frac{2}{\rho} \left[c \sum_{k=0}^{m-1} \lambda_k^2 \frac{f(x_k) - f(x_{k+1})}{\|x_k - R_C(v_k)\|^2} \right]^{-1} . \tag{6.41}$$

REMARK 6.29 *We can weaken the inequality (6.39) using the second bound of Lemma 6.26. Then estimate (6.39) changes to*

$$f(x_m) - f(x^*) \le \left[c \sum_{k=0}^{m-1} \frac{f(x_k) - f(x_{k+1})}{\|\nabla f(x_k)\|^2} \right]^{-1} . \tag{6.42}$$

THEOREM 6.30 *Assume that f is continuously differentiable and*

$$\|\nabla f(x) - \nabla f(y)\| \leq L\|x-y\|^{\omega} \quad (x, y \in C_0) \tag{6.43}$$

holds for a given $0 < \omega \leq 1$. If the sequence $\{x_k\}$ is such that

$$0 < \|x_k - x_{k+1}\|^{\omega} \leq \frac{\alpha_k}{L}\|\nabla f(x_k)\|, \tag{6.44}$$

where

$$\alpha_k = \frac{\nabla f(x_k)^T (x_k - x_{k+1})}{\|\nabla f(x_k)\| \|x_k - x_{k+1}\|} > 0, \tag{6.45}$$

then

$$f(x_k) - f(x_{k+1}) \geq \frac{1}{2}\nabla f(x_k)^T (x_k - x_{k+1}) > 0. \tag{6.46}$$

Proof. A simple calculation yields

$$\begin{aligned} f(x_k) - f(x_{k+1}) &= \nabla f(x_k)^T (x_k - x_{k+1}) \\ &+ \textstyle\int_0^1 \left[\nabla f(x_{k+1} + \tau(x_k - x_{k+1})) - \nabla f(x_k)\right]^T (x_k - x_{k+1})\, d\tau \\ &\geq \nabla f(x_k)^T (x_k - x_{k+1}) - L\|x_k - x_{k+1}\|^{1+\omega} \textstyle\int_0^1 (1-\tau)\, d\tau \\ &= \nabla f(x_k)^T (x_k - x_{k+1}) - \frac{L}{2}\|x_k - x_{k+1}\|^{1+\omega}. \end{aligned} \tag{6.47}$$

It follows from conditions (6.44) and (6.45) that

$$\begin{aligned} f(x_k) - f(x_{k+1}) &\geq \alpha_k \|\nabla f(x_k)\| \|x_k - x_{k+1}\| - \frac{L}{2}\|x_k - x_{k+1}\|^{1+\omega} \\ &\geq \frac{\alpha_k}{2}\|\nabla f(x_k)\| \|x_k - x_{k+1}\| \\ &= \frac{1}{2}\nabla f(x_k)^T (x_k - x_{k+1}) > 0. \end{aligned}$$

■

Condition (6.46) implies condition *(P2)*. The Hölder condition (6.43) implies the boundedness of $\nabla f(x)$ on C_0. Theorems 6.27 and 6.30 imply

COROLLARY 6.31 *Under the conditions of Theorems 6.27 and 6.30*

$$f(x_m) - f(x^*) < \left[\frac{c}{2}\sum_{k=0}^{m-1} \lambda_k^2 \frac{\nabla f(x_k)^T (x_k - x_{k+1})}{\|x_k - R_C(v_k)\|^2}\right]^{-1}. \tag{6.48}$$

THEOREM 6.32 *(Karmanov). Assume that conditions (P1)-(P4) hold and f is strongly convex on C. Then*

$$f(x_m) - f(x^*) \leq (f(x_0) - f(x^*)) \exp\left(-\rho \sum_{k=0}^{m-1} \frac{f(x_k) - f(x_{k+1})}{\|\nabla f(x_k)\|^2}\right) \tag{6.49}$$

and

$$\|x_m - x^*\|^2 \le \frac{2}{\rho}\left(f(x_0) - f(x^*)\right)\exp\left(-\rho\sum_{k=0}^{m-1}\frac{f(x_k) - f(x_{k+1})}{\|\nabla f(x_k)\|^2}\right). \quad (6.50)$$

Proof. It follows from Theorem 6.21 that $\|\nabla f(x_k)\|^2 \ge \rho\mu_k$. Hence

$$\mu_k - \mu_{k+1} = \frac{f(x_k) - f(x_{k+1})}{\|\nabla f(x_k)\|^2}\|\nabla f(x_k)\|^2 \ge \rho\frac{f(x_k) - f(x_{k+1})}{\|\nabla f(x_k)\|^2}\mu_k,$$

which implies that

$$\mu_{k+1} \le \left(1 - \rho\frac{f(x_k) - f(x_{k+1})}{\|\nabla f(x_k)\|^2}\right)\mu_k. \quad (6.51)$$

We can assume that $\mu_k > 0$ (if $\mu_k = 0$, then $x_k = x^*$ and the process stops). This implies

$$0 \le \rho\frac{f(x_k) - f(x_{k+1})}{\|\nabla f(x_k)\|^2} < 1.$$

From inequality (6.51) we obtain

$$\begin{aligned}\mu_m &\le \mu_0\prod_{k=0}^{m-1}\left(1 - \rho\frac{f(x_k) - f(x_{k+1})}{\|\nabla f(x_k)\|^2}\right)\\ &\le \mu_0\exp\left(-\rho\sum_{k=0}^{m-1}\frac{f(x_k) - f(x_{k+1})}{\|\nabla f(x_k)\|^2}\right).\end{aligned}$$

The rest of the proof follows from Theorem 6.21. ■

Although the presented convergence theory is applicable to unconstrained optimization algorithms [218] we concentrate on the gradient projection method (6.35).

We assume that ∇f satisfies the Lipschitz condition

$$\|\nabla f(x) - \nabla f(y)\| \le L\|x - y\| \quad (x, y \in C_0). \quad (6.52)$$

Let $x_{k+1}^{\beta} = x_k - \beta d_k$. Using inequality (6.47) and Lemma 6.26 we obtain that

$$\begin{aligned}f(x_k) - f\left(x_{k+1}^{\beta}\right) &\ge \beta\nabla f(x_k)^T\left(x_k - R_C(v_k)\right) - \frac{L}{2}\beta^2\|x_k - R_C(v_k)\|^2\\ &\ge \frac{\beta}{\lambda_k}\|x_k - R_C(v_k)\|^2 - \frac{L}{2}\beta^2\|x_k - R_C(v_k)\|^2\\ &= \left(\frac{\beta}{\lambda_k} - \frac{L}{2}\beta^2\right)\|x_k - R_C(v_k)\|^2.\end{aligned}$$

In view of Theorem 6.27 the convergence of the gradient projection method (6.35) depends on the behavior of the quantities

$$\lambda_k^2 \left(\frac{\beta_k}{\lambda_k} - \frac{L}{2} \beta_k^2 \right). \tag{6.53}$$

Next we show the convergence of four particular methods. In each case we prove that inequality

$$\|f(x_k) - f(x_{k+1})\| \geq \varepsilon \|x_k - R_C(v_k)\|^2$$

holds with a positive constant ε. Assuming that $0 < \lambda' \leq \lambda_k$ we obtain the estimate

$$f(x_m) - f(x^*) \leq \left[c \sum_{k=0}^{m-1} (\lambda')^2 \varepsilon \right]^{-1} \to 0 \quad (m \to \infty)$$

from Theorem 6.27. Note that the speed of convergence is $O(1/m)$.
Case 1. $0 < \lambda' \leq \lambda_k \leq \lambda'' < +\infty$, *$\beta_k$ is selected by the criterion*

$$f(x_{k+1}) = \min_{0 \leq \beta \leq 1} f\left(x_{k+1}^{\beta} \right). \tag{6.54}$$

In this case

$$\begin{aligned} f(x_k) - f(x_{k+1}) &\geq f(x_k) - f\left(x_{k+1}^{\beta} \right) \geq \left(\frac{\beta}{\lambda_k} - \frac{L}{2} \beta^2 \right) \|x_k - R_C(v_k)\|^2 \\ &\geq \left(\frac{\beta}{\lambda''} - \frac{L}{2} \beta^2 \right) \|x_k - R_C(v_k)\|^2. \end{aligned}$$

Since this inequality holds for arbitrary $0 \leq \beta \leq 1$, we can take $\beta = \beta'$, where $0 < \beta' < \min\{1, 2/(L\lambda'')\}$. Then

$$f(x_k) - f(x_{k+1}) \geq \varepsilon_1 \|x_k - R_C(v_k)\|^2 \quad (\varepsilon_1 > 0).$$

Case 2. $0 < \lambda' \leq \lambda_k \leq \lambda'' < 2/L$, $0 < \beta_0 \leq \beta_k \leq 1$.
In this case

$$\begin{aligned} f(x_k) - f(x_{k+1}) &\geq \beta_k \left(\frac{1}{\lambda''} - \frac{L}{2} \beta_k \right) \|x_k - R_C(v_k)\|^2 \\ &\geq \beta_0 \left(\frac{1}{\lambda''} - \frac{L}{2} \right) \|x_k - R_C(v_k)\|^2 \\ &= \varepsilon_2 \|x_k - R_C(v_k)\|^2 \quad (\varepsilon_2 > 0). \end{aligned}$$

Case 3. $\beta_k \equiv 1$ *(method (6.24)),* λ_k *is selected by the criterion*

$$f(R_C(x_k - \lambda_k \nabla f(x_k))) = \min_{\lambda' \leq \lambda \leq \lambda''} f(R_C(v_k(\lambda))), \tag{6.55}$$

where

$$0 < \lambda' \leq \lambda'' \leq \frac{2}{2\varepsilon + L}, \quad \varepsilon > 0 \tag{6.56}$$

and $v_k(\lambda) = x_k - \lambda \nabla f(x_k)$.

Since $\beta_k = 1$ and $\lambda_k \leq \lambda''$,

$$\begin{aligned} f(x_k) - f(x_{k+1}) &= f(x_k) - f(R_C(v_k)) \\ &\geq \left(\frac{1}{\lambda''} - \frac{L}{2}\right) \|x_k - R_C(v_k)\|^2 \\ &= \varepsilon_3 \|x_k - R_C(v_k)\|^2 \quad (\varepsilon_3 > 0). \end{aligned}$$

Case 4. $0 < \lambda' \leq \lambda_k \leq \lambda''$,

$$\beta_k = \min\left\{1, \gamma_k \frac{\nabla f(x_k)^T (x_k - R_C(v_k))}{\|x_k - R_C(v_k)\|^2}\right\}, \tag{6.57}$$

$$0 < \delta_1 \leq \gamma_k \leq \frac{2 - \delta_2}{L}, \quad \delta_2 > 0. \tag{6.58}$$

If $\beta_k = 1$, then

$$f(x_k) - f(x_{k+1}) \geq \varepsilon_3 \|x_k - R_C(v_k)\|^2.$$

If $\beta_k = \gamma_k \nabla f(x_k)^T (x_k - R_C(v_k)) / \|x_k - R_C(v_k)\|^2$, then Lemma 6.26 and Theorem 6.30 imply

$$\begin{aligned} f(x_k) - f(x_{k+1}) &\geq \frac{\gamma_k}{2} \frac{\left[\nabla f(x_k)^T (x_k - R_C(v_k))\right]^2}{\|x_k - R_C(v_k)\|^2} \\ &\geq \frac{\gamma_k}{2\lambda_k^2} \|x_k - R_C(v_k)\|^2 \\ &\geq \widetilde{\varepsilon}_4 \|x_k - R_C(v_k)\|^2, \end{aligned}$$

where $\widetilde{\varepsilon}_4 = \delta_1 / \left[2(\lambda'')^2\right] > 0$. The choice $\varepsilon_4 = \min\{\varepsilon_3, \widetilde{\varepsilon}_4\}$ gives the requested lower bound for $f(x_k) - f(x_{k+1})$.

The application of the gradient projection method depends on the calculation of the projection R_C, which might be difficult if C is not a linear subspace.

If C is given by simple bound constraints, that is

$$C = \{x \in \mathbb{R}^n \mid a_i \leq x_i \leq b_i, \, i = 1, \ldots, n\} \tag{6.59}$$

or

$$C = \{x \in \mathbb{R}^n \mid 0 \le x_i, \, i = 1, \ldots, n\}, \tag{6.60}$$

then

$$p = R_C(v), \quad p_i = \begin{cases} b_i, & \text{if } v_i \ge b_i \\ v_i, & \text{if } a_i < v_i < b_i \\ a_i, & \text{if } v_i \le a_i \end{cases} \quad (i = 1, \ldots, n) \tag{6.61}$$

in the first case and

$$p = R_C(v), \quad p_i = \begin{cases} v_i, & \text{if } v_i \ge 0 \\ 0, & \text{if } v_i < 0 \end{cases} \quad (i = 1, \ldots, n) \tag{6.62}$$

in the second case. If C is a hyperplane of the form

$$C = \left\{x \in \mathbb{R}^n \mid a^T x = b\right\} \quad (a \in \mathbb{R}^n, a \ne 0), \tag{6.63}$$

then

$$R_C(v) = v - \frac{a^T v - b}{\|a\|^2} a, \quad d(v, C) = \frac{\left|a^T v - b\right|}{\|a\|} \quad (v \in \mathbb{R}^n). \tag{6.64}$$

Let $\alpha^+ = \max\{0, \alpha\}$ for $\alpha \in \mathbb{R}$. If C is a half-space given by

$$C = \left\{x \in \mathbb{R}^n \mid a^T x \le b\right\} \quad (a \in \mathbb{R}^n, \; a \ne 0), \tag{6.65}$$

then

$$R_C(v) = v - \frac{\left(a^T v - b\right)^+}{\|a\|^2} a, \quad d(v, C) = \frac{\left(a^T v - b\right)^+}{\|a\|} \quad (v \in \mathbb{R}^n). \tag{6.66}$$

If C is a hyperslab of the form

$$C = \left\{x \in \mathbb{R}^n \mid c \le a^T x \le b\right\} \quad (a \in \mathbb{R}^n, a \ne 0),$$

then for every $v \in \mathbb{R}^n$,

$$R_C(v) = v - \frac{\left(a^T v - b\right)^+ - \left(c - a^T v\right)^+}{\|a\|^2} a \tag{6.67}$$

and

$$d(v, C) = \frac{\left|\left(a^T v - b\right)^+ - \left(c - a^T v\right)^+\right|}{\|a\|}. \tag{6.68}$$

If $C = \{x \in \mathbb{R}^n \mid \|x - x_0\| \le r < \infty\}$, then

$$R_C(v) = \begin{cases} x_0 + \frac{r(v - x_0)}{\|v - x_0\|}, & \text{if } \|v - x_0\| > r, \\ \\ v, & \text{if } \|v - x_0\| \le r, \end{cases} \tag{6.69}$$

(see e.g. [27]). Further characterizations of R_C, when C is an affine manifold of the form $C = \{x \in \mathbb{R}^n \mid A^T x = d\}$ ($A \in \mathbb{R}^{n \times m}$) or a closed convex cone, are given in [73], [250], [213] and [200].

The choice of λ_k (β_k) also affects the efficiency of implementation. For the selection of λ_k and the implementation of the gradient projection method we refer to Bertsekas [44], [45], Dunn [112], Scales [309] and Kelley [222].

3. Methods for convex feasibility problems

It is a very common problem in different fields of mathematics and applications to find a point in the intersection of convex sets. Its precise formulation is the following.

Suppose that

$$C_1, \ldots, C_m \subset \mathbb{R}^n \tag{6.70}$$

are closed convex subsets with nonempty intersection $C = \cap_{i=1}^m C_i$. The convex feasibility problem is to find a point

$$x \in C. \tag{6.71}$$

Applications of the convex feasibility problem can be found in [77], [32] and the references therein.

If the functions $f_i : \mathbb{R}^n \to \mathbb{R}$ are continuous and convex, then

$$C_i = \{x \in \mathbb{R}^n \mid f_i(x) \leq 0\}$$

and $C = \cap_{i=1}^m C_i$ are closed convex sets. Belousov [38] characterizes C, when the functions f_i are convex multivariate polynomials. Any system of equations

$$f_i(x) = 0 \quad (f_i : \mathbb{R}^n \to \mathbb{R},\ i = 1, \ldots, m) \tag{6.72}$$

with convex f_i's can be written as a convex feasibility problem by observing that

$$f_i(x) = 0 \Leftrightarrow f_i(x) \leq 0 \wedge f_i(x) \geq 0. \tag{6.73}$$

Since we deal with methods of Kaczmarz and Cimmino type, the convergence results of this section can be compared with those of Chapters 4 and 5.

Assume now that C_i's are half-spaces of the form

$$C_i = \{x \in \mathbb{R}^n \mid a_i^T x \leq b_i\} \quad (i = 1, \ldots, m). \tag{6.74}$$

Thus

$$C = \{x \in \mathbb{R}^n \mid A^T x \leq b\} \tag{6.75}$$

is a convex polyhedron, where $A = [a_1, \ldots, a_m]$ and $b = [b_1, \ldots, b_m]^T$. In this case the convex feasibility problem is simply called the linear feasibility problem.

Agmon [9], Motzkin and Schoenberg [261] developed the first projection method for solving the linear feasibility problem.

If the point $x_j \notin C$, then it is orthogonally projected to the half-spaces C_i not containing x_j. We select the projection which is furthest away from x_j. Hence

$$x_{j+1} = x_j - \frac{a_i^T x_j - b_i}{\|a_i\|^2} a_i \quad (j \geq 0), \tag{6.76}$$

where $i = i(j)$ is selected by the criterion

$$d(x_j, C_i) = \frac{a_i^T x_j - b_i}{\|a_i\|} = \max\left\{\frac{a_l^T x_j - b_l}{\|a_l\|} \mid a_l^T x_j - b_l > 0\right\} = \max_{x_j \notin C_l} d(x_j, C_l). \tag{6.77}$$

This algorithm is called the method of orthogonal projection. The relaxed version of the algorithm is given by

$$x_{j+1} = x_j - \mu \frac{a_i^T x_j - b_i}{\|a_i\|^2} a_i \quad (j \geq 0), \tag{6.78}$$

where $0 < \mu \leq 2$ and i is selected by the above criterion. For $\mu = 1$ we obtain the method of orthogonal projection. If $\mu = 2$, then x_{j+1} is the reflection of x_j on the hyperplane $a_i^T x = b_i$. A somewhat modified version of the algorithm can be found in Censor [77]. We also note that the Agmon, Motzkin, Schoenberg algorithm is very similar to the Kaczmarz method except for the selection of C_i.

Tompkins [342] suggested the following generalization of the Kaczmarz method for solving the nonlinear system of inequalities

$$f_i(x) \leq 0 \quad (i = 1, \ldots, m). \tag{6.79}$$

The Tompkins, Kaczmarz algorithm can be written in the form

$$x_{j+1} = x_j - \frac{f_i(x_j)}{\|\nabla f_i(x_j)\|^2} \nabla f_i(x_j) \quad (i \equiv j \pmod m + 1). \tag{6.80}$$

For the general convex feasibility problem Bregman [54] developed the method of sequential projections. Let $x_0 \in \mathbb{R}^n$ be arbitrary and

$$x_{j+1} = R_{C_i}(x_j) \quad (i \equiv j \pmod m + 1). \tag{6.81}$$

This method is also a generalization of the Kaczmarz algorithm .

THEOREM 6.33 *(Bregman). For any $x_0 \in \mathbb{R}^n$ the sequence $\{x_j\}$ generated by algorithm (6.81) converges to a solution $x^* \in C$.*

Proof. Let $z \in C$ be arbitrary. Since $\|x_{j+1} - z\| \leq \|x_j - z\|$ by Corollary 6.8, the sequence $\{x_j\}$ is bounded. Consider the sequence $\{x_{km+1}\}$ and select a subsequence $\{x_{k_l m+1}\}$ that converges to some point x^*. Since $x_{km+1} \in C_1$ and C_1 is closed, $x^* \in C_1$. Also by Corollary 6.8 we obtain the inequality

$$\left\|x_{k_l m+1} - x_{k_l m+2}\right\|^2 \le \left\|x_{k_l m+1} - z\right\|^2 - \left\|x_{k_l m+2} - z\right\|^2.$$

Using that $\lim_{j\to\infty} \|x_j - z\|$ exists we obtain $\|x_{k_l m+1} - x_{k_l m+2}\| \to 0$. It follows that $x_{k_l m+2} \to x^*$. Relation $x_{k_l m+2} \in C_2$ implies that $x^* \in C_2$. We similarly obtain that $x^* \in C_i$ $(i = 3, \dots, m)$. Hence $x^* \in C$. We now prove that the whole sequence $\{x_j\}$ converges to x^*. For any $\varepsilon > 0$ there is an integer $k_0(\varepsilon) > 0$ such that $\|x_{k_l m+1} - x^*\| < \varepsilon$ for all $k_l > k_0(\varepsilon)$. Since $\|x_{j+1} - z\| \le \|x_j - z\|$ holds with arbitrary $z \in C$, we have

$$\|x_n - x^*\| \le \|x_{k_l m+1} - x^*\| < \varepsilon$$

for $n \ge k_l m + 1 > k_0(\varepsilon) m + 1$. Hence $x_n \to x^*$. ∎

Bregman [54] proved the theorem for Hilbert space and weak convergence. Here we presented a simplified proof.

Assume now that $i = i(j)$ is selected by the criterion

$$d\left(x_j, C_{i(j)}\right) = \max_{1\le l\le m} d(x_j, C_l). \tag{6.82}$$

Hence the method of sequential projections takes the form

$$x_{j+1} = R_{C_{i(j)}}(x_j) \quad (j \ge 0). \tag{6.83}$$

THEOREM 6.34 *(Bregman). Algorithm (6.83) is convergent, that is, for any $x_0 \in \mathbb{R}^n$ the sequence $\{x_j\}$ converges to some $x^* \in C$.*

Proof. Let $z \in C$ be arbitrary. Then Corollary 6.8 implies that

$$\|x_{j+1} - z\| = \left\|R_{C_{i(j)}}(x_j) - z\right\| \le \|x_j - z\|.$$

Hence $\{x_j\}$ is bounded and there is a subsequence $\{x_{j_k}\}$ converging to some x^*. Since $\|x_j - z\| \to a \ge 0$ and $\|x_j - x_{j+1}\|^2 \le \|x_j - z\|^2 - \|x_{j+1} - z\|^2$, we obtain $\|x_j - x_{j+1}\| \to 0$. The inequality $d(x_j, C_i) \le \|x_j - x_{j+1}\|$ implies that $d(x_j, C_i) \to 0$ for all $i = 1, \dots, m$. Let $\{y_k\} \subset C_i$ be such that $\|x_{j_k} - y_k\| \to 0$. The sequence $\{y_k\}$ is bounded. Let $y_{k_l} \to y^*$. Since $y^* \in C_i$ and $\left\|x_{j_{k_l}} - y_{k_l}\right\| \to 0$, we have $x_{j_{k_l}} \to y^*$ and $x^* = y^* \in C_i$. This holds for every $i = 1, \dots, m$ implying that $x^* \in C$. The relation $x_n \to x^*$ can now be proved as in the previous theorem. ∎

Consider the linear feasibility problem (6.74). Then $d(x_j, C_i) = 0$ for $x_j \in C_i$ and

$$d(x_j, C_i) = \frac{a_i^T x_j - b_i}{\|a\|} > 0$$

for $x_j \notin C_i$. Hence for $x_j \notin C$ algorithm (6.83) becomes

$$x_{j+1} = x_j - \frac{a_i^T x_j - b_i}{\|a\|^2} a_i \quad (j \geq 0), \tag{6.84}$$

where $i = i(j)$ is given by

$$d(x_j, C_i) = \max_l d(x_j, C_l) = \max_{x_j \notin C_l} d(x_j, C_l). \tag{6.85}$$

Hence we obtained algorithm (6.76) as a special case of (6.83).

Bregman [54] also points out how the method of successive projections can be used to solve linear and quadratic programming problems.

There are several generalizations of Bregman's sequential projection method. Gurin, Polyak and Raik [173] developed the relaxed projection method

$$x_{j+1} = x_j + \lambda_j (R_{C_i}(x_j) - x_j) \quad (j \geq 0), \tag{6.86}$$

where $0 < \lambda_j \leq 2$. For $\lambda_j \equiv 1$ we obtain algorithm (6.83). The index $i = i(j)$ may be selected either by

$$i \equiv j \pmod m + 1 \tag{6.87}$$

or

$$d(x_j, C_i) = \max_l d(x_j, C_l). \tag{6.88}$$

Define

$$\phi(x_j) = \max_l d(x_j, C_l).$$

Then any selection $i = i(j)$ is allowed if

$$d\left(x_j, C_{i(j)}\right) \to 0 \Rightarrow \phi(x_j) \to 0 \tag{6.89}$$

holds. We will show that selections (6.87) and (6.88) satisfy this condition.

THEOREM 6.35 *(Gurin, Polyak, Raik). Assume that algorithm (6.86) satisfies conditions (6.89) and*

$$0 < \varepsilon_1 \leq \lambda_j \leq 2 - \varepsilon_2, \quad \varepsilon_2 > 0. \tag{6.90}$$

Then for any $x_0 \in \mathbb{R}^n$ the sequence $\{x_j\}$ converges to some $x^* \in C$.

Proof. We first show that

$$\|x_{j+1} - z\| \leq \|x_j - z\| \quad (j \geq 0) \tag{6.91}$$

holds for any selection $i = i(j)$ and for any $z \in C$. By definition and Theorem 6.7 we obtain

$$\begin{aligned} \|x_{j+1} - z\|^2 &= \|x_j + \lambda_j (R_{C_i}(x_j) - x_j) - z\|^2 \\ &= \|x_j - z\|^2 + \lambda_j^2 \|R_{C_i}(x_j) - x_j\|^2 \\ &\quad + 2\lambda_j (x_j - z)^T (R_{C_i}(x_j) - x_j) \\ &= \|x_j - z\|^2 + \left(\lambda_j^2 - 2\lambda_j\right) \|R_{C_i}(x_j) - x_j\|^2 \\ &\quad + 2\lambda_j (R_{C_i}(x_j) - z)^T (R_{C_i}(x_j) - x_j) \\ &\leq \|x_j - z\|^2 + \left(\lambda_j^2 - 2\lambda_j\right) \|R_{C_i}(x_j) - x_j\|^2 \\ &\leq \|x_j - z\|^2. \end{aligned}$$

For $z = R_C(x_j)$ inequality (6.91) implies

$$d(x_j, C) = \|x_j - R_C(x_j)\| \geq \|x_{j+1} - R_C(x_j)\|$$

$$\geq \|x_{j+1} - R_C(x_{j+1})\| = d(x_{j+1}, C).$$

Hence $\{d(x_j, C)\}$ is monotone decreasing and there exists

$$\lim_{j\to\infty} d(x_j, C) = \alpha \geq 0.$$

Furthermore

$$d^2(x_j, C) - d^2(x_{j+1}, C) \geq \|x_j - R_C(x_j)\|^2 - \|x_{j+1} - R_C(x_j)\|^2$$

$$= \|x_j - R_C(x_j)\|^2 - \left\|x_j + \lambda_j \left(R_{C_{i(j)}}(x_j) - x_j\right) - R_C(x_j)\right\|^2$$

$$= -\lambda_j^2 \left\|R_{C_{i(j)}}(x_j) - x_j\right\|^2 - 2\lambda_j (x_j - R_C(x_j))^T \left(R_{C_{i(j)}}(x_j) - x_j\right)$$

$$= \lambda_j (2 - \lambda_j) \left\|R_{C_{i(j)}}(x_j) - x_j\right\|^2$$

$$-2\lambda_j \left(R_{C_{i(j)}}(x_j) - R_C(x_j)\right)^T \left(R_{C_{i(j)}}(x_j) - x_j\right)$$

$$\geq \lambda_j (2 - \lambda_j) \left\|R_{C_{i(j)}}(x_j) - x_j\right\|^2$$

$$\geq \varepsilon_1 \varepsilon_2 \left\|R_{C_{i(j)}}(x_j) - x_j\right\|^2 = \varepsilon_1 \varepsilon_2 d^2\left(x_j, C_{i(j)}\right).$$

Since $\{d(x_j, C)\}$ is convergent, we obtain

$$d\left(x_j, C_{i(j)}\right) \to 0 \tag{6.92}$$

and $\phi(x_j) \to 0$. We now prove that

$$\lim_{j\to\infty} d(x_j, C) = 0. \tag{6.93}$$

Assume that there exists a subsequence $\{x_{j_k}\}$ such that $d(x_{j_k}, C) \geq \alpha > 0$. Since $\{x_{j_k}\}$ is bounded, there is a subsequence $\left\{x_{j_{k_i}}\right\}$ such that $x_{j_{k_i}} \to x^*$ as $i \to \infty$. For any $1 \leq l \leq m$, the inequality $d\left(x_{j_{k_i}}, C_l\right) \leq \phi\left(x_{j_{k_l}}\right)$ implies that $d\left(x_{j_{k_i}}, C_l\right) \to 0$. Consequently, $d(x^*, C_l) = 0$ for all $l = 1, \ldots, m$. Hence $x^* \in C$ and $d(x^*, C) = 0$ which contradicts the assumption $d(x_{j_k}, C) \geq \alpha > 0$. Select now a subsequence $\{x_{j_k}\}$ such that $x_{j_k} \to x^* \in C$. Then $\|x_{j_k} - x^*\| < \varepsilon$ for all $j_k \geq j(\varepsilon)$. Setting $z = x^*$ in (6.91) we obtain that

$$\|x_{j+1} - x^*\| \leq \|x_{j_k} - x^*\| < \varepsilon \quad (j \geq j_k \geq j(\varepsilon)).$$

Hence $x_{j+1} \to x^*$. ■

We prove that selections (6.87) and (6.88) satisfy condition (6.89) provided that (6.90) holds. It is obvious that relation (6.92) implies condition (6.89) for selection (6.88). Consider now the cyclic selection (6.87). For any $\varepsilon > 0$ there is an integer $j(\varepsilon) > 0$ such that

$$d\left(x_j, C_{i(j)}\right) \leq \frac{\varepsilon}{2m} \quad (j \geq j(\varepsilon)).$$

Then

$$\|x_{j+1} - x_j\| = \lambda_j \left\|R_{C_{i(j)}}(x_j) - x_j\right\| = \lambda_j d\left(x_j, C_{i(j)}\right) \leq \frac{\varepsilon}{m}.$$

Let $j \geq j(\varepsilon)$ and $1 \leq l \leq m$. There is a nonnegative integer $k < m$ such that $i(j+k) = l$. Hence

$$\begin{aligned} d(x_j, C_l) &\leq \|x_j - R_{C_l}(x_{j+k})\| = \left\|x_j - R_{C_{i(j+k)}}(x_{j+k})\right\| \\ &\leq \sum_{s=1}^{k} \|x_{j+s-1} - x_{j+s}\| + \left\|x_{j+k} - R_{C_{i(j+k)}}(x_{j+k})\right\| \\ &\leq k\frac{\varepsilon}{m} + \frac{\varepsilon}{2m} < \varepsilon. \end{aligned}$$

Since this holds for all $1 \leq l \leq m$ we obtained that

$$\phi(x_j) = \max_l d(x_j, C_l) < \varepsilon \quad (j \geq j(\varepsilon)).$$

Thus we proved that condition (6.89) also holds for selection (6.87).

Gurin, Polyak and Raik proved several variants of the above convergence theorem [173]. An interesting generalization of their method is given by Pierra [281].

The following extension of Bregman's algorithm (6.83) is due to Bregman himself [55].

Let $S \subset \mathbb{R}^n$ be a convex set such that $S \cap C \neq \emptyset$. Assume that function $D : S \times S \to \mathbb{R}$ has the following properties:
(i) $D(x, y) \geq 0$, $D(x, y) = 0 \Leftrightarrow x = y$.
(ii) For any $y \in S$ and $1 \leq i \leq m$ there exists a point $x = P_i^D(y) \in C_i \cap S$ such that

$$D(x, y) = \min_{z \in C_i \cap S} D(z, y).$$

The point x is called a D-projection of y on the set C_i.
(iii) For every $1 \leq i \leq m$ and $y \in S$ the function

$$G_i(z) = D(z, y) - D\left(z, P_i^D(y)\right)$$

is convex on $C_i \cap S$.
(iv) There exists $D'_x(x, y)$ and $D'_x(y, y) = 0$.
(v) For any $z \in C \cap S$ and $L \in \mathbb{R}$ the level set

$$T = \{x \in S \mid D(z, x) \leq L\}$$

is compact.
(vi) If $D(x_j, y_j) \to 0$, $y_j \to y^* \in \overline{S}$ and the sequences $\{x_j\}$ belong to a compact set, then $x_j \to y^*$.

Later we show examples of D-functions. Bregman's D-projection method is defined as

$$x_0 \in S, \tag{6.94}$$

$$x_{j+1} = P_{i(j)}^D(x_j) \quad (j \geq 0). \tag{6.95}$$

The selection function $i = i(j)$ will be specified later.

LEMMA 6.36 *Let $z \in C_i \cap S$. Then for any $y \in S$,*

$$D\left(P_i^D(y), y\right) \leq D(z, y) - D\left(z, P_i^D(y)\right). \tag{6.96}$$

Proof. It follows from assumption (iii) that

$$D\left(\lambda z + (1-\lambda) P_i^D(y), y\right) - D\left(\lambda z + (1-\lambda) P_i^D(y), P_i^D(y)\right)$$
$$\leq \lambda\left(D(z, y) - D\left(z, P_i^D(y)\right)\right) + (1-\lambda) D\left(P_i^D(y), y\right)$$

holds for all $\lambda \in [0, 1]$. For $\lambda > 0$ we obtain

$$D(z, y) - D\left(z, P_i^D(y)\right) - D\left(P_i^D(y), y\right)$$
$$\geq \frac{D\left(\lambda z+(1-\lambda)P_i^D(y),y\right)-D\left(P_i^D(y),y\right)}{\lambda} - \frac{D\left(\lambda z+(1-\lambda)P_i^D(y),P_i^D(y)\right)}{\lambda}.$$

Since $\lambda z + (1-\lambda) P_i^D(y) \in C_i \cap S$, assumption (ii) implies that the first fraction of the right-hand side is nonnegative. If $\lambda \to 0$, the second fraction tends to 0 because of condition (iv). Hence by letting $\lambda \to 0$ we obtain the inequality $D(z,y) - D\left(z, P_i^D(y)\right) - D\left(P_i^D(y), y\right) \geq 0$. ■

LEMMA 6.37 *For any selection $i(j)$, algorithm (6.94), (6.95) has the following properties:*
(a) The sequences $\{x_j\}$ belong to a compact set.
(b) For any $z \in C$ there exists $\lim_{j\to\infty} D(z, x_j)$.
(c) $D(x_{j+1}, x_j) \to 0$ as $j \to \infty$.

Proof. Let $z \in C \cap S$. By Lemma 6.36

$$D(x_{j+1}, x_j) \leq D(z, x_j) - D(z, z_{j+1}). \tag{6.97}$$

Since $D(x_{j+1}, x_j) \geq 0$, it follows that

$$D(z, z_{j+1}) \leq D(z, x_j).$$

Hence $\lim_{j\to\infty} D(z, x_j)$ exists and $D(x_{j+1}, x_j) \to 0$. Since $\{x_j\} \subset T = \{x \in S \mid D(z,x) \leq D(z, x_0)\}$ and T is compact by assumption (v), we proved the lemma. ■

THEOREM 6.38 *(Bregman). If the selection function $i = i(j)$ of algorithm (6.94), (6.95) is given by $i \equiv j \pmod m + 1$, then any accumulation point x^* of the sequence $\{x_j\}$ belongs to C.*

Proof. If x^* is an accumulation point of $\{x_j\}$, then there exist $1 \leq i \leq m$ and a subsequence $\{x_{j_k}\}_{k=0}^{\infty} \subset \{x_j\}$ such that $\{x_{j_k}\}_{k=0}^{\infty} \subset C_i$ and $x_{j_k} \to x^* \in C_i$. For simplicity, we can assume that $i = 1$. Then $\{x_{j_k+1}\} \subset C_2$. We can select a convergent subsequence, also denoted by $\{x_{j_k+1}\}$, that $x_{j_k+1} \to x_2^* \in C_2$. Since $\{x_{j_k+2}\} \subset C_3$, a subsequence (also denoted by) $\{x_{j_k+2}\}$ exists such that $x_{j_k+2} \to x_3^* \in C_3$. Continuing this way we find a sequence $\{x_{j_k}\}$ such that

$$x_{j_k} \to x^* \in C_1, \quad x_{j_k+1} \to x_2^* \in C_2, \quad \ldots, x_{j_k+m} \to x_m^* \in C_m.$$

It follows from Lemma 6.37 that $D(x_{j_k+1}, x_{j_k}) \to 0$. Condition (vi) implies that $x_2^* = \lim_k x_{j_k+1} = \lim_k x_{j_k} = x^*$. Hence $x^* \in C_2$. Similarly we obtain that $x_3^* = \cdots = x_m^* = x^*$ and $x^* \in C$. ■

Bregman [55] proved the same convergence result for the selection function $i = i(j)$ defined by

$$\min_{x \in C_{i(j)}} D(x, x_j) = \max_{1 \leq l \leq m} \min_{x \in C_l} D(x, x_j). \tag{6.98}$$

Proposition 6.39 *There is only one accumulation point if any of the following conditions is satisfied:*
(α) The set S is closed and for any $z_1, z_2 \in C \cap S$ the function $H(y) = D(z_1, y) - D(z_2, y)$ is continuous on S.
(β) Function $D(x, y)$ is defined for $x \in \overline{S}$, and $D(y^*, y_j) \to 0$ holds for $y_j \to y^* \in S$.

Proof. Assume that (α) holds, $x_{j_k} \to x^* \in C$ and $x_{j_l} \to x^{**} \in C$. Lemma 6.37 implies that

$$\lim_j H(x_j) = \lim_j (D(x^*, x_j) - D(x^{**}, x_j))$$

exists. For the subsequence $\{x_{j_k}\}$, $\lim_k H(x_{j_k}) = -D(x^{**}, x^*) \leq 0$. For the other subsequence $\{x_{j_l}\}$, $\lim_l H(x_{j_l}) = D(x^*, x^{**}) \geq 0$. Consequently, $D(x^*, x^{**}) = D(x^{**}, x^*) = 0$ and $x^* = x^{**}$. Assume that condition (β) is true. Then

$$\lim_j D(x^*, x_j) = \lim_l D(x^*, x_{j_l}) = \lim_k D(x^*, x_{j_k}) = 0.$$

Condition (vi) then implies that $x^* = x^{**}$. ■

Bregman [55] gave the following examples of D-functions the first of which shows that algorithm (6.94), (6.95) is indeed a generalization of algorithm (6.83).

1. $S = \mathbb{R}^n$ and
$$D(x, y) = \|x - y\|_2^2 . \tag{6.99}$$

It is easy to verify that D satisfies conditions *(i)-(vi)* and (α) and that the D-projection $P_i^D(x)$ coincides with the orthogonal projection $R_{C_i}(x)$ ($1 \leq i \leq m$).

2. Let $f : S \to \mathbb{R}$ be strictly convex and differentiable on the convex set $S \subset \mathbb{R}^n$. Then

$$D(x, y) = f(x) - f(y) - \nabla f(y)^T (x - y) \tag{6.100}$$

is a D-function satisfying conditions *(i)-(vi)* and (α).

3. $S = \mathbb{R}^n$ and
$$D(x, y) = (x - y)^T C (x - y), \tag{6.101}$$

where $C \in \mathbb{R}^{n \times n}$ is symmetric and positive definite. It satisfies conditions *(i)-(vi)* and (α).

Censor and Lent [82] derived different characterizations of D-projections, especially on hyperplanes. Applications of Bregman's D-projection algorithm to convex optimization are given by Bregman [55], Censor and Lent [82], and De Pierro and Iusem [99]. The best source on Bregman's D-projection algorithm is Censor and Zenios [83].

We close the section with the Cimmino algorithm extended to the convex feasibility problem.

Consider first the linear feasibility problem (6.74), (6.75). Let the weights $0 < w_i \leq 1$ $(i = 1, \ldots, m)$ be given such that $\sum_{i=1}^{m} w_i = 1$. The reflection of x_j on $C_i = \left\{x \in \mathbb{R}^n \mid a_i^T x \leq b_i\right\}$ is given by

$$\delta_i = x_j - 2\frac{\left(a_i^T x_j - b_i\right)^+}{\|a_i\|^2} a_i \quad (i = 1, \ldots, m).$$

Observe that $\delta_i = x_j$ for $x_j \in C_i$. The weighted sum of the reflected points gives the formula

$$x_{j+1} = x_j - 2\sum_{i=1}^{m} w_i \frac{\left(a_i^T x_j - b_i\right)^+}{\|a_i\|^2} a_i = x_j - 2\sum_{x_j \notin C_i} w_i \frac{a_i^T x_j - b_i}{\|a_i\|^2} a_i.$$

A simple modification of this formula gives the algorithm of Censor and Elfving [81]:

$$x_{j+1} = x_j - \frac{2}{\mu_j}\sum_{x_j \notin C_i} w_i \frac{a_i^T x_j - b_i}{\|a_i\|^2} a_i \quad \left(\mu_j = \sum_{x_j \notin C_i} w_i,\ j \geq 0\right). \tag{6.102}$$

Assuming that C is nonempty, Censor and Elfving [81] proved the convergence of algorithm (6.102) for any $x_0 \in \mathbb{R}^n$. A different convergence proof was given by De Pierro and Iusem [98].

We now derive the Cimmino algorithm for the general convex feasibility problem (6.70), (6.71). The reflection of x_j on the closed convex set C_i is given by $\delta_i = 2R_{C_i}(x_j) - x_j$. The weighted sum of the reflected points has the form

$$x_{j+1} = 2\sum_{i=1}^{m} w_i R_{C_i}(x_j) - x_j = x_j + 2\left(\sum_{i=1}^{m} w_i R_{C_i}(x_j) - x_j\right).$$

Setting $P(x) = \sum_{i=1}^{m} w_i R_{C_i}(x)$ we can write

$$x_{j+1} = x_j + 2\left(P(x_j) - x_j\right). \tag{6.103}$$

Introducing the relaxation parameter μ we obtain the nonlinear Cimmino algorithm

$$x_{j+1} = x_j + \mu\left(P(x_j) - x_j\right) \quad (\mu > 0,\ j \geq 0). \tag{6.104}$$

Iusem and De Pierro [213] proved convergence for the modified algorithm

$$x_{j+1} = x_j + \mu(x_j)\left(P(x_j) - x_j\right) \quad (j \geq 0), \tag{6.105}$$

where

$$\mu(x) = \begin{cases} \left(\sum_{x \notin C_i} w_i\right)^{-1}, & \text{if } |\{i \mid x \notin C_i\}| \geq 2 \\ 1, & \text{otherwise} \end{cases} \tag{6.106}$$

and $\mu(x) \geq 1$.

We prove the convergence of algorithm (6.104) for a fixed μ in the range $0 < \mu \leq 1$. Introducing

$$P_\mu(x) = x + \mu(P(x) - x) = (1-\mu)x + \mu P(x) \quad (\mu > 0) \tag{6.107}$$

we can write algorithm (6.104) in the form

$$x_{j+1} = P_\mu(x_j) \quad (j \geq 0). \tag{6.108}$$

LEMMA 6.40 *Let $D \subset \mathbb{R}^n$ be closed and convex. If $V : D \to D$ is firmly contractive, then the following properties hold for all $x, y \in D$:*
(i) $\|V(x) - V(y)\| \leq \|x - y\|$.
(ii) $\|V(x) - V(y)\|^2 + \|(x - V(x)) - (y - V(y))\|^2 \leq \|x - y\|^2$.
(iii) $\|V(x) - V(y)\| = \|x - y\| \Rightarrow V(x) - V(y) = x - y$.

Proof. (i) See the proof of Theorem 6.9. (ii) Using the definition of firm contractivity we obtain

$$\begin{aligned} \|(x - V(x)) - (y - V(y))\|^2 &= \|x - y\|^2 + \|V(x) - V(y)\|^2 \\ &\quad - 2(x-y)^T (V(x) - V(y)) \\ &\leq \|x - y\|^2 - \|V(x) - V(y)\|^2, \end{aligned}$$

which clearly implies (ii). (iii) Under the assumption, (ii) implies

$$0 \leq \|(x - V(x)) - (y - V(y))\|^2 \leq 0$$

and $x - V(x) = y - V(y)$. ■

Let $D \subset \mathbb{R}^n$ be a closed convex set. The fixed point set of a mapping $V : D \to D$ will given by

$$F(V) = \{x \in D \mid V(x) = x\}. \tag{6.109}$$

LEMMA 6.41 *Let $D \subset \mathbb{R}^n$ be closed and convex. Assume that $V : D \to D$ is firmly contractive and $F(V) \neq \emptyset$. Then for any $x_0 \in D$, the sequence $x_{j+1} = V(x_j)$ $(j \geq 0)$ converges to a fixed point $x^* \in F(V)$.*

Proof. For any $z \in F(V)$,

$$\|x_{j+1} - z\| = \|V(x_j) - V(z)\| \leq \|x_j - z\| \quad (j \geq 0).$$

Hence the sequence $\{x_j\}$ is bounded and there exists a subsequence $x_{j_k} \to x^*$. Since $\{\|x_j - z\|\}$ is monotone decreasing, $\|x_j - z\| \to \alpha(z) \geq 0$. Thus $\|x_{j_k} - z\| \to \|x^* - z\| = \alpha(z)$ and

$$\|x_{j_k+1} - z\| = \|V(x_{j_k}) - z\| \to \|V(x^*) - z\| = \alpha(z).$$

It follows that $\|V(x^*) - z\| = \|V(x^*) - V(z)\| = \|x^* - z\|$, which implies that $V(x^*) - V(z) = x^* - z$ and $V(x^*) = x^*$. Consequently $x^* \in F(V)$. Setting $z = x^*$ we obtain that $\|x_j - x^*\| \to \alpha(x^*) = 0$. ■

If D is bounded, then $F(V)$ has at least one element by the Schauder fixed point theorem. It can be proved that $F(V)$ is convex (see, e.g., [271]). In our application D will not be necessarily bounded. We will show however that $F(V) = C$.

LEMMA 6.42 *Let $D \subset \mathbb{R}^n$ be closed and convex. Assume that the mappings $V_i : D \to D$ are firmly contractive ($i = 1, \ldots, m$), $w_i > 0$ and $\sum_{i=1}^m w_i = 1$. Define $V(x)$ by the convex combination of V_i's:*

$$V(x) = \sum_{i=1}^{m} w_i V_i(x). \tag{6.110}$$

Then
(i) $V : D \to D$ is firmly contractive.
(ii) $\|V(x) - V(y)\| = \|x - y\| \Rightarrow V_i(x) - V_i(y) = x - y$ ($i = 1, \ldots, m$).

Proof. (i) By definition we have

$$\begin{aligned}(V(x) - V(y))^T (x - y) &= \sum_{i=1}^{m} w_i (V_i(x) - V_i(y))^T (x - y) \\ &\geq \sum_{i=1}^{m} w_i \|V_i(x) - V_i(y)\|^2 .\end{aligned}$$

The Jensen inequality (see [284], [38]) implies that

$$\sum_{i=1}^{m} w_i \|V_i(x) - V_i(y)\|^2 \geq \left(\sum_{i=1}^{m} w_i \|V_i(x) - V_i(y)\| \right)^2 .$$

Since

$$\|V(x) - V(y)\| \leq \sum_{i=1}^{m} w_i \|V_i(x) - V_i(y)\|$$

we obtain

$$(V(x) - V(y))^T (x - y) \geq \|V(x) - V(y)\|^2 ,$$

which was to be proved.
(ii) If $\|V(x) - V(y)\| = \|x - y\|$, then

$$\|x - y\| = \|V(x) - V(y)\| \leq \sum_{i=1}^{m} w_i \|V_i(x) - V_i(y)\| \leq \|x - y\|$$

implies that

$$\sum_{i=1}^{m} w_i \|V_i(x) - V_i(y)\| = \|x - y\|. \qquad (6.111)$$

Because of Lemma 6.40 (property (i)) the equality holds if and only if

$$\|V_i(x) - V_i(y)\| = \|x - y\| \quad (i = 1, \ldots, m).$$

Property (iii) of the same lemma implies that $V_i(x) - V_i(y) = x - y$ for all i. ■

Let us define the mapping

$$V_\mu(x) = (1 - \mu)x + \mu V(x). \qquad (6.112)$$

Since $\mathrm{Id}(x) = x$ is firmly contractive, V_μ is also firmly contractive for $0 < \mu \leq 1$. For any $\mu \neq 0$, $F(V_\mu) = F(V)$.

COROLLARY 6.43 *Let $0 < \mu \leq 1$. For any $x_0 \in D$, the sequence $x_{j+1} = V_\mu(x_j)$ $(j \geq 0)$ converges to a fixed point x of mapping V provided that $F(V) \neq \emptyset$.*

Proof. It follows from Lemmas 6.41 and 6.42. ■

LEMMA 6.44 *Under the assumptions of Lemma 6.42*

$$F(V) = \cap_{i=1}^{m} F(V_i) \qquad (6.113)$$

provided that $\cap_{i=1}^{m} F(V_i) \neq \emptyset$.

Proof. If $x \in \cap_{i=1}^{m} F(V_i)$, then $V_i(x) = x$ $(i = 1, \ldots, m)$ implies $V(x) = x$ and $F(V) \supset \cap_{i=1}^{m} F(V_i)$. In order to prove the reverse inclusion we define the distance function

$$f(x) = \sum_{i=1}^{m} w_i \|V_i(x) - x\|^2 \quad (x \in D). \qquad (6.114)$$

It is obvious that $f(x) = 0$ if and only if $x \in \cap_{i=1}^{m} F(V_i)$. It follows from property (ii) of Lemma 6.40 that for any $y \in F(V_i)$

$$\|V_i(x) - x\|^2 \leq \|x - y\|^2 - \|V_i(x) - y\|^2 \quad (x \in D).$$

Taking $y \in \cap_{i=1}^{m} F(V_i)$ we obtain the estimate

$$f(x) \le \|x - y\|^2 - \sum_{i=1}^{m} w_i \|V_i(x) - y\|^2 \quad (x \in D). \tag{6.115}$$

If $x \in F(V)$, then $\|V(x) - V(y)\| = \|x - y\|$ implies $V_i(x) - V_i(y) = V_i(x) - y = x - y$ for all i. Hence the estimate implies $0 \le f(x) \le 0$ and $x \in \cap_{i=1}^{m} F(V_i)$. ∎

THEOREM 6.45 *Consider the convex feasibility problem (6.70), (6.71). If $0 < \mu \le 1$, then for any $x_0 \in \mathbb{R}^n$ the sequence $\{x_j\}$ generated by (6.108) converges to a feasible point $x^* \in C$.*

Proof. Set $V_i(x) = R_{C_i}(x)$ for $i = 1, \dots, m$. Then $V(x) = P(x)$, $V_\mu(x) = P_\mu(x)$, $F(V_i) = C_i$ and $F(V) = \cap_{i=1}^{m} C_i = C$. The rest follows from Corollary 6.43. ∎

Here the distance function (6.114) becomes

$$f(x) = \sum_{i=1}^{m} w_i d^2(x, C_i) \quad (x \in \mathbb{R}^n), \tag{6.116}$$

which is convex. The results can be extended to Hilbert spaces, but instead of Lemma 6.41 we need the fixed point theorems of Browder [63], [64], [61], [62] and Opial [269] (see also [224], [37], [183], [70], [229], [91], [32]).

In the proof of Theorems 6.23, 6.33, 6.34, 6.35 and Lemma 6.41 we exploited the special properties of the Fejér sequences. The notion of Fejér sequences was introduced by Motzkin and Schoenberg [261].

DEFINITION 6.46 *Let $K \subset \mathbb{R}^n$ be a closed convex nonempty set. The sequence $\{x_j\}$ is a Fejér sequence with respect to K if*

$$\|x_{j+1} - z\| \le \|x_j - z\| \quad (j \ge 0) \tag{6.117}$$

holds for all $z \in K$.

The most important properties of the Fejér sequences are collected in [32].

General iterative schemes of projection methods for solving the convex feasibility problem are given by Schott [313], Flam and Zowe [123], and Bauschke and Borwein [32]. Censor [77], [78] gives surveys on the algorithms and applications. Bauschke and Borwein [32] give a unified theory of the projection methods in real Hilbert spaces. Related generalizations can be found in [225], [226] and [79].

Finally we note that the method of alternating projections studied in Chapter 7 is also related to the convex feasibility problem.

Chapter 7

PROJECTION METHODS FOR LINEAR EQUATIONS IN HILBERT SPACES

Projections and projection methods in abstract vector spaces show similarities and differences when compared to the finite-dimensional case. The difference stems from the change from finite- to infinite-dimensional vector spaces. Since a deeper study requires a strong background in functional analysis we restrict the subject to results that can be understood easily. We concentrate on the alternating projection method that is behind the many algorithms we studied in earlier chapters. For easier understanding we summarize the necessary basic concepts and results from functional analysis. For deeper details the reader can consult with any of the books [177], [6], [301], [236], [217], [113], [128], [27] and the references therein.

1. Introduction

We first summarize the most important concepts and results we need. Then we study the projections of Hilbert spaces. The section concludes with a survey on angles between subspaces in Hilbert spaces.

DEFINITION 7.1 *The set $X \neq \emptyset$ is called a metric space if there is defined a function d with domain $X \times X$ and range in $\mathbb{R}$ such that for all $x, y, z \in X$,*
(i) $d(x,y) \geq 0$, $d(x,y) = 0 \Leftrightarrow x = y$;
(ii) $d(x,y) = d(y,x)$;
(iii) $d(x,y) \leq d(x,z) + d(z,y)$.
d is called the metric or distance function of X.

DEFINITION 7.2 *A linear vector space X over the field $\mathbb{F}$ ($\mathbb{F} = \mathbb{R}$ or $\mathbb{F} = \mathbb{C}$) is called a normed linear space if for every $x \in X$, there is associated a real number $\|x\|$, the norm of the vector x, such that*
(i) $\|x\| \geq 0$ ($x \in X$), $\|x\| = 0 \Leftrightarrow x = 0$;

(ii) $\|\lambda x\| = |\lambda| \, \|x\|$ $(\lambda \in \mathbb{F})$;
(iii) $\|x + y\| \leq \|x\| + \|y\|$ $(x, y \in X)$.

A normed linear space X is a metric space with the distance function

$$d(x, y) = \|x - y\|. \tag{7.1}$$

DEFINITION 7.3 *A sequence of elements $\{x_n\}$ of a metric space X is said to converge to an element $x \in X$ if*

$$\lim_{n \to \infty} d(x_n, x) = 0.$$

The limit is unique and we may write

$$\lim_{n \to \infty} x_n = x. \tag{7.2}$$

In a normed linear space the above definition is equivalent to

$$\lim_{n \to \infty} \|x_n - x\| = 0. \tag{7.3}$$

Hence it is called convergence in norm.

Let X be a metric space and $S \subseteq X$. The closure $\overline{S}$ of S is defined as the set of all limits of convergent sequences of S. Obviously $S \subseteq \overline{S}$. If $S = \overline{S}$, S is called closed. A set S is dense in X if $\overline{S} = X$. X is separable if it has a countable dense set.

DEFINITION 7.4 *A sequence of elements of X, $\{x_n\}$ is called a Cauchy sequence, if for every $\varepsilon > 0$ there is an integer $n(\varepsilon)$ such that $d(x_m, x_n) < \varepsilon$ for all $m, n > n(\varepsilon)$. A space X is called complete if every Cauchy sequence has a limit in X.*

Specifically, if X is a complete normed linear space, and if for any $\varepsilon > 0$, we can find $n(\varepsilon) > 0$ such that

$$\|x_m - x_n\| < \varepsilon \quad (m, n > n(\varepsilon)),$$

then there is an element $x \in X$ such that

$$\lim_{n \to \infty} \|x - x_n\| = 0.$$

DEFINITION 7.5 *A complete normed linear space is called a Banach space.*

DEFINITION 7.6 *Let X, Y be linear vector spaces over the same field $\mathbb{F}$. A mapping $T : X \to Y$ defined on a linear subspace $\mathcal{D}(T)$ of X is said to be linear, if*

$$T(\alpha x_1 + \beta x_2) = \alpha T(x_1) + \beta T(x_2) \quad (x_1, x_2 \in \mathcal{D}(T), \alpha, \beta \in \mathbb{F}).$$

If $Y = \mathbb{F}$, the linear operator T is called a linear functional over X.

DEFINITION 7.7 *Let X, Y be normed linear spaces. Then a mapping $T : X \to Y$ is continuous at a point $x_0 \in X$ if for every $\varepsilon > 0$ there exists $\delta(\varepsilon) > 0$ such that $\|T(x) - T(x_0)\| < \varepsilon$ holds for $\|x - x_0\| \leq \delta(\varepsilon)$. T is said to be continuous (on X) if it continuous at every point $x_0 \in X$.*

THEOREM 7.8 *Let X, Y be normed linear spaces. If the linear operator $T : X \to Y$ is continuous at 0, then it is continuous everywhere.*

DEFINITION 7.9 *Let X, Y be normed linear spaces. The linear operator $T : X \to Y$ is said to be bounded if there exists a positive constant β such that*

$$\|Tx\| \leq \beta \|x\| \quad (x \in X). \tag{7.4}$$

The number β is a bound of operator T.

THEOREM 7.10 *Let X, Y be normed linear spaces. The linear operator $T : X \to Y$ is continuous if and only if it is bounded.*

The norm of operator T is defined by

$$\|T\| = \sup_{\|x\| \leq 1} \|Tx\| = \sup_{\|x\|=1} \|Tx\|. \tag{7.5}$$

The oblique projections of infinite-dimensional linear vector spaces are defined exactly the same way as we did in Chapter 2.

DEFINITION 7.11 *Let X be a linear vector space. A linear operator $P : X \to X$ will be called a projection if*

$$P^2 = P.$$

THEOREM 7.12 *Let X be a linear vector space and $P : X \to X$ a projection. Then*
(i) $I - P$ is a projection;
(ii) $\mathcal{R}(I - P) = \mathcal{N}(P)$, $\mathcal{N}(I - P) = \mathcal{R}(P)$;
(iii) $X = \mathcal{N}(P) \oplus \mathcal{R}(P)$.

Proof. (i) $(I - P)^2 = I - 2P + P^2 = I - P$.
(ii) If $x \in \mathcal{R}(I - P)$, then $x = (I - P)y$, where $y \in X$. Hence $Px = 0$. Conversely, for $x \in \mathcal{N}(P)$ we have $(I - P)x = x$. Then also $\mathcal{N}(I - P) = \mathcal{R}(I - (I - P)) = \mathcal{R}(P)$.
(iii) For $x \in X$ we have $x = Px + (I - P)x$, that is $X = \mathcal{R}(P) + \mathcal{N}(P)$. If $y \in \mathcal{R}(P) \cap \mathcal{N}(P)$, then $y = Px$, $Py = 0$, where $x \in X$. Thus $0 = Py = P^2x = Px = y$. ∎

The geometric interpretation of the projection operators is also the same. Let X be a linear vector space, M and N subspaces of X such that $M+N=X$ and $M\cap N=\{0\}$ ($X=M\oplus N$). Then for any $x\in X$ there is a unique decomposition

$$x=a+b\quad (a\in M,\, b\in N)\,. \tag{7.6}$$

The projections $P_{M,N}$ and $P_{N,M}$ are defined by $P_{M,N}x=a$ and $P_{N,M}x=b$, respectively. It is clear that both $P_{M,N}$ and $P_{N,M}$ are idempotent linear operators

$$P_{M,N}^2=P_{M,N},\quad P_{N,M}^2=P_{N,M}$$

and

$$P_{M,N}+P_{N,M}=I. \tag{7.7}$$

It is also clear that $\mathcal{N}(P_{M,N})=N$, $\mathcal{R}(P_{M,N})=M$, $\mathcal{N}(P_{N,M})=M$ and $\mathcal{R}(P_{N,M})=N$.

In finite dimensional spaces the projections are bounded. Provided that the space is not finite dimensional, unbounded idempotents exist and are relatively easy to construct in any normed linear space (for such a construction, see [72]).

If the projection is bounded in a normed linear space, then its range is closed.

THEOREM 7.13 *Let X be a normed space and $P:X\to X$ a bounded (continuous) projection. Then*
(i) $\mathcal{R}(P)=\overline{\mathcal{R}(P)}$;
(ii) $\|P\|\geq 1$ for $P\neq 0$.

Proof. (i) Let $y\in\overline{\mathcal{R}(P)}$, that is, $y=\lim_{n\to\infty}Px_n$, $x_n\in X$. Then $Px=\lim_{n\to\infty}P^2x_n=\lim_{n\to\infty}Px=y$, that is $y\in\mathcal{R}(P)$. (ii) Choose $y=Px\neq 0$. Then $Py=y$ and $\|Py\|=\|y\|$. ■

DEFINITION 7.14 *Let X and Y be arbitrary sets. The graph of any mapping $T:X\to Y$ is defined by*

$$G(T)=\{(x,Tx)\mid x\in\mathcal{D}(T)\}\subset X\times Y.$$

If X and Y are normed linear spaces over the same field, then $X\times Y$ is a normed linear space with operations

$$(x_1,y_1)+(x_2,y_2)=(x_1+x_2,y_1+y_2)\,,$$
$$\lambda(x_1,y_1)=(\lambda x_1,\lambda y_1)$$

and norm

$$\|(x_1,y_1)\|=\|x_1\|+\|y_1\|\,.$$

DEFINITION 7.15 *Let X and Y be normed linear spaces. The operator $T : \mathcal{D}(T) \subset X \to Y$ is said to be closed if $G(T)$ is closed in $X \times Y$.*

An alternative definition for a closed operator T is the following. Let $x_n \in \mathcal{D}(T)$, $x_n \to x$ and $Tx_n \to y$. Then $x \in \mathcal{D}(T)$ and $Tx = y$.

THEOREM 7.16 *(The closed graph theorem). Let X and Y be Banach spaces. Let $T : X \to Y$ be a closed linear operator. If $\mathcal{D}(T) = X$, then T is bounded (continuous).*

THEOREM 7.17 *Let X be a Banach space, M and N closed subspaces such that $X = M \oplus N$. Then there exists a bounded projection $P : X \to X$ such that $\mathcal{R}(P) = M$ and $\mathcal{N}(P) = N$.*

Proof. Define P by $Px = m$, where $x = m + n$, $m \in M$, $n \in N$. It is clear that P is linear, idempotent, $\mathcal{R}(P) = M$ and $\mathcal{N}(P) = N$. We show that P is a closed operator. Assume that $x_n \to x$, $Px_n \to y$. Then $Px_n \in M$ and the closedness of M implies that $y \in M$. Hence $Py = y$. Since $P(x_n - Px_n) = 0$, $x_n - Px_n \in N$. But N is also closed. Hence $x - y \in N$, $Px = Py$ and finally $Px = y$ follow. Thus P is closed and is bounded by the closed graph theorem. ∎

DEFINITION 7.18 *Let X be a normed linear space. The sequence $\{e_n\} \subset X$ is said to be the Schauder basis of X, if every $x \in X$ can be uniquely represented in the form*

$$x = \sum_{i=1}^{\infty} c_i e_i, \tag{7.8}$$

where the coefficients c_i are scalars.

Concerning the Schauder basis we refer to [113] and [236]. The following result can be proved (see. e.g., [236]).

THEOREM 7.19 *If X is a Banach space with a Schauder basis $\{e_n\} \subset X$ and $P_n x = \sum_{i=1}^{n} c_i e_i$. Then $P_n : X \to X$ is a bounded linear operator and $\|P_n\| \leq C$ $(n \geq 1)$.*

For $x \in X$ let $R_n x = \sum_{i=n+1}^{\infty} c_i e_i$ and $f_i(x) = c_i$ $(i \geq 1)$. It follows from the theorem that R_n is a uniformly bounded linear operator and f_i's are bounded linear functionals. The linearity is a consequence of the unique representation $x = \sum_{i=1}^{\infty} c_i e_i$. The boundedness follows from the relations $R_n = I - P_n$ and

$$|f_i(x)| = |c_i| = \frac{\|P_i x - P_{i-1} x\|}{\|e_i\|} \leq \frac{2C}{\|e_i\|} \|x\|.$$

It is easy to recognize that P_n is a projection. Since $f_i(e_j) = \delta_{ij}$, where δ_{ij} is the Kronecker delta,

$$P_n^2 x = \sum_{i=1}^{n} c_i (P_n x) e_i = \sum_{i=1}^{n} c_i \left(\sum_{j=1}^{n} c_j (x) e_j \right) e_j$$

$$= \sum_{i=1}^{n} \sum_{j=1}^{n} c_j (x) c_i (e_j) e_j = P_n x.$$

Also $R_n = I - P_n$ is the complementary projection.

Later we use the following result known as the Banach-Steinhaus theorem and also as the principle of uniform boundedness.

THEOREM 7.20 *Let X and Y be Banach spaces and let $\{A_n\}$ be a sequence of linear operators from X into Y. If the sequence $\{A_n\}$ is bounded at every point, i.e., $\sup_n \|A_n x\| < \infty$ $(x \in X)$, then there exists a bound $M > 0$ such that $\|A_n\| \leq M$ $(n \geq 1)$.*

The concept of orthogonal projections requires the notion of orthogonality which is readily provided in the Hilbert spaces. From now on we study the projections of Hilbert spaces (for Banach space projections, see the references [310], [19], [220], [227], [228], [345]).

DEFINITION 7.21 *A linear vector space X over the field $\mathbb{F}$ ($\mathbb{F} = \mathbb{R}$ or $\mathbb{F} = \mathbb{C}$) will be called an inner product space, if there is a map $\langle \cdot, \cdot \rangle : X \times X \to F$ with the properties*
(i) $\langle x, x \rangle \geq 0$ $(x \in X)$, $\langle x, x \rangle = 0 \Leftrightarrow x = 0$;
(ii) $\langle x + y, z \rangle = \langle x, z \rangle + \langle y, z \rangle$ $(x, y, z \in X)$;
(iii) $\langle \lambda x, y \rangle = \lambda \langle x, y \rangle$ $(\lambda \in \mathbb{F},\ x, y \in X)$;
(iv) $\langle x, y \rangle = \overline{\langle y, x \rangle}$ $(x, y \in X)$.
The quantity $\langle x, y \rangle$ is called the inner product of x and y.

If $\mathbb{F} = \mathbb{R}$, then condition (iv) becomes $\langle x, y \rangle = \langle y, x \rangle$.

THEOREM 7.22 *If X is an inner product space, then $\|x\| = \sqrt{\langle x, x \rangle}$ is a norm on X and X becomes a normed linear space.*

THEOREM 7.23 *(Cauchy-Schwarz inequality). In any inner product space X,*

$$|\langle x, y \rangle| \leq \|x\| \, \|y\|, \tag{7.9}$$

where $\|x\| = \sqrt{\langle x, x \rangle}$. The equality holds if and only if x and y are dependent.

In real inner product spaces we can define the angle between two vectors.

DEFINITION 7.24 *The angle θ between elements x, y in a real inner product space X is defined by*

$$\cos\theta = \frac{\langle x, y\rangle}{\|x\|\,\|y\|} \quad (0 \le \theta \le \pi)\,. \tag{7.10}$$

DEFINITION 7.25 *Let X be an inner product space, $x, y \in X$. Then $x \perp y$ if and only if $\langle x, y\rangle = 0$.*

DEFINITION 7.26 *Let X be an inner product space. A vector x is orthogonal to a subset M of X, in symbols $x \perp M$, if $x \perp y$ for all $y \in M$.*

DEFINITION 7.27 *The subspaces M and N of H are orthogonal, that is $M \perp N$, if $x \perp y$ for all $x \in M$ and $y \in N$.*

DEFINITION 7.28 *Let X be a linear vector space. A subset C of X is convex, if for all $x_1, x_2 \in C$, $\lambda x_1 + (1-\lambda)\, x_2 \in C$ $(0 \le \lambda \le 1)$ also holds.*

DEFINITION 7.29 *If the inner product space X is complete in the norm $\|x\| = \sqrt{\langle x, x\rangle}$, then it will called a Hilbert space.*

The vector spaces $\mathbb{R}^m$ and $\mathbb{C}^m$ are finite-dimensional Hilbert spaces with the natural inner products $\langle x, y\rangle = y^T x$ and $\langle x, y\rangle = y^H x$, respectively.

THEOREM 7.30 *(Parallelogram law). For any two elements x, y in a Hilbert space H,*

$$\|x+y\|^2 + \|x-y\|^2 = 2\,\|x\|^2 + 2\,\|y\|^2\,. \tag{7.11}$$

THEOREM 7.31 *(The Pythagorean theorem). If $x \perp y$, then*

$$\|x+y\|^2 = \|x\|^2 + \|y\|^2\,. \tag{7.12}$$

THEOREM 7.32 *In a Hilbert space the inner product is continuous.*

Proof. If $\|x_n - x\| \to 0$ and $\|y_n - y\| \to 0$ for $n \to \infty$, then

$$\begin{aligned} |\langle x, y\rangle - \langle x_n, y_n\rangle| &\le |\langle x, y - y_n\rangle| + |\langle x - x_n, y_n\rangle| \\ &\le \|x\|\,\|y - y_n\| + \|x - x_n\|\,\|y_n\|\,. \end{aligned}$$

The bound of the right side tends to 0 as $n \to \infty$. ■

The linear functionals of a Hilbert space have a simple representation, which is extremely important in applications.

THEOREM 7.33 *(Riesz). Let H be a Hilbert space and $f : H \to \mathbb{F}$ $(\mathbb{F} = \mathbb{R}, \mathbb{C})$ any bounded (continuous) linear functional. There exists one and only one element $z \in H$ such that*

$$f(x) = \langle x, z\rangle \quad (x \in H)\,.$$

Moreover, $\|f\| = \|z\|$.

Proof. The null space $\mathcal{N}(f) = \{x \in H \mid f(x) = 0\}$ is closed in H. If $\mathcal{N}(f) = H$, then $f(x) = \langle x, 0 \rangle = 0$ for every $x \in H$ and $z = 0$. Assume that $\mathcal{N}(f) \neq H$. Choose a nonzero element $u \in \mathcal{N}^{\perp}(f)$. Then for all $x \in H$,

$$f(f(x)u - f(u)x) = f(x)f(u) - f(u)f(x) = 0.$$

Hence $f(x)u - f(u)x \in \mathcal{N}(f)$. As

$$0 = \langle f(x)u - f(u)x, u \rangle = f(x)\langle u, u \rangle - f(u)\langle x, u \rangle,$$

one has

$$f(x) = \left\langle x, \frac{\overline{f(u)u}}{\|u\|^2} \right\rangle = \langle x, z \rangle,$$

where $z = \overline{f(u)u}/\|u\|^2$. The element is uniquely determined, for if there is $v \in H$ such that $\langle x, z \rangle = \langle x, v \rangle$ for all $x \in H$, then $v = z$. As $|f(x)| \leq \|x\| \|z\|$ for all $x \in H$, the inequality $\|f\| \leq \|z\|$ holds. If $z \neq 0$, then the identity

$$f\left(\frac{z}{\|z\|^2}\right) = \left\langle \frac{z}{\|z\|^2}, z \right\rangle = \|z\|$$

implies that $\|f\| = \|z\|$. ∎

In Hilbert spaces we can define different types of convergence.

DEFINITION 7.34 *A sequence* $\{x_n\} \subset H$ *is said to converge weakly to* x *if*

$$\lim_{n \to \infty} \langle x_n, z \rangle = \langle x, z \rangle \tag{7.13}$$

for all $z \in H$.

THEOREM 7.35 *(The weak compactness property). Every bounded sequence of elements in a Hilbert space contains a weakly convergent subsequence.*

DEFINITION 7.36 *A sequence* $\{x_n\} \subset H$ *is said to converge strongly to* x *if*

$$\lim_{n \to \infty} \|x_n - x\| = 0. \tag{7.14}$$

Strong convergence clearly implies the weak convergence.

THEOREM 7.37 *Suppose* x_n *converges weakly to* x *and in addition* $\|x_n\| \to \|x\|$. *Then* x_n *converges strongly to* x.

Proof.

$$\|x_n - x\|^2 = \|x_n\|^2 + \|x\|^2 - \langle x_n, x \rangle - \langle x, x_n \rangle \to 2\|x\|^2 - 2\langle x, x \rangle = 0.$$

■

THEOREM 7.38 *In finite dimensional Hilbert spaces the strong and weak convergence are equivalent.*

THEOREM 7.39 *Let M be a closed linear subspace of H. If $x_n \in M$ and $x_n \to x$ weakly, then $x \in M$.*

For the proof of the last two claims, see e.g., [236], [220].

DEFINITION 7.40 *An operator sequence $\{T_n\}$ is said to converge weakly to T if $T_n x \to Tx$ weakly for all $x \in H$.*

DEFINITION 7.41 *An operator sequence $\{T_n\}$ is said to converge strongly to T if $T_n x \to Tx$ strongly for all $x \in H$.*

DEFINITION 7.42 *An operator sequence $\{T_n\}$ is said to converge to T in norm if $\|T_n - T\| \to 0$.*

1.1 Projections of Hilbert spaces

We prove the following important result.

THEOREM 7.43 *Every closed convex set in a Hilbert space has a unique element of minimal norm.*

Proof. Let C be a closed convex set and $d = \inf_{x \in C} \|x\|$. There exists a sequence $\{x_n\} \subset C$ such that $d = \lim_n \|x_n\|$. By the parallelogram law,

$$\left\|\frac{x_n - x_m}{2}\right\|^2 = \frac{1}{2}\left(\|x_n\|^2 + \|x_m\|^2\right) - \left\|\frac{1}{2}x_n + \frac{1}{2}x_m\right\|^2.$$

Since $\frac{1}{2}x_n + \frac{1}{2}x_m \in C$, the right side of the equation is less than or equal to

$$\frac{1}{2}\left(\|x_n\|^2 + \|x_m\|^2\right) - d^2.$$

This bound tends to 0 for $m, n \to \infty$. Hence $\{x_n\}$ is a Cauchy sequence. Since C is closed and H is complete, there is an element $z \in C$ such that $\lim_n x_n = z$. Furthermore the inequality $|\|x_n\| - \|z\|| \leq \|x_n - z\|$ implies that $\|z\| = d$. Assume that there exist two elements z_1, z_2 in C with norm d. Then by calling again the parallelogram law we get

$$\left\|\frac{z_1 - z_2}{2}\right\| = d^2 - \left\|\frac{z_1 + z_2}{2}\right\|^2 \leq 0,$$

so that $z_1 = z_2$. ■

COROLLARY 7.44 *Let C be a closed convex set in H. For any x in H there is a unique element in C which is closest to x; that is, there is a unique element z in C such that*

$$\|x - z\| = d(x, C) = \inf_{y \in C} \|x - y\| . \tag{7.15}$$

Proof. Observe that the set $x - C = \{x - y \mid y \in C\}$ is also convex and closed. ∎

THEOREM 7.45 *Let C be a closed convex set in H. For any x in H, z is the unique element in C which is closest (in norm) to x if and only if*

$$Re \langle x - z, z - y \rangle \geq 0 \quad (y \in C) . \tag{7.16}$$

DEFINITION 7.46 *Let C be a closed convex set of H. For each x in H define $R_C(x)$ as the closest point to x in C.*

The mapping $R_C(x)$ is called the projection of x onto C. R_C is not necessarily linear but is continuous [27]. Note that R_C leaves C invariant.

LEMMA 7.47 *If C is a closed convex set of a real Hilbert space H, then R_C is contractive in the sense that*

$$\|R_C(x) - R_C(y)\| \leq \|x - y\| \quad (x, y \in H) . \tag{7.17}$$

For the proof, see Chapter 6 or [64]. The differential properties of the projection R_C were studied among others by Zarontonello [367], Holmes [202], Fitzpatrick and Phelps [122] (see also Chapter 6 and the references therein). Zarontonello [367] is a comprehensive survey on metric projections R_C in Hilbert spaces.

DEFINITION 7.48 *A cone is a set with the property that if an element x belongs to it, so does αx for every $\alpha \geq 0$. A convex cone is a cone that is also convex.*

Note that C is a convex cone if whenever x_1 and x_2 belong to it, so do $\alpha x_1 + \beta x_2$ for every $\alpha, \beta \geq 0$.

COROLLARY 7.49 *Suppose that C is a closed convex cone. Let z denote the projection of x on C. Then*

$$Re \langle x - z, y \rangle \leq 0 \quad (y \in C) , \quad Re \langle x - z, z \rangle = 0. \tag{7.18}$$

Conversely, if any element z in C satisfies these two equations, then it is the projection of x on C.

COROLLARY 7.50 *Let M be a closed linear subspace of H. Then for each x in H there is a unique element in M which is closest to x, being the projection of x onto M, denoted $P_M x$, and is characterized by the fact that*

$$\langle x - P_M x, m \rangle = 0 \quad (m \in M). \tag{7.19}$$

Also P_M is linear.

Proof. Let m_0 be the unique element in M closest to x. If there exists an $m \in M$ such that $\langle x - m_0, m \rangle \neq 0$, then we can assume that $\|m\| = 1$. Let $m_1 = m_0 + \langle x - m_0, m \rangle\, m \in M$. Then

$$\begin{aligned} \|x - m_1\|^2 &= \|(x - m_0) - \langle x - m_0, m \rangle\, m\|^2 \\ &= \|x - m_0\|^2 - |\langle x - m_0, m \rangle|^2 < \|x - m_0\|^2 . \end{aligned}$$

This contradicts the assumption that m_0 is the unique closest element to x in M. Conversely, if $x - m_0 \perp M$, then $x - m_0 \perp m_0 - m \in M$ $(m \in M)$ and

$$\|x - m\|^2 = \|(x - m_0) + (m_0 - m)\|^2 = \|x - m_0\|^2 + \|m_0 - m\|^2 .$$

Thus $\|x - m_0\|^2 \leq \|x - m\|^2$ for any $m \in M$. It is clear that $P_M(\alpha x) = \alpha P_M x$. For any $x, y \in H$, $P_M x$ and $P_M y$ satisfy the relations

$$\langle x - P_M x, m \rangle = 0, \quad \langle y - P_M y, m \rangle = 0 \quad (x \in M).$$

It follows that

$$\langle x - P_M x, m \rangle + \langle y - P_M y, m \rangle = \langle (x + y) - (P_M x + P_M y), m \rangle = 0 \quad (m \in M).$$

Hence $P_M(x + y) = P_M x + P_M y$. ∎

COROLLARY 7.51 *If M and N are closed subspaces of a Hilbert space H such that $M \subset N$ and $M \neq N$, then there exists a nonzero vector z in N such that $z \perp M$.*

Proof. There is a point $x \in N$ and $x \notin M$. Then $z = x - P_M x \in N$ and $z \perp M$. ∎

DEFINITION 7.52 *The orthogonal complement of a subset M of H is defined by*

$$M^{\perp} = \{x \in H \mid x \perp M\}. \tag{7.20}$$

THEOREM 7.53 *(i) If M is a subset of H, then $M^{\perp}$ is a closed subspace of H.*
(ii) If M is a subset of H, then $M \subset M^{\perp\perp}$.
(iii) If M and N are subsets of H such that $M \subset N$, then $M^{\perp} \supset N^{\perp}$.

(iv) If M is a subset of H, then $M^{\perp} = M^{\perp\perp\perp}$.
(v) If M is a closed subspace of H, then $M = M^{\perp\perp}$.

Proof. (i) Let $x, y \in M^{\perp}$. Then $\langle x, m\rangle = 0$ and $\langle y, m\rangle = 0$ imply $\langle \alpha x + \beta y, m\rangle = 0$ for all $m \in M$. Hence $\alpha x + \beta y \in M^{\perp}$. The fact that $M^{\perp}$ is closed follows from the continuity of the inner product.
(ii) If $x \in M$ and $y \in M^{\perp}$, then $x \perp y$, so that $x \perp M^{\perp}$. Hence $x \in M^{\perp\perp}$.
(iii) $x \in N^{\perp} \Leftrightarrow \langle x, y\rangle = 0$ $(y \in N)$. $M \subset N$ implies that $x \in M^{\perp}$.
(iv) Property (ii) implies that $M^{\perp} \subset M^{\perp\perp\perp}$. Applying (iii) to the relation $M \subset M^{\perp\perp}$, we get $M^{\perp} \supset M^{\perp\perp\perp}$. Hence (iv) holds.
(v) Property (ii) gives $M \subset M^{\perp\perp}$. If M were a proper subset of $M^{\perp\perp}$, then, by the last corollary, $M^{\perp\perp}$ would have a nonzero vector orthogonal to M. Since this would contradict the relation $M^{\perp} \cap M^{\perp\perp} = \{0\}$, the proof is complete. ■

If M is a subspace, then $M \cap M^{\perp} = \{0\}$. If $x \in M \cap M^{\perp}$, then $\langle x, x\rangle = \|x\|^2 = 0 \Leftrightarrow x = 0$.

DEFINITION 7.54 *The closed linear subspace generated by a set of elements M is the subspace $\mathcal{L}(M)$ with the property that every closed subspace containing M contains $\mathcal{L}(M)$.*

PROPOSITION 7.55 $\mathcal{L}(M) = M^{\perp\perp}$.

Proof. The inclusion $M \subset M^{\perp\perp}$ implies $\mathcal{L}(M) \subset M^{\perp\perp}$. Assume the existence of an element $x \in M^{\perp\perp}$ such that $x \notin \mathcal{L}(M)$. Then $P_{\mathcal{L}(M)}x \in \mathcal{L}(M)$ and $x - P_{\mathcal{L}(M)}x \perp M$. Since $x \perp M^{\perp}$, relation $x \perp x - P_{\mathcal{L}(M)}x$ also holds. Hence $0 = \langle x - P_{\mathcal{L}(M)}x, x\rangle = \left\|\left(I - P_{\mathcal{L}(M)}\right)x\right\|^2$, which implies that $x = P_{\mathcal{L}(M)}x$ and $x \in \mathcal{L}(M)$. This is a contradiction. ■

Particularly, if $M \subset H$ is a linear subspace, then $\overline{M} = M^{\perp\perp}$.

The sum $M+N$ of two closed subspaces M and N is not necessarily closed. For details, see Halmos [177], [181]. The following results give conditions for $M + N$ to be closed.

THEOREM 7.56 *If M and N are orthogonal closed subspaces, then $M + N$ is closed.*

Proof. Assume that $\{z_n\} \subset M + N$ is such that $z_n \to z \in H$ and $z_n = x_n + y_n$ $(x_n \in M,\ y_n \in N)$ for all n. The Pythagorean theorem implies that for every n and m,

$$\|z_n - z_m\|^2 = \|x_n - x_m\|^2 + \|y_n - y_m\|^2 .$$

Therefore both sequences $\{x_n\}$ and $\{y_n\}$ are Cauchy sequences. If $x_n \to x$ and $y_n \to y$, then $x \in M$ and $y \in N$. It follows from the continuity of addition that $z_n \to x + y$. Hence $z \in M + N$. ■

THEOREM 7.57 *If M and N are closed subspaces and $M+N$ is closed, then*

$$(M \cap N)^{\perp} = M^{\perp} + N^{\perp}. \tag{7.21}$$

In particular, $M^{\perp}+N^{\perp}$ is closed. Consequently, $M+N$ is closed if and only if $M^{\perp}+N^{\perp}$ is closed.

Proof. The result can be found, for example in Kato [220], Theorem 4.8. The last two claims of the theorem follow from the first one using the facts that $(M \cap N)^{\perp}$ is always closed, $M^{\perp\perp} = M$, and $N^{\perp\perp} = N$. ■

The following result is a direct consequence of Corollary 7.50.

THEOREM 7.58 *(The projection theorem). If M is a closed subspace of H, then $M + M^{\perp} = H$.*

Proof. If $M + M^{\perp} = N$, then N is a closed subspace. Since $M \subset N$ and $M^{\perp} \subset N$, it follows that $N^{\perp} \subset M^{\perp}$ and $N^{\perp} \subset M^{\perp\perp}$. Hence $N^{\perp} = \{0\}$ and $N = N^{\perp\perp} = \{0\}^{\perp} = H$. ■

Let M, N and Y be linear subspaces of a Hilbert space H. If $Y = M+N$ and $M \cap N = \{0\}$, then we call $M+N$ the direct sum of M and N. We write $M+N$ also in the form $M \oplus N$. If $M \perp N$, then we write $N = Y \ominus M$ and call N the orthogonal complement of M in Y. Similarly we write $M = Y \ominus N$ and call M the orthogonal complement of N in Y.

Notice that the projection theorem states that if M is a closed linear subspace, then $H = M \oplus M^{\perp}$, $H \ominus M = M^{\perp}$ and $H \ominus M^{\perp} = M$. If M and N are closed linear subspaces and $N \subset M$, then $M \cap N^{\perp} = M \ominus N$ is the set of vectors $x \in M$ orthogonal to N.

REMARK 7.59 *It is customary that $M \oplus N$ denotes a direct sum with orthogonal complements. Keeping the consistency of notation we do not assume that $H = M \oplus N$ implies $M \perp N$.*

THEOREM 7.60 *Let X, Y be Hilbert spaces. If $A : X \to Y$ is a bounded linear operator, then there exists a unique linear operator $A^* : Y \to X$, called the adjoint of A, such that $\langle Ax, y\rangle = \langle x, A^*y\rangle$ for all $x \in X$ and $y \in Y$; A^* is such that $\|A^*\| = \|A\|$.*

Proof. Choose $y \in Y$. The mapping $x \to \langle Ax, y\rangle$ is a bounded linear functional on X. The Riesz theorem implies that there exists a unique element $z \in X$ such that $\langle Ax, y\rangle = \langle x, z\rangle$ for all $x \in X$. Let us define $A^*y = z$. Then $A^* : Y \to X$ and $\langle Ax, y\rangle = \langle x, A^*y\rangle$. Let $S : Y \to X$ be such that

$$\langle x, Sy\rangle = \langle x, A^*y\rangle \quad (x \in X,\, y \in Y).$$

Then $Sy = A^*y$ for all $y \in Y$, that is $S = A^*$. The operator A^* is linear since

$$\langle x, A^*(\alpha y_1 + \beta y_2)\rangle = \overline{\alpha}\langle Ax, y_1\rangle + \overline{\beta}\langle Ax, y_2\rangle = \langle x, \alpha A^*y_1 + \beta A^*y_2\rangle$$

for $x \in X$, $y_1, y_2 \in Y$ and $\alpha, \beta \in \mathbb{F}$. For $y \in Y$ one has

$$\|A^* y\|^2 = \langle A^* y, A^* y\rangle = \langle A(A^* y), y\rangle \le \|A(A^* y)\| \|y\| \le \|A\| \|A^* y\| \|y\|,$$

which yields $\|A^* y\| \le \|A\| \|y\|$ and $\|A^*\| \le \|A\|$. As $(A^*)^* = A$ the rest follows. ■

It is easy to prove that

$$(A+B)^* = A^* + B^*, \tag{7.22}$$

$$(\lambda A)^* = \overline{\lambda} A^* \quad (\lambda \in F). \tag{7.23}$$

THEOREM 7.61 *Let X, Y be Hilbert spaces and $A : X \to Y$ a bounded linear operator. Then $\mathcal{N}(A^*) = \mathcal{R}^\perp(A)$ and $\overline{\mathcal{R}(A^*)} = \mathcal{N}^\perp(A)$.*

Proof. $y \perp \mathcal{R}(A) \Leftrightarrow \langle Ax, y\rangle = 0$ for all $x \in X$. $y \in \mathcal{N}(A^*) \Leftrightarrow \langle x, A^* y\rangle = 0$ for all $x \in X$. But $\langle Ax, y\rangle = \langle x, A^* y\rangle$. Similarly we have $\mathcal{N}(A) = \mathcal{R}^\perp(A^*)$. Taking the orthogonal complements we get $\mathcal{N}^\perp(A) = \mathcal{R}^{\perp\perp}(A^*) = \overline{\mathcal{R}(A^*)}$. ■

REMARK 7.62 *If $A : H \to H$ is a bounded linear operator, then $H = \overline{\mathcal{R}(A)} \oplus \mathcal{N}(A^*)$.*

THEOREM 7.63 *Let X, Y be Hilbert spaces and $A : X \to Y$ a bounded linear operator. If $x_n \in X$ and $x_n \to x$ weakly, then $Ax_n \to Ax$ weakly.*

Proof. By definition $x_n \to x$ weakly if and only if $\langle x_n, z\rangle \to \langle x, z\rangle$ for all $z \in X$. Since for all $y \in Y$, $\langle Ax_n, y\rangle = \langle x_n, A^* y\rangle \to \langle x, A^* y\rangle = \langle Az, y\rangle$, the claim is proved. ■

THEOREM 7.64 *If $P : H \to H$ is a bounded projection, then so is P^*.*

Proof. The statement follows from the relations

$$\langle Px, y\rangle = \langle P^2 x, y\rangle = \langle Px, P^* y\rangle = \langle x, P^*(P^* y)\rangle = \left\langle x, (P^*)^2 y\right\rangle = \langle x, P^* y\rangle$$

which hold for any $x, y \in H$. ■

REMARK 7.65 *As $\mathcal{R}(P)$ is closed, $\mathcal{R}(P^*) = \mathcal{N}^\perp(P)$ and $\mathcal{N}(P^*) = \mathcal{R}^\perp(P)$.*

An operator A is Hermitian (self-adjoint) if $A = A^*$. A Hermitian operator A is said to be positive semidefinite, that is $A \ge 0$, if $\langle Ax, x\rangle \ge 0$ for every vector x. A Hermitian operator A is said to be negative semidefinite, if $-A$ is positive semidefinite. The Hermitian operator A is said to be positive definite if $\langle Ax, x\rangle > 0$ for every vector $x \ne 0$.

We define a partial ordering in the set of Hermitian operators by writing $A \le B$ whenever $B - A$ is positive . A sequence of Hermitian operators $\{A_i\}$

is said to be monotone increasing (decreasing) if $A_i \le A_{i+1}$ $(A_i \ge A_{i+1})$ for all i.

DEFINITION 7.66 *A subspace M is invariant under an operator A if $AM \subset M$. A subspace M reduces an operator A if both M and $M^\perp$ are invariant under A.*

THEOREM 7.67 *A necessary and sufficient condition that a subspace M be invariant under an operator A is that $M^\perp$ be invariant under A^*.*

Proof. By symmetry it is enough to prove that the condition is necessary. Assume that M is invariant under A. Let $x \in M$ and $y \in M^\perp$. Then $Ax \in M$ and

$$\langle x, A^*y\rangle = \langle Ax, y\rangle = 0.$$

Thus $A^*y \in M^\perp$, and consequently $M^\perp$ is invariant under A^*. ∎

COROLLARY 7.68 *A necessary and sufficient condition that a subspace M reduce an operator A is that it should be invariant under both A and A^*.*

The following definition already appeared implicitly in Corollary 7.50 and in the projection theorem.

DEFINITION 7.69 *Let M be a closed subspace of a Hilbert space H. For every $x \in H$ let $x = m + n$, where $m \in M$ and $n \in M^\perp$. The element m is called the orthogonal projection of x upon M and will be denoted by $P_M x$; P_M is called the orthogonal projection operator or the projector upon M.*

It is clear that $I - P_M$ is the orthogonal projection on $M^\perp$. Thus we have the relation $I = P_M + P_{M^\perp}$. We emphasize again that $P_M x$ is the best approximation or nearest point to x from M:

$$\|x - P_M x\| = d(x, M),$$

and $x - P_M x$ is orthogonal to M.

The following results are well known (see, e.g., Halmos [177], [6], or [104]).

THEOREM 7.70 *The orthogonal projection P_M on a closed subspace M of H is idempotent ($P_M^2 = P_M$) and Hermitian ($P_M^* = P_M$) operator; if $M \ne \{0\}$, then $\|P_M\| = 1$.*

Proof. Let $x_i = y_i + z_i$ (for $i = 1, 2$), where $y_i \in M$, $z_i \in M^\perp$. Then $\lambda_1 x_1 + \lambda_2 x_2 = (\lambda_1 y_1 + \lambda_2 y_2) + (\lambda_1 z_1 + \lambda_2 z_2)$ and

$$(\lambda_1 y_1 + \lambda_2 y_2) \in M, \ \ (\lambda_1 z_1 + \lambda_2 z_2) \in M^\perp.$$

It follows that

$$P_M(\lambda_1 x_1 + \lambda_2 x_2) = \lambda_1 y_1 + \lambda_2 y_2 = \lambda_1 P_M x_1 + \lambda_2 P_M x_2.$$

Thus P_M is a linear operator. Next, $P_M^2 x_1 = P_M(P_M x_1) = P_M y_1 = y_1 = Px_1$, so that $P_M^2 = P_M$. Since $\|x_1\|^2 = \|y_1\|^2 + \|z_1\|^2 \geq \|y_1\|^2 = \|P_M x_1\|^2$, it also follows that $\|P_M\| \leq 1$. If $P_M \neq 0$, there is a point $x \neq 0$ in M. Thus, $\|P_M x\| = \|x\|$. This gives $\|P_M\| = 1$. It remains to show that P_M is Hermitian. This is proved as follows:

$$\langle P_M x_1, x_2 \rangle = \langle y_1, x_2 \rangle = \langle y_1, y_2 \rangle = \langle x_1, y_2 \rangle = \langle x_1, P_M x_2 \rangle .$$

■

The converse of the theorem is also true.

THEOREM 7.71 *A Hermitian operator P with $P^2 = P$ is an orthogonal projection.*

Proof. Let $M = P(H)$. M is a linear subspace of H. It is also closed. Indeed, if $y_n = Px_n \to z$, then $Py_n = P^2 x_n = Px_n = y_n$. Hence $z = \lim y_n = \lim Py_n = Pz \in M$. Next, since P is Hermitian and idempotent,

$$\langle x - Px, Py \rangle = \langle Px - P^2 x, y \rangle = 0 \quad (y \in M),$$

so that $x - Px$ lies in $M^\perp$. Therefore $x = Px + (x - Px)$ is the unique decomposition of x as a sum $y + z$ with $y \in M$ and $z \in M^\perp$. This shows that P is the orthogonal projection on $M = P(H)$. ■

THEOREM 7.72 *A bounded projection P of a Hilbert space H is orthogonal if and only if $\|P\| \leq 1$.*

Proof. The necessity is obvious. Let $P^2 = P$, $\|P\| \leq 1$. If $\|P\| < 1$, then $P = 0$. For $\|P\| = 1$ we prove $\mathcal{N}^\perp(P) \subset \mathcal{R}(P)$. If $x \in \mathcal{N}^\perp(P)$, then $\langle x, x - Px \rangle = 0$, $Px = x + (Px - x)$, $\|x\|^2 \geq \|Px\|^2 = \|x\|^2 + \|Px - x\|^2$, which implies $x = Px \in \mathcal{R}(P)$. Similarly, the inclusion $\mathcal{N}^\perp(P^*) \subset \mathcal{R}(P^*)$ holds. Theorems 7.61 and 7.64 yield

$$\mathcal{N}^\perp(P) \subset \mathcal{R}(P) = \mathcal{N}^\perp(P^*) \subset \mathcal{R}(P^*) = \mathcal{N}^\perp(P).$$

Hence $\mathcal{R}(P) = \mathcal{N}^\perp(P)$ and P is orthogonal. ■

For any orthogonal projection P, $0 \leq P \leq I$ holds in the partial ordering of Hermitian operators.

THEOREM 7.73 *If P is the orthogonal projection on a closed subspace M of H and if x is a vector such that $\|Px\| = \|x\|$, then $Px = x$ (and therefore $x \in M$).*

Proof. For any $x \in H$, $x = Px + (x - Px)$, $Px \in M$ and $x - Px \in M^\perp$. It follows that $\|x\|^2 = \|Px\|^2 + \|x - Px\|^2$. The assumed relation $\|Px\| = \|x\|$ implies $\|x - Px\| = 0$. ■

THEOREM 7.74 *A closed subspace $M \subset H$ with orthogonal projection P is invariant under an operator A if and only if $AP = PAP$.*

Proof. If $AP = PAP$ and $x \in M$, then $Ax = APx = PAPx \in M$. If M is invariant under A, then $APx \in M$ and therefore, $APx = PAPx$ for every $x \in H$. ■

THEOREM 7.75 *A closed subspace $M \subset H$ with orthogonal projection P reduces an operator A if and only if $AP = PA$.*

Proof. If $AP = PA$, then multiplying by P on the right and on the left, we obtain that $AP = PAP = PA$. Taking the adjoints we get that $PA^* = PA^*P = A^*P$. Since $AP = PAP$ and $A^*P = PA^*P$ hold, M is invariant under both A and A^* by the previous theorem. The rest follows from Corollary 7.68. ■

DEFINITION 7.76 *Two orthogonal projections P and Q are orthogonal to each other, in symbols $P \perp Q$, if $PQ = 0$.*

Since $(PQ)^* = Q^*P^* = QP$, $PQ = 0$ if and only if $QP = 0$.

THEOREM 7.77 *Let M and N be closed subspaces of H. Then $P_M \perp P_N$ if and only if $M \perp N$.*

Proof. Let $P_M P_N = 0$ and let $x \in M$, $y \in N$. Then

$$\langle x, y\rangle = \langle P_M x, P_N y\rangle = \langle P_N P_M x, y\rangle = 0.$$

Thus $M \perp N$. Suppose conversely that $M \perp N$. For any $x \in H$, $P_N x \in N$ and hence $P_N x \perp M$. It follows that $P_M(P_N x) = 0$. Thus $P_M P_N = 0$. ■

THEOREM 7.78 *Let M and N be closed subspaces of H. The sum $P_M + P_N$ is a projection if and only if $P_M P_N = 0$. In that case, $P_M + P_N = P_{M\oplus N}$.*

Proof. If $P = P_M + P_N$ is a projection, then $P^2 = P$. Hence $P_M P_N + P_N P_M = 0$. Multiplying both sides by P_M, we get

$$P_M P_N + P_M P_N P_M = 0$$

and

$$P_M P_N P_M + P_N P_M = 0.$$

By subtraction we obtain that $P_M P_N - P_N P_M = 0$. Hence $P_M P_N = P_N P_M$. Conversely, if $P_M P_N = 0$, then $P_N P_M = 0$. It follows that $P^2 = P$. Since P is also Hermitian, it is an orthogonal projection. Finally, $Px = P_M x + P_N x$ varies over $M \oplus N$ as x varies in H. Therefore, $P = P_{M\oplus N}$. ■

Since $P_M P_N = 0 \Leftrightarrow P_M \perp P_N \Leftrightarrow M \perp N$, $M \oplus N$ is an orthogonal direct sum.

THEOREM 7.79 *Let M and N be closed subspaces of H. The product $P_M P_N$ is an orthogonal projection if and only if P_M and P_N commute, that is, $P_M P_N = P_N P_M$. In that case $P_M P_N = P_{M \cap N}$.*

Proof. If $P = P_M P_N$ is an orthogonal projection, then $P^* = P$. Hence

$$P_M P_N = (P_M P_N)^* = P_N^* P_M^* = P_N P_M.$$

Conversely, let $P_M P_N = P_N P_M = P$. Then $P^* = P$, so that P is Hermitian. Also

$$P^2 = P_M P_N P_M P_N = P_M^2 P_N^2 = P_M P_N = P.$$

Thus P is a projection. Further, $Px = P_M (P_N x) = P_N (P_M x)$. Therefore, $Px \in M \cap N$. On the other hand, if $x \in M \cap N$, then $Px = P_M (P_N x) = P_M x = x$. Thus $P = P_{M \cap N}$. ■

THEOREM 7.80 *If M and N are closed subspaces of H, then the following statements are equivalent:*
(i) $P_M \leq P_N$.
(ii) $\|P_M x\| \leq \|P_N x\|$ for every vector $x \in H$.
(iii) $M \subset N$.
(iv) $P_N P_M = P_M$.
(v) $P_M P_N = P_M$.

Proof. (i) holds if and only if $\langle (P_N - P_M) x, x \rangle \geq 0$ for any $x \in H$. Thus

$$\|P_N x\|^2 = \langle P_N x, x \rangle \geq \langle P_M x, x \rangle = \|P_M x\|^2,$$

so that (ii) holds. If $M \not\subseteq N$, then there is a point $x_0 \in M$, $x_0 \notin N$. Let $x_0 = y_0 + z_0$ where $y_0 \in N$, $z_0 \in N^\perp$, $z_0 \neq 0$. Then

$$\|P_M x_0\|^2 = \|y_0\|^2 + \|z_0\|^2 > \|y_0\|^2 = \|P_N x_0\|^2.$$

This contradicts (ii). If $M \subset N$, then $P_M x \in M$ for any $x \in H$. Hence $P_N P_M x = P_M x$, and (iv) follows. Since $P_M^* = (P_N P_M)^* = P_M P_N = P_M$, condition (iv) implies (v). Similarly, (v) implies (iv). It remains to show that (v) implies (i). Since

$$\begin{aligned}\langle (P_N - P_M) x, x \rangle &= \langle (P_N - P_M) P_N x, x \rangle = \langle P_N (I - P_M) P_N x, x \rangle \\ &= \langle (I - P_M) P_N x, P_N x \rangle \geq 0,\end{aligned}$$

we proved the theorem. ■

THEOREM 7.81 *Let M and N be closed subspaces of H. The difference $P = P_N - P_M$ is a projection if and only if $M \subset N$ ($P_M \leq P_N$). If this is the case, then $P = P_{N \ominus M}$.*

Proof. If P is a projection, then so is $I - P = (I - P_N) + P_M$. $I - P_N$ is also a projection. Applying Theorem 7.78 we get, $(I - P_N) P_M = 0$, that is, $P_M = P_N P_M$. Hence by Theorem 7.80, $M \subset N$. Suppose conversely that $M \subset N$. Then by Theorem 7.80, $(I - P_N) P_M = 0$. Hence, by Theorem 7.78, $(I - P_N) + P_M = I - P = P_{N^\perp \oplus M}$ is a projection. Thus $P = P_{(N^\perp \oplus M)^\perp} = P_{N \cap M^\perp} = P_{N \ominus M}$. ■

LEMMA 7.82 *Let M and N be closed subspaces of H. If P_M and P_N commute, then*

$$P_{\overline{M+N}} = P_M + P_N - P_M P_N. \tag{7.24}$$

Consequently, if $M \subset N^\perp$, then $M + N$ is closed and

$$P_{M+N} = P_M + P_N. \tag{7.25}$$

Proof. Let $P = P_M + P_N - P_M P_N$. Since P_M and P_N commute, $I - P_M$ and P_N also commute. Theorem 7.79 implies that $(I - P_M) P_N = P_{M^\perp \cap N}$. P_M is orthogonal to $(I - P_M) P_N$. Hence by Theorem 7.78 we obtain that $P = P_M + (I - P_M) P_N = P_{M \oplus (M^\perp \cap N)}$. P is symmetric in P_M and P_N. Therefore, we can also write that $P = P_{N \oplus (N^\perp \cap M)}$. Since

$$M, N \subset M \oplus \left(M^\perp \cap N\right) = N \oplus \left(N^\perp \cap M\right) \subset \overline{M + N}$$

we obtain that $P = P_{\overline{M+N}}$. Assume that $M \subset N^\perp$. Then by Theorem 7.77 $P_M P_N = P_N P_M = 0$, so P_M and P_N commute, and the first relation implies that

$$P_{\overline{M+N}} = P_M + P_N.$$

It remains to verify that $M + N$ is closed. It suffices to verify that $\overline{M + N} \subset M + N$. But if $x \in \overline{M + N}$, then the last relation implies that $z = P_{\overline{M+N}} x = P_M x + P_N x \in M + N$. ■

LEMMA 7.83 *The following statements are equivalent:*
(i) P_M commutes with P_N.
(ii) $P_{M^\perp}$ commutes with P_N.
(iii) P_M commutes with $P_{N^\perp}$.
(iv) $P_{M^\perp}$ commutes with $P_{N^\perp}$.
(v) $M = M \cap N + M \cap N^\perp$.

Proof. The equivalence of the first four claims obviously follows from the facts that $P_{M^\perp} = I - P_M$ and $P_{N^\perp} = I - P_N$.

(i)⇒(v) If P_M and P_N commute, so do P_M and $P_{N^\perp}$. Hence by Theorem 7.79,

$$P_{M\cap N} + P_{M\cap N^\perp} = P_M P_N + P_M P_{N^\perp} = P_M\left(P_N + P_{N^\perp}\right) = P_M.$$

Thus $M \subset M\cap N + M\cap N^\perp$. Since the reverse inclusion is obvious, (v) holds.
(v)⇒(i) If (v) holds, then the last statement of Lemma 7.82 implies that

$$P_M = P_{M\cap N + M\cap N^\perp} = P_{M\cap N} + P_{M\cap N^\perp},$$

and hence

$$P_M P_N = P_N P_{M\cap N} + P_N P_{M\cap N^\perp} = P_{M\cap N}.$$

By Theorem 7.79, (i) holds. ■

If $A : H \to H$ has a closed range, define A^+ by

$$A^+\big|_{\mathcal{R}(A)} = \left(A|_{\mathcal{R}(A)}\right)^{-1}, \quad A^+\big|_{\mathcal{N}(A)} = 0,$$

where $A|_M$ denotes the restriction of A to M.

THEOREM 7.84 *(Anderson, Schreiber). Let M and N be closed subspaces of H. If $M+N$ is closed, then*

$$P_{M\cap N} = 2P_M\left(P_M + P_N\right)^+ P_N. \tag{7.26}$$

The operator $A : B = A(A+B)^+ B$ is called the parallel sum. For its properties and related theory we refer to [15], [14], [17] and [18].

Next we generalize Theorem 7.78.

THEOREM 7.85 *The sum*

$$Q = P_{M_1} + P_{M_2} + \cdots + P_{M_n}. \tag{7.27}$$

of orthogonal projections is an orthogonal projection if and only if $P_{M_i} \perp P_{M_j}$ ($M_i \perp M_j$, $i \neq j$). In that case

$$Q = P_{M_1 \oplus M_2 \oplus \cdots \oplus M_n}. \tag{7.28}$$

Proof. If $M_i \perp M_j$ for all $i \neq j$, then $P_{M_i}P_{M_j} = 0$ $(i \neq j)$. Thus $Q^2 = Q$. Since $Q^* = Q$, Q is an orthogonal projection. The rest is obvious. We prove now the necessity. Since for any $x \in H$, $j \neq k$,

$$\|x\|^2 \geq \langle Qx, x\rangle = \sum_{i=1}^{n} \langle P_{M_i}x, x\rangle \geq \left\langle P_{M_j}x, x\right\rangle + \langle P_{M_k}x, x\rangle,$$

we get

$$\left\|P_{M_j}x\right\|^2 + \|P_{M_k}x\|^2 \leq \|x\|^2.$$

Let $x = P_{M_k}y$. Then

$$\left\|P_{M_j}P_{M_k}y\right\|^2 + \|P_{M_k}y\|^2 \leq \|P_{M_k}y\|^2$$

implies that

$$\left\|P_{M_j}P_{M_k}y\right\| = 0$$

holds for any $y \in H$. Hence $P_{M_j}P_{M_k} = 0$ showing the orthogonality of P_{M_j} (M_j) and P_{M_k} (M_k). ∎

Let $N = \sum_{i=1}^{n} M_i$. Since $M_i \cap M_j = \{0\}$ $(i \neq j)$ we have the unique decomposition of any $x \in N$ in the form

$$x = \sum_{i=1}^{n} x_i, \quad (x_i \in M_i;\, \langle x_i, x_j\rangle = 0,\, i \neq j)\,. \tag{7.29}$$

If $Q = I$, that is

$$I = P_{M_1} + P_{M_2} + \cdots + P_{M_n}, \tag{7.30}$$

we speak of this as a resolution of the identity.

In Section 2.3.2 we investigated the sum of projections. In Hilbert space, Stampfli [324] proved the following results for projections, which are not necessarily Hermitian.

THEOREM 7.86 *(Stampfli). Let H be a separable and infinite dimensional Hilbert space.*
(i) If A is any Hermitian linear operator on H, then there exist projections P_1, P_2, P_3, P_4 such that $P_1 + P_2 + P_3 + P_4 = A$.
(ii) If $S \neq 0$ is a Hermitian operator on H with $S \leq 0$, then there exist no projections P_1, P_2 such that $P_1 + P_2 = S$.
(iii) If T is a linear operator on H, then T is the sum of eight projections.

The orthogonal projections of a Hilbert space are dependent on the inner product $\langle .,. \rangle$ defined in the Hilbert space H. Let A be a positive definite Hermitian operator in H and define the inner product

$$\langle x, y\rangle_A = \langle x, Ay\rangle \quad (x, y \in H)\,. \tag{7.31}$$

Let V be an arbitrary but fixed closed subspace of H. Let $P(A)$ denote the projection of H onto V, orthogonal with respect to the inner product $\langle .,. \rangle_A$. Pasternak-Winiarski [275] proved that the orthogonal projection $P(A)$ is an analytic function of the inner product (or A). He also gave an explicit formula for its Taylor expansion. We note that Wimmer [360] proved the Lipschitz continuity of oblique projections in finite dimension (see Section 2.7).

The sequences of orthogonal projections have special convergence properties.

THEOREM 7.87 *If the sequence $\{P_{M_n}\}$ of orthogonal projections is monotone, then it strongly converges to some orthogonal projection P_M. If the sequence of projections is decreasing, then $M = \cap_{i=1}^{\infty} M_i$. If the sequence is monotone increasing, then $M = \cup_{i=1}^{\infty} M_i$.*

COROLLARY 7.88 *Let $\{P_n\}$ be a sequence of orthogonal projections such that $P_i P_j = 0$ for all $i \neq j$. The series*

$$Px = \sum_{i=1}^{\infty} P_i x \tag{7.32}$$

is convergent for any $x \in H$, and the operator P is a projection.

THEOREM 7.89 *If the sequence $\{P_n\}$ of orthogonal projections weakly converge to the orthogonal projection P, then it converges strongly to P.*

For the proof we refer to Kantorovich, Akilov [217], Achieser, Glasmann [6] or Halmos [181].

The following result of Buckholtz [71], [72] establishes interesting connections between orthogonal and oblique projections.

THEOREM 7.90 *(Buckholtz). Let M and N be closed subspaces of a Hilbert space H, and let P_M and P_N denote the orthogonal projections of H onto these subspaces. The following are equivalent:*
(i) H is the direct sum of M and N.
(ii) There exists a bounded linear projection Q with range M and null space N.
(iii) The operator $P_M - P_N$ is invertible.
(iv) $\|P_M + P_N - I\| < 1$.

Proof. If (i) holds, then every $x \in H$ has a unique decomposition of the form $x = m + n$, $m \in M$, $n \in N$. Define Q by $Qx = m$. By Theorem 7.17, Q is a bounded projection with range M and null space N. Therefore, (i) implies (ii). The easy way to prove that (ii)⇒(iii) is to exhibit an inverse for $P_M - P_N$. We use the fact that, if Q is a projection with range M and null space N, then Q^* is a projection with range $N^{\perp}$ and null space $M^{\perp}$. Furthermore $I - Q$ is a projection with range N and null space M. Finally, $I - Q^*$ is projection with range $M^{\perp}$ and $N^{\perp}$. Therefore,

$$P_M Q = Q, \quad P_N Q^* = 0, \quad P_N (I - Q) = I - Q, \quad P_M (I - Q^*) = 0.$$

Combining these results yields that

$$(Q + Q^* - I)(P_M - P_N) = I.$$

These operators are Hermitian. Taking the adjoint yields

$$(P_M - P_K)(Q + Q^* - I) = I.$$

Therefore,

$$(P_M - P_N)^{-1} = Q + Q^* - I. \tag{7.33}$$

To establish the implication (iii)⇒(iv) we use the identity

$$\|(P_M - P_N)x\|^2 + \|(P_M + P_N - I)x\|^2 = \|x\|^2.$$

Set $\mu = \inf\{\|(P_M - P_N)x\| \mid \|x\| = 1\}$. We have $\mu^2 + \|(P_M + P_N - I)\|^2 = 1$. Since $P_M - P_N$ is invertible, $\mu > 0$ and (iv) follows. It remains to prove (iv)⇒(i). Certainly (iv) implies the invertibility of $P_M + P_N$. The identity

$$x = P_M (P_M + P_N)^{-1} x + P_N (P_M + P_N)^{-1} x$$

ensures that $H = M + N$. Note that (iv) also requires M and N to be disjoint except for 0. Therefore H is a direct sum of M and N, and the proof of the theorem is complete. ■

Buckholtz also discovered the following inversion formulas

$$(\alpha P_M + \beta P_N)^{-1} = \frac{1}{\beta}(I - Q - Q^*) + \left(\frac{1}{\alpha} + \frac{1}{\beta}\right) QQ^* \tag{7.34}$$

and

$$(\alpha Q + \beta (I - Q^*))^{-1} = \frac{1}{\alpha} P_M + \frac{1}{\beta} P_N - \left(\frac{1}{\alpha} + \frac{1}{\beta}\right) P_N P_M. \tag{7.35}$$

THEOREM 7.91 *(Ljance). Let Q be a bounded linear projection with range M and null space N. If $Q \neq 0$, then*

$$\|Q\|^2 = \frac{1}{1 - \|P_N P_M\|^2}, \tag{7.36}$$

where P_M and P_N denote the orthogonal projections onto M and N.

Proof. For $x \in H$ write $x = x_1 + x_2$, where $x_1 = P_N x$ and $x_2 = (I - P_N)x$. Then $Qx = Qx_2$ and $\|x_2\| \leq \|x\|$. Consequently the norm of Q is the same as the norm of the restriction of Q to $N^\perp$. For $x \in N^\perp$ write $Qx = x + (Qx - x)$ and observe that $(Qx - x)$ is the orthogonal projection of Qx onto N. Therefore $Qx - x = P_N Qx = P_N P_M Qx$, since $Qx \in M$. From $x \in N^\perp$ and $(Qx - x) \in N$ we obtain $\|Qx\|^2 = \|x\|^2 + \|Qx - x\|^2 = \|x\|^2 + \|P_N P_M Qx\|^2$. Dividing by $\|Qx\|^2$ and some manipulation yields the identity

$$\frac{\|Qx\|}{\|x\|} = \left(1 - (\|P_N P_M Qx\| / \|Qx\|)^2\right)^{-1/2}.$$

It follows from this that $\|Q\| \leq \left(1 - \|P_N P_M\|^2\right)^{-1/2}$. To obtain equality rather than inequality, two observations are needed: restricting the operator Q to $N^{\perp}$ does not change its range, and restricting the operator $P_N P_M$ to M does not diminish its norm. This completes the proof. ■

Here we used the proof of Buckholtz [72]. For other proofs, see [295] and [89]. For the finite-dimensional case, see Section 2.5).

Further characterizations of Hilbert space projections can be found in [258], [318], [20], [89], [90].

In certain cases we can give simple representations of projections. Let

$$\{\phi_i\}_{i=1}^n, \{\gamma_i\}_{i=1}^n \subset H$$

be biorthogonal, that is, let

$$\langle \phi_i, \gamma_j \rangle = \begin{cases} 1, & i = j, \\ 0, & i \neq j. \end{cases} \tag{7.37}$$

In case of biorthogonality, both $\{\phi_i\}_{i=1}^n$ and $\{\gamma_i\}_{i=1}^n$ are linearly independent. It is easy to see that

$$Px = \sum_{i=1}^{n} \langle x, \gamma_i \rangle \phi_i \tag{7.38}$$

is a projection. Also, relations

$$\langle Px, y \rangle = \sum_{i=1}^{n} \langle x, \gamma_i \rangle \langle \phi_i, y \rangle = \left\langle x, \sum_{i=1}^{n} \overline{\langle \phi_i, y \rangle} \gamma_i \right\rangle = \langle x, P^* y \rangle$$

imply that

$$P^* y = \sum_{i=1}^{n} \overline{\langle \phi_i, y \rangle} \gamma_i. \tag{7.39}$$

If $\{\phi_i\}_{i=1}^n \subset H$ is an orthonormal system, then

$$Px = \sum_{i=1}^{n} \langle x, \phi_i \rangle \phi_i \tag{7.40}$$

is the orthogonal projection on $\mathcal{R}(\phi_1, \ldots, \phi_n)$. Notice that the $\langle x, \phi_i \rangle$'s are the Fourier coefficients of x. These representations of projections are related to the representations of Theorems 2.25 and 2.26.

THEOREM 7.92 *Let $\{\phi_i\}_{i=1}^\infty$ be an arbitrary orthonormal sequence in H. Then for any $x \in H$, $y = \sum_{i=1}^\infty \langle x, \phi_i \rangle \phi_i$ is the orthogonal projection of x on $\overline{\mathcal{L}(\{\phi_i\}_{i=1}^\infty)}$.*

For the above representations, see e.g., [95] or [128]. Similar representations can be given for Banach space projections by using the concept of conjugate spaces. For details we refer to [220], [217] or [345]. The biorthogonal systems of Hilbert spaces are intensively studied in [335].

1.2 Angles between subspaces

We investigate different characterizations of the relative position of two subspaces of a Hilbert space. The results will be used in estimating the convergence speed of certain projection methods. Halmos [179], [180] points out that the characterization of the "generic" position of two subspaces is not an easy task. The need for such characterizations induced, however, a number of different, and sometimes independent developments (for references, see Davis and Kahan [94]). Here we deal with the gap and the angle concepts of Friedrichs and Dixmier. For the Hilbert space analogue of the theory introduced in Section 1.7 we refer to Davis and Kahan [94].

We first deal with the gap concept using Achieser and Glasmann [6] and Krasnosel'skii et al.[230]. The study of angle concepts follows Deutsch [104].

Given a pair of closed subspaces M and N, their respective orthogonal projections P_M and P_N can provide information on the relative position of the subspaces. Theorems 7.77, 7.80, 7.90 and Lemma 7.83 gave a kind of characterization in certain special cases. The concept of gap is measuring the "distance" of the two subspaces.

DEFINITION 7.93 *Let M and N be closed subspaces of a Hilbert space H, The gap (aperture, opening) between the two subspaces is defined by*

$$\Theta(M,N) = \|P_M - P_N\|, \tag{7.41}$$

where P_M and P_N are the orthogonal projections on M and N, respectively.

It is clear from the definition that

$$\Theta(M,N) \geq 0, \quad \Theta(M,N) = \Theta(N,M) \tag{7.42}$$

and

$$\Theta(M,N) = \Theta\left(M^{\perp}, N^{\perp}\right). \tag{7.43}$$

Using the identity

$$P_M - P_N = P_M(I - P_N) - (I - P_M)P_N$$

we obtain

$$(P_M - P_N)x = P_M(I - P_N)x - (I - P_M)P_N x,$$

where the vectors $P_M(I-P_N)x$ and $(I-P_M)P_Nx$ are orthogonal. Therefore,

$$\begin{aligned}\|(P_M-P_N)x\|^2 &= \|P_M(I-P_N)x\|^2 + \|(I-P_M)P_Nx\|^2 \\ &\le \|(I-P_N)x\|^2 + \|P_Nx\|^2 = \|x\|^2.\end{aligned}$$

This proves that

$$\Theta(M,N) \le 1. \tag{7.44}$$

REMARK 7.94 *It is easy to see that $\Theta(\cdot,\cdot)$ is a metric on the closed subspaces of H. In fact, $\theta(M,N) \ge 0$ and $\Theta(M,N) = 0$ if and only if $P_M = P_N$, that is $M = N$. Also $\theta(M,N) = \Theta(N,M)$ and for any closed subspace $Z \subset H$, we have*

$$\theta(M,N) = \|P_M - P_N\| \le \|P_M - P_Z\| + \|P_Z - P_N\| = \Theta(M,Z) + \Theta(Z,N). \tag{7.45}$$

REMARK 7.95 *Assume that $M \subset N$ and $M \ne N$. Then $P_M - P_N = P_{N\ominus M}$ and $\Theta(M,N) = 1$. If $N = M^\perp$, then $P_M - P_N = P_M - (I-P_M) = I$ and $\Theta(M,M^\perp) = 1$.*

THEOREM 7.96 *If $\Theta(M,N) < 1$, then* $\dim(M) = \dim(N)$.

An equivalent definition of gap is given by

$$\Theta(M,N) = \max\left(\sup_{x\in N,\,\|x\|=1}\|(I-P_M)x\|,\ \sup_{x\in M,\,\|x\|=1}\|(I-P_N)x\|\right). \tag{7.46}$$

Since $\|(I-P_M)x\| = d(x,M)$ representation (7.46) can be written in the form

$$\Theta(M,N) = \max\left\{\sup_{x\in N,\,\|x\|=1} d(x,M),\ \sup_{x\in M,\,\|x\|=1} d(x,N)\right\}. \tag{7.47}$$

This is the formula that Krein, Krasnosel'skii and Milman used to extend the gap concept to Banach spaces (for details and references, see [230], [42], [220]). We prove now the equivalence of (7.46) and (7.41). By definition

$$\begin{aligned}\Theta(M,N) &= \sup_{x\in H}\frac{\|(P_M-P_N)x\|}{\|x\|} \\ &= \sup_{x\in H}\frac{\sqrt{\|P_M(I-P_N)x\|^2 + \|(I-P_M)P_Nx\|^2}}{\|x\|}.\end{aligned} \tag{7.48}$$

If x is restricted to $N \subset H$, then

$$\begin{aligned}\Theta(M,N) &\geq \sup_{x\in N} \frac{\sqrt{\|P_M(I-P_N)x\|^2+\|(I-P_M)P_N x\|^2}}{\|x\|}\\ &= \sup_{x\in N} \frac{\|(I-P_M)x\|}{\|x\|} = \rho_1.\end{aligned}$$

Similarly we can prove that

$$\Theta(M,N) \geq \sup_{x\in M} \frac{\|(I-P_N)x\|}{\|x\|} = \rho_2.$$

Hence

$$\Theta(M,N) \geq \max\{\rho_1,\rho_2\}.$$

We have the estimates

$$\|(I-P_M)P_N x\|^2 \leq \rho_1^2 \|P_N x\|^2, \tag{7.49}$$

$$\begin{aligned}\|P_M(I-P_N)x\|^2 &= \langle P_M(I-P_N)x, P_M(I-P_N)x\rangle\\ &= \langle P_M(I-P_N)x, (I-P_N)x\rangle\\ &= \langle (I-P_N)P_M(I-P_N)x, (I-P_N)x\rangle\\ &\leq \|(I-P_N)P_M(I-P_N)x\| \, \|(I-P_N)x\|,\end{aligned}$$

$$\|P_M(I-P_N)x\|^2 \leq \rho_2 \|P_M(I-P_N)x\| \, \|(I-P_N)x\|$$

and

$$\|P_M(I-P_N)x\| \leq \rho_2 \|(I-P_N)x\|. \tag{7.50}$$

It follows from (7.49) and (7.50) that

$$\begin{aligned}\|P_M(I-P_N)x\|^2 + \|(I-P_M)P_N x\|^2 &\leq \rho_2^2 \|(I-P_N)x\|^2 + \rho_1^2 \|P_N x\|^2\\ &\leq \max\{\rho_1^2,\rho_2^2\} \|x\|^2.\end{aligned}$$

Hence $\Theta(M,N) \leq \max\{\rho_1,\rho_2\}$ follows, which together with the lower bound $\Theta(M,N) \geq \max\{\rho_1,\rho_2\}$ proves the equivalence of the two definitions.

If the gap between two n-dimensional subspaces is less than 1, then they have orthogonal bases which are close to each other in a certain sense [230].

LEMMA 7.97 *Let M and N be n-dimensional subspaces of a Hilbert space H with gap $\Theta(M,N) < 1$. Then M and N have orthogonal bases $u_1,\dots,u_n$ and $v_1,\dots,v_n$, respectively, such that*

$$\langle u_i, v_j\rangle = \beta_i \delta_{ij} \quad (i,j = 1,\dots,n),$$

where δ_{ij} is the Kronecker delta, and the constants β_i satisfy the inequalities

$$\left(1-\Theta^2(M,N)\right)^{1/2} \leq \beta_i \leq 1 \quad (i = 1,\dots,n).$$

For other results on the gap in finite dimension we refer to Section 2.5. We now define the angle concepts of Friedrichs and Dixmier.

DEFINITION 7.98 *(Friedrichs, 1937). The angle between the subspaces M and N of a Hilbert space H is the angle $\alpha(M,N)$ in $[0,\pi/2]$ whose cosine is given by*

$$c(M,N) = \sup\Big\{|\langle x,y\rangle| \mid x\in M\cap(M\cap N)^{\perp},\ \|x\|\le 1,\ y\in N\cap(M\cap N)^{\perp},\ \|y\|\le 1\Big\} \tag{7.51}$$

DEFINITION 7.99 *(Dixmier, 1949). The minimal angle between the subspaces M and N is the angle $\alpha_0(M,N)$ in $[0,\pi/2]$ whose cosine is defined by*

$$c_0(M,N) = \sup\{|\langle x,y\rangle| \mid x\in M,\ \|x\|\le 1,\ y\in N,\ \|y\|\le 1\}. \tag{7.52}$$

The two definitions are different except for the case $M\cap N=\{0\}$ when they clearly agree. The following results are quoted from Deutsch [104].

LEMMA 7.100 *(i)* $0\le c(M,N)\le c_0(M,N)\le 1$.
(ii) (a) $c(M,N)=c(N,M)$.
(b) $c_0(M,N)=c_0(N,M)$.
(iii) $c(M,N)=c_0\left(M\cap(M\cap N)^{\perp},N\cap(M\cap N)^{\perp}\right)$.
(iv) If $M\cap N=\{0\}$, then $c(M,N)=c_0(M,N)$ and $\alpha(M,N)=\alpha_0(M,N)$.
(v) If $M\cap N\ne\{0\}$, then $c_0(M,N)=1$ and $\alpha_0(M,N)=0$.

LEMMA 7.101 *The following properties hold:*
(i) $c(M,N)=c_0\left(M,N\cap(M\cap N)^{\perp}\right)=c_0\left(M\cap(M\cap N)^{\perp},N\right)$.
(ii) (a) $|\langle x,y\rangle|\le c_0(M,N)\,\|x\|\,\|y\|$ for all $x\in M$, $y\in N$.
(b) $|\langle x,y\rangle|\le c(M,N)\,\|x\|\,\|y\|$ for all $x\in M$, $y\in N$, and at least one of x or y is in $(M\cap N)^{\perp}$.
(iii) $c_0(M,N)=\|P_MP_N\|=\|P_MP_NP_M\|^{1/2}$.
(iv) $c(M,N)=\|P_MP_N-P_{M\cap N}\|=\left\|P_MP_NP_{(M\cap N)^{\perp}}\right\|$.
(v) $c(M,N)=\left\|P_MP_{(M\cap N)^{\perp}}P_NP_{(M\cap N)^{\perp}}\right\|$.
(vi) $c_0(M,N)=0$ if and only if $M\perp N$ (i.e. $M\subset N^{\perp}$).
(vii) $c(M,N)=0$ if and only if P_M and P_N commute.

THEOREM 7.102 *The following statements are equivalent.*
(i) $c_0(M,N)<1$.
(ii) $M\cap N=\{0\}$ and $M+N$ is closed.
(iii) There exists a constant $\rho>0$ such that

$$\|x+y\|\ge\rho\,\|y\| \tag{7.53}$$

for all $x \in M$ *and* $y \in N$.
(iv) $\inf\{d(y, M) \mid y \in N, \|y\| = 1\} > 0$.
(v) $\inf\{d(x, N) \mid x \in M, \|x\| = 1\} > 0$.

Using Lemma 7.57 and Theorem 7.102, we can derive

THEOREM 7.103 *The following statements are equivalent.*
(i) $c(M, N) < 1$.
(ii) $M \cap (M \cap N)^{\perp} + N \cap (M \cap N)^{\perp}$ *is closed.*
(iii) $M + N$ *is closed.*
(iv) $M^{\perp} + N^{\perp}$ *is closed.*

LEMMA 7.104 *(Deutsch, Hundal). Let* M, N, *and* S *be closed subspaces of* H *such that* $S \subset M \cap N$. *Then*

$$c(M, N) = c\left(M \cap S^{\perp}, N \cap S^{\perp}\right). \tag{7.54}$$

The angle between the orthogonal complements of the two subspaces is the same as the angle between the subspaces. However this is true for the minimal angle only with restrictions.

LEMMA 7.105 *(Hundal). If* $c_0(M, N) < 1$, *then for any closed subspace* X *which contains* $M + N$, *we have*

$$c_0(M, N) \leq c_0\left(M^{\perp} \cap X, N^{\perp} \cap X\right). \tag{7.55}$$

We will write $H = M \oplus N$ if $H = M + N$ and $M \cap N = \{0\}$.

THEOREM 7.106 *(Krein, Krasnosel'skii, Milman). If* $H = M \oplus N$, *then*

$$c_0(M, N) = c_0\left(M^{\perp}, N^{\perp}\right). \tag{7.56}$$

THEOREM 7.107

$$c(M, N) = c\left(M^{\perp}, N^{\perp}\right). \tag{7.57}$$

We close this section with a result on the norm of oblique projections, which is complementary to Theorem 7.90.

THEOREM 7.108 *(Buckholtz). Suppose that the Hilbert space* H *is the direct sum of closed subspaces* M *and* N. *Let* θ *be the minimal angle between* M *and* N, *let* Q *be the projection with range* M *and null space* N. *Then*

$$\|Q\| = \|I - Q\| = \|I - Q^*\| = \|Q^*\| = \csc(\theta).$$

Proof. From Theorem 7.91 we have

$$\|Q\| = \left(1 - \|P_N P_M\|^2\right)^{-1/2} = \left(1 - \cos^2\theta\right)^{-1/2} = \csc(\theta).$$

The order of operators is unimportant. $P_M P_N$ and $P_N P_M$ have the same norm because they are adjoint. Since $I - Q$ is a projection with range N and null space M we also have the corresponding result that $\|I - Q\| = \|Q\| = \csc(\theta)$. Because they are adjoint of $I - Q$ and Q, we also have $\|I - Q^*\| = \|Q^*\| = \csc(\theta)$. ■

For similar results in finite-dimensional spaces, see Section 2.5.

2. General convergence theorems

Let X and Y be (complex or real) Banach spaces. Consider the equation

$$Ax = y, \tag{7.58}$$

where $A : X \to Y$ is a linear operator. The *projection method* for the solution of this equation is defined as follows. Let $\{X_k\}$ and $\{Y_k\}$ be two sequences of subspaces,

$$X_k \subset \mathcal{D}(A) \subset X, \quad Y_k \subset Y \quad (k = 1, 2, \ldots),$$

and let $P_k : Y \to Y_k$ be linear projection operators, that is

$$P_k^2 = P_k,\ P_k Y = Y_k \quad (k = 1, 2, \ldots).$$

Now replace equation $Ax = y$ by the approximation

$$P_k(Ax_k - y) = 0 \quad (x_k \in X_k). \tag{7.59}$$

Note that the approximate solution belongs to X_k.

If $X = Y$, $X_k = Y_k$ ($k = 1, 2, \ldots$), the projection method is called the Galerkin method. The following variants of the projection method (7.59) are the best known ones.

Let X and Y be Hilbert spaces and let $\{\phi_j\} \subset \mathcal{D}(A) \subset X$ and $\{\psi_j\} \subset Y$ be complete sequences in X and Y, respectively. Let $X_k = \mathcal{R}(\phi_1, \ldots, \phi_k)$ and $Y_k = \mathcal{R}(\psi_1, \ldots, \psi_k)$.

LEMMA 7.109 *Let $P_k : Y \to Y_k$ be the orthogonal projection. Then equations*

$$P_k z = 0$$

and

$$\langle z, \psi_j \rangle = 0 \quad (j = 1, \ldots, k)$$

are equivalent.

Proof. Assume that $P_k z = 0$. As $P_k \psi_j = \psi_j$ we have $\langle z, \psi_j \rangle = \langle z, P_k \psi_j \rangle = \langle P_k z, \psi_j \rangle = 0$ ($j = 1, \ldots, k$). Suppose that $\langle z, \psi_j \rangle = 0$ ($j = 1, \ldots, k$) holds.

Then $P_k z = \sum_{j=1}^k \alpha_j \psi_j$ and

$$\|P_k z\|^2 = \left\langle P_k z, \sum_{j=1}^k \alpha_j \psi_j \right\rangle = \sum_{j=1}^k \alpha_j \langle P_k z, \psi_j \rangle$$
$$= \sum_{j=1}^k \alpha_j \langle z, P_k \psi_j \rangle = \sum_{j=1}^k \alpha_j \langle z, \psi_j \rangle = 0,$$

which implies $P_k z = 0$. ■

Assume that $P_k : Y \to Y_k$ is the orthogonal projection. Then by the lemma $P_k (Ax_k - y) = 0$ holds if and only if

$$\langle Ax_k - y, \psi_j \rangle = 0 \quad (j = 1, \ldots, k). \tag{7.60}$$

Taking

$$x_k = \sum_{j=1}^k \xi_j \phi_j, \tag{7.61}$$

we can write this condition as

$$\sum_{j=1}^k \langle A\phi_j, \psi_i \rangle \xi_j = \langle y, \psi_i \rangle \quad (i = 1, \ldots, k), \tag{7.62}$$

which is a linear system for the coefficients ξ_j. Hence the solution x_k is defined by the orthogonality condition

$$Ax_k - y \perp \psi_j \quad (j = 1, \ldots, k). \tag{7.63}$$

This version of projection methods is called the Galerkin-Petrov method. Note that the subspaces X_k and Y_k in the Galerkin-Petrov method are finite dimensional and

$$X_k \subset X_{k+1}, \quad Y_k \subset Y_{k+1} \quad (k = 1, 2, \ldots). \tag{7.64}$$

There are various ways to choose $\{\phi_j\}$ and $\{\psi_j\}$ in the Galerkin-Petrov method. If

$$\psi_i = A\phi_i \quad (i = 1, 2, \ldots), \tag{7.65}$$

the Galerkin-Petrov method is known as the *least squares method.* It can be proved that u_k minimizes the functional $\|Ax_k - y\|^2$ in $X_k = \mathcal{R}(\phi_1, \ldots, \phi_k)$, if and only if x_k is defined by the least squares method. If $X = Y$ and $\psi_i = \phi_i$ $(i = 1, 2, \ldots)$, then the projection method is called the Bubnov-Galerkin method.

The following convergence result or its variants can be found in several books (see, e.g., [230], [237], [214] or [363]).

DEFINITION 7.110 *A sequence of subspaces* $\{X_n\}_{n=1}^{\infty}$ *is said to be ultimately dense in a Banach space* X, *if for every* $z \in X$,

$$\lim_{n\to\infty} d(z, X_n) = 0,$$

where

$$d(z, X_n) = \inf_{u\in X_n} \|z - u\|.$$

THEOREM 7.111 *Let the domain* $\mathcal{D}(A)$ *of operator* A *be dense in* X *and the range* $\mathcal{R}(A)$ *dense in* Y, *and assume that* A *is one-to-one on* $\mathcal{D}(A)$. *Assume that the subspaces* AX_n *and* Y_n *are closed in* Y. *Finally, assume that the projections* P_n *are uniformly bounded:*

$$\|P_n\| \le c \quad (n = 1, 2, \ldots). \tag{7.66}$$

For any $y \in Y$ *and sufficiently large* n ($n \ge n_0$, *say*) *there exists a unique solution of equation (7.59) and* $\lim_{n\to\infty} \|Ax_n - y\| = 0$ *if and only if the following conditions hold:*
(i) The sequence of subspaces $\{AX_n\}$ *is ultimately dense in* Y.
(ii) For $n \ge n_0$, *the operator* P_n *is a bijective mapping of* AX_n *onto* Y_n.
(iii) $\tau = \underline{\lim}_{n\to\infty}\tau_n > 0$, *where*

$$\tau_n = \inf_{z_n\in AX_n, \|z_n\|=1} \|P_n z_n\|. \tag{7.67}$$

If conditions (i)-(iii) are satisfied, the rate of convergence is given by the inequality

$$d(y, AX_n) \le \|Ax_n - y\| \le \left(1 + \frac{c}{\tau_n}\right) d(y, AX_n). \tag{7.68}$$

Proof. Substituting $Ax_n = y_n$, equation (7.59) becomes

$$P_n y_n = P_n y \quad (y_n \in AX_n). \tag{7.69}$$

Assume that (i)-(iii) hold. Let $P_n|_{AX_n}$ denote the restriction of P_n to the subspace AX_n. For $n \ge n_0$ it follows from (ii) that $P_n|_{AX_n}$ is a bijective mapping of AX_n onto Y_n, and it has a bounded inverse $\left(P_n|_{AX_n}\right)^{-1}$ mapping Y_n onto AX_n. Hence there is a unique element y_n satisfying (7.69):

$$y_n = \left(P_n|_{AX_n}\right)^{-1} P_n y.$$

But then $x_n = A^{-1}y_n$ is the unique element satisfying (7.59). It follows from condition (iii) that $\left\|\left(P_n|_{AX_n}\right)^{-1}\right\| \le 1/\tau_n \le 1/\tau$. Then condition (7.66) implies

$$\left\|\left(P_n|_{AX_n}\right)^{-1} P_n\right\| \le \left\|\left(P_n|_{AX_n}\right)^{-1}\right\| \|P_n\| \le \frac{c}{\tau_n} \le \frac{c}{\tau} < \infty,$$

that is the norm of the operators $\left(P_n|_{AX_n}\right)^{-1} P_n$ is uniformly bounded. If $z_n \in AX_n$, then $\left(P_n|_{AX_n}\right)^{-1} P_n z_n = z_n$, and so

$$\begin{aligned} Ax_n - y = y_n - y &= \left(P_n|_{AX_n}\right)^{-1} P_n y - y \\ &= \left(P_n|_{AX_n}\right)^{-1} P_n (y - z_n) - (y - z_n) \end{aligned} \tag{7.70}$$

and

$$\|Ax_n - y\| \leq \left(1 + \frac{c}{\tau}\right) \|y - z_n\|.$$

Since $z_n \in AX_n$ is arbitrary, it follows that the residual $\|Ax_n - y\|$ tends to zero and (7.68) holds. We now prove the necessity. Suppose that for every $y \in Y$ and $n \geq n_0$ the approximation x_n is uniquely determined by (7.59) or (7.69), and $\lim_{n\to\infty} \|Ax_n - y\| = 0$. Since

$$\lim_{n\to\infty} \|Ax_n - y\| = \lim_{n\to\infty} \|y_n - y\| = 0 \quad (\forall y \in Y),$$

where $y_n \in AX_n$, condition (i) is satisfied. Furthermore, for $n \geq n_0$ equation (7.59) has a unique solution. Consequently, equation (7.69) has also a unique solution. Hence condition (ii) is also satisfied. To show (iii) it suffices to prove that the norms $\left\|\widetilde{P}_n^{-1}\right\| = 1/\tau_n$ are uniformly bounded for $n \geq n_0$. We have $y_n = \left(P_n|_{AX_n}\right)^{-1} P_n y$. Thus for any $y \in Y$,

$$\left(P_n|_{AX_n}\right)^{-1} P_n y \to y \quad (n \to \infty).$$

The Banach-Steinhaus theorem implies that the norms $\left\|\left(P_n|_{AX_n}\right)^{-1} P_n\right\|$ are uniformly bounded:

$$\left\|\left(P_n|_{AX_n}\right)^{-1} P_n\right\| \leq c' \quad (n \geq n_0).$$

In particular, for any $y_n \in Y_n$,

$$\left\|\left(P_n|_{AX_n}\right)^{-1} y_n\right\| = \left\|\left(P_n|_{AX_n}\right)^{-1} P_n y_n\right\| \leq c' \|y_n\| \quad (n \geq n_0),$$

and therefore $\left\|\left(P_n|_{AX_n}\right)^{-1}\right\| \leq c'$. ■

REMARK 7.112 *For* $\dim(X_n) = \dim(Y_n) < \infty$, *condition (ii) follows from (iii).*

REMARK 7.113 *Condition (7.66) is a serious restriction even for finite dimensional subspaces* Y_n. *For example, if* $Y = C[0,1]$, Y_n *is the set of polynomials of degree* $\leq n$, *then any projection* P_n *mapping* C *onto* Y_n *has norm*

$$\|P_n\| \geq c \ln n \quad (c = const > 0)$$

(see, e.g., [95]).

For Hilbert spaces the above theorem can be refined with the use of the gap between two subspaces. We need the following result ([230], [301]).

LEMMA 7.114 *Let M_1 and M_2 be closed subspaces of a Hilbert space H, P_1 and P_2 the corresponding orthogonal projections. The operator P_2 is a bijective mapping of M_1 onto M_2 if and only if $\Theta(M_1, M_2) < 1$. If $\theta = \Theta(M_1, M_2)$ and $\tau = \inf_{z \in M_1,\, \|z\|=1} \|P_2 z\|$, then*

$$\theta^2 + \tau^2 = 1. \tag{7.71}$$

Proof. Assume that $\|P_1 - P_2\| = \theta < 1$. Then the operator $I + (P_1 - P_2)$ maps H onto itself:

$$(I + P_1 - P_2) H = H.$$

Applying P_2 to both sides we get $P_2 P_1 H = P_2 H$, or $P_2 M_1 = M_2$. Hence P_2 maps M_1 onto M_2. It is not difficult to see that P_2 is also bijective on M_1. If $P_2 x_0 = 0$ and $x_0 \in M_1$, $x_0 \neq 0$, we have

$$\|x_0\| = \|(I - P_2) x_0\| = \|(I - P_2) P_1 x_0\| = \|(P_1 - P_2) x_0\| < \|x_0\|,$$

which is a contradiction. This proves the sufficiency. Assume that P_2 is a bijective mapping of M_1 onto M_2. The restriction $P_2|_{M_1}$ of P_2 to M_1 is invertible, $\left(P_2|_{M_1}\right)^{-1}$ maps M_2 onto M_1, it is bounded, and $\left\|\left(P_2|_{M_1}\right)^{-1}\right\| = 1/\tau$. Then

$$\|P_2 x\| \geq \tau \|x\| \quad (x \in M_1),$$

and for any $\varepsilon > 0$ there exists an element $x_\varepsilon \in M_1$ such that

$$\|P_2 x_\varepsilon\| \leq (\tau + \varepsilon) \|x_\varepsilon\|.$$

Thus

$$\|(I - P_2) x\|^2 = \|x\|^2 - \|P_2 x\|^2 \leq \left(1 - \tau^2\right) \|x\|^2 \quad (x \in M_1),$$

$$\|(I - P_2) x_\varepsilon\| \geq [1 - (\tau + \varepsilon)]^2 \|x_\varepsilon\|^2,$$

that is

$$\sup_{x \in M_1,\, \|x\|=1} \|(I - P_2) x\|^2 = 1 - \tau^2. \tag{7.72}$$

The adjoint of $P_2|_{M_1}$ is $P_1|_{M_2}$, the restriction of projection P_1 to M_2. In fact, for any $x \in M_1$ and $y \in M_2$,

$$\begin{aligned} \left\langle P_2|_{M_1} x, y \right\rangle_{M_2} &= \langle P_2 x, y \rangle = \langle x, P_2 y \rangle = \langle x, y \rangle \\ &= \langle P_1 x, y \rangle = \langle x, P_1 y \rangle = \left\langle x, P_1|_{M_2} y \right\rangle_{M_1}. \end{aligned}$$

Consequently, the adjoint $P_1|_{M_2}$ of $P_2|_{M_1}$ is also invertible:

$$\left\|\left(P_1|_{M_2}\right)^{-1}\right\| = \left\|\left(P_2|_{M_1}\right)^{-1}\right\| = 1/\tau.$$

This is equivalent with

$$\inf_{z\in M_2,\, \|z\|=1} \|P_1 z\| = \tau.$$

Replacing M_1 and M_2 by each other and repeating the same arguments we obtain the following analogue of (7.72):

$$\sup_{z\in M_2,\, \|z\|=1} \|(I - P_1)\, z\|^2 = 1 - \tau^2.$$

Together with (7.72) this leads to

$$\Theta\,(M_1, M_2) = \sqrt{1-\tau^2} < 1.$$

This proves (7.71) and completes the proof. ∎

We now return to equation $Ax = y$ and the projection method

$$P_k\,(Ax_k - y) = 0 \quad (x_k \in X_k).$$

We assume that Y is a Hilbert space and P_k is the orthogonal projection onto Y_k. Using the last lemma we can reformulate the general convergence theorem.

THEOREM 7.115 *Let the domain $\mathcal{D}\,(A)$ of operator A be dense in the Banach space X and let the range $\mathcal{R}\,(A)$ be dense in the Hilbert space Y. Assume that the subspaces AX_n and Y_n are closed in Y. Let Q_n and P_n be the corresponding orthogonal projections. For any $y \in Y$ and sufficiently large n ($n \geq n_0$, say) there exists a unique solution of equation (7.59) and* $\lim_{n\to\infty} \|Ax_n - y\| = 0$ *if and only if the following conditions hold:*
(i) For any $y \in Y$, $Q_n y \to y$ in norms as $n \to \infty$.
(ii) $\overline{\lim}_{n\to\infty}\Theta\,(AX_n, Y_n) < 1$.
If the conditions (i) and (ii) are satisfied, the rate of convergence is given by the inequality

$$\|(I - Q_n)\, y\| \leq \|Ax_n - y\| \leq \frac{1}{\sqrt{1-\theta_n^2}}\, \|(I - Q_n)\, y\|\,, \qquad (7.73)$$

where $\theta_n = \Theta\,(AX_n, Y_n)$.

Proof. We have to prove only the bound (7.73) which does not follow from (7.68). If $z_n \in AX_n$, then $\left(P_n|_{AX_n}\right)^{-1} P_n z_n = z_n$, and so

$$\begin{aligned} Ax_n - y &= y_n - y = \left(P_n|_{AX_n}\right)^{-1} P_n y - y \\ &= \left(P_n|_{AX_n}\right)^{-1} P_n\,(y - z_n) - (y - z_n)\,. \end{aligned} \qquad (7.74)$$

We start from (7.70), setting $z_n = Q_n y$:

$$Ax_n - y = \left(P_n|_{AX_n}\right)^{-1} P_n (I - Q_n) y - (I - Q_n) y.$$

Since $\left(P_n|_{AX_n}\right)^{-1} P_n (I - Q_n) y \in AX_n$ and $(I - Q_n) y \in (AX_n)^{\perp}$,

$$\begin{aligned} \|Ax_n - y\|^2 &= \left\|\left(P_n|_{AX_n}\right)^{-1} P_n (I - Q_n) y\right\|^2 + \|(I - Q_n) y\|^2 \\ &\leq \left(\frac{1}{\tau_n^2} \|P_n (I - Q_n)\|^2 + 1\right) \|(I - Q_n) y\|^2, \end{aligned}$$

where $\tau_n = \inf_{z \in AX_n, \|z\|=1} \|P_n z\|$. Relations (7.48) and (7.71) imply

$$\|P_n (I - Q_n)\| = \|(I - Q_n) P_n\| \leq \theta_n$$

and $\tau_n^2 = 1 - \theta_n^2$, respectively. Thus (7.73) follows. ■

REMARK 7.116 *In the least squares method* $Y_n = AX_n$ *and* $\Theta(AX_n, Y_n) = 0$. *Hence condition (ii) is clearly satisfied.*

For the application of the above theory in solving operator equations we refer to the very rich special literature, especially to Krasnosel'skii et al. [230], Gochberg, Feldman [159], Prößdorf, Silberman [293] and Luchka [240]. The papers [24], [277] and [278] are also of interest.

3. The method of alternating projections

The method of alternating projections can be defined as follows. Let M_1, $M_2, \ldots, M_k$ be closed subspaces of the Hilbert space H, $M = \cap_{i=1}^k M_i$, and $P_i = P_{M_i}$ $(i = 1, \ldots, k)$. The orthogonal projection P_M is called the intersection projection. We seek for $P_M x$, the best approximation in the intersection M to any $x \in H$, by projecting a point cyclically through the individual subspaces. Thus the method of alternating projection is defined by

$$\begin{aligned} x_0 &= x, \\ x_{j+1} &= P_i x_j \quad (i \equiv j \pmod{k} + 1). \end{aligned} \tag{7.75}$$

Chapters 4, 6 contain several applications of the alternating projection method. For other applications, which range from probability to computer tomography, we refer to Deutsch [103] (see, also [259]). Deutsch [102], [104], [105] also extensively studied the convergence of the alternating projection method. We first deal with the convergence of the alternating projection method. The speed estimates will be given in a separate section.

3.1 Convergence results

Considering the subsequence $\{x_{nk}\}_{n=0}^{\infty}$, we see that

$$x_{nk} = P_k P_{k-1} \cdots P_1 x_{(n-1)k} = (P_k P_{k-1} \cdots P_1)^n x_0. \tag{7.76}$$

It is easy to see that if $x_{nk} \to P_M x$ $(n \to \infty)$, then $x_j \to P_M x$ $(n \to \infty)$ also holds. The relations $x_{j+1} = P_i x_j$ and $P_M x = P_i P_M x$ imply $x_{j+1} - P_M x = P_i (x_j - P_M x)$ and

$$\|x_{j+1} - P_M x\| \leq \|x_j - P_M x\| \leq \cdots \leq \|x_{nk} - P_M x\| \tag{7.77}$$

for some $n \geq 0$. Thus the convergence of the alternating projection method depends on the convergence of the sequence $\{(P_k P_{k-1} \ldots P_1)^n x - P_M x\}_{n=0}^{\infty}$.

Von Neumann [266] proved that for $k = 2$,

$$\lim_{n \to \infty} (P_2 P_1)^n x = P_M x \quad (x \in H). \tag{7.78}$$

An alternative formulation of the result is obtained by using the fact that

$$M_1^{\perp} \cap M_2^{\perp} = \left(\overline{M_1 + M_2}\right)^{\perp}$$

and replacing M_1 by $M_1^{\perp}$ and M_2 by $M_2^{\perp}$. Hence

$$\lim_{n \to \infty} [(I - P_2)(I - P_1)]^n = \left(I - P_{\overline{M_1 + M_2}}\right) x \quad (x \in H). \tag{7.79}$$

Variants of the von Neumann result appeared in Aronszajn [22], Nakano [262], Wiener [358], and Browder [60]. It is worth mentioning that Stiles (see [103]) and Bosznay [50] generalized the result for Banach spaces.

Browder [60], Prager [292], Halperin [182], and Pyle [297] extended the result for $k \geq 2$. Further extensions of the von Neumann theorem involve contractive or nonexpansive mappings.

Definition 7.117 *A mapping $\pi : \mathbb{N} \to A \subset \mathbb{N}$ will be called a (random) selection. The selection π is said to be admissible, if for each integer $j \in A$, there exists a positive integer $m(j)$ such that for any n,*

$$\{\pi(n), \pi(n+1), \ldots, \pi(n + m(j) - 1)\}$$

must contain the integer j. A selection π is called periodic, if $\pi(n+m) = \pi(n)$ for some m and all n. The selection π is said to be quasi-periodic, if there is an m such that for any n,

$$\{\pi(n), \pi(n+1), \ldots, \pi(n+m-1)\}$$

contains each $j \in A$ at least once.

It is easy to see that for any A, finite or infinite, there are admissible selections. If $A = \{1,2\}$, then for an admissible selection π, the sequence $\{\pi(j)\}_{j=1}^{\infty}$ consists of alternating 1's and 2's. It is also easy to see that for finite A, an admissible selection π is also quasi-periodic.

THEOREM 7.118 *(Browder, 1958). Let $\{M_i\}_{i\in A}$ be a sequence of closed subspaces in a Hilbert space H, $M = \cap_{i\in A} M_i$, $P_i = P_{M_i}$ $(i \in A)$, $\pi : \mathbb{N} \to A$ an admissible selection, and $R_k = P_{\pi(k)}P_{\pi(k-1)}\cdots P_{\pi(2)}P_{\pi(1)}$. Then for any $x \in H$, $R_k x$ converges weakly in H to $P_M x$ as $k \to \infty$.*

Prager [292] developed the following results.

THEOREM 7.119 *(Prager, 1960). Let A be a bounded linear operator in the Hilbert space H and satisfy the following conditions:*
(i) $A = A^*$.
(ii) $A \geq 0$.
(iii) $Ax = x \Rightarrow x = 0$.
(iv) $\|A\| \leq 1$.
Then $A^n \to 0$ strongly, that is for every element $x \in H$, $\lim_{n\to\infty} \|A^n x\| = 0$.

Proof. For any $k \geq 1$, $A^k \geq 0$ and $\|A^k\| \leq 1$. Also we have

$$\left|\left\langle A^k x, x\right\rangle\right| \leq \left\|A^k x\right\| \|x\| \leq \|x\|^2 .$$

Hence

$$\left\langle \left(I - A^k\right) x, x, \right\rangle = \|x\|^2 - \left\langle A^k x, x\right\rangle \geq 0,$$

that is $I - A^k$ is also positive. Let $x \in H$ be arbitrary. For $n > m$ we prove

$$\langle A^m x, x\rangle \geq \langle A^n x, x\rangle \Leftrightarrow \langle A^m \left(I - A^{n-m}\right) x, x\rangle \geq 0.$$

Since A^m and $I - A^{n-m}$ are both positive and commutative, $A^m\left(I - A^{n-m}\right)$ is also positive (see, e.g., [301]). Hence $\langle A^n x, x\rangle$ is monotone decreasing and bounded from below. Let $d = \lim_{n\to\infty} \langle A^n x, x\rangle$. Consider now

$$\begin{aligned}\left\|A^k x - A^l x\right\| &= \left\langle A^k x - A^l x, A^k x - A^l x\right\rangle \\ &= \left\langle A^k x, A^k x\right\rangle - 2\left\langle A^k x, A^l x\right\rangle + \left\langle A^l x, A^l x\right\rangle \\ &= \left\langle A^{2k} x, x\right\rangle - 2\left\langle A^{k+l} x, x\right\rangle + \left\langle A^{2l} x, x\right\rangle .\end{aligned}$$

Given $\varepsilon > 0$ there is a number $n(\varepsilon)$ such that for any $j > n(\varepsilon)$, $\langle A^j x, x\rangle \leq d + \varepsilon$. Hence for $k, l > n(\varepsilon)$, we have $\langle A^{2k} x, x\rangle \leq d + \varepsilon$, $\langle A^{2l} x, x\rangle \leq d + \varepsilon$, $\langle A^{k+l} x, x\rangle \geq d$ and $\|A^k x - A^l x\| \leq 2\varepsilon$. Consequently $A^n x$ is a Cauchy sequence and has a limit $y = \lim_{n\to\infty} A^n x$. But $y = Ay$ which implies $y = 0$. ■

THEOREM 7.120 *(Prager, 1960). Let $M_1, M_2, \ldots, M_k$ be closed subspaces of the Hilbert space H, $M = \cap_{i=1}^k M_i$, and $P_i = P_{M_i}$ $(i = 1, \ldots, k)$. For each $x \in H$,*

$$\lim_{n\to\infty} \|(P_1 P_2 \cdots P_{k-1} P_k P_{k-1} \cdots P_2 P_1)^n x - P_M x\| = 0. \tag{7.80}$$

Proof. It is easy to see that operator $Q = P_1 P_2 \cdots P_{k-1} P_k P_{k-1} \cdots P_2 P_1$ satisfies conditions 1,2 and 4 of Theorem 7.119. In order to verify condition 3 observe that for $x \notin \cap_{i=1}^k M_i$ we have $\|P_1 P_2 \cdots P_{k-1} P_k P_{k-1} \cdots P_2 P_1 x\| < \|x\|$. Hence the equality $P_1 P_2 \cdots P_{k-1} P_k P_{k-1} \cdots P_2 P_1 x = x$ is impossible. Assume now that $M \neq \{0\}$. Decompose the subspaces M_i in the form $M_i = M + (M_i \ominus M)$. Since $M_i \ominus M \subset M^\perp$, Lemma 7.82 implies that $P_{M_i} = P_M + P_{M_i \ominus M}$. Since M is orthogonal to every subspace $M_i \ominus M$, any product of the form $P_M P_{M_i \ominus M}$ or $P_{M_i \ominus M} P_M$ is 0. Hence

$$Qx = P_M x + P_{M_1 \ominus M} \cdots P_{M_{k-1} \ominus M} P_{M_k \ominus M} P_{M_{k-1} \ominus M} \cdots P_{M_1 \ominus M} x$$

and

$$Q^n x = P_M x + \left(P_{M_1 \ominus M} \cdots P_{M_{k-1} \ominus M} P_{M_k \ominus M} P_{M_{k-1} \ominus M} \cdots P_{M_1 \ominus M}\right)^n x.$$

It is also true that $\cap_{i=1}^k (M_i \ominus M) = \{0\}$. Therefore

$$\lim_{n\to\infty} \left(P_{M_1 \ominus M} \cdots P_{M_{k-1} \ominus M} P_{M_k \ominus M} P_{M_{k-1} \ominus M} \cdots P_{M_1 \ominus M}\right)^n x = 0$$

and Theorem 7.119 implies

$$\lim_{n\to\infty} Q^n x = P_M x.$$

■

THEOREM 7.121 *(Prager, 1960). Let $M_1, M_2, \ldots, M_k$ be closed subspaces of the Hilbert space H, $M = \cap_{i=1}^k M_i$, and $P_i = P_{M_i}$ $(i = 1, \ldots, k)$, $\pi : \mathbb{N} \to \{1, \ldots, k\}$ a quasi-periodic selection, and $R_k = P_{\pi(k)} P_{\pi(k-1)} \cdots P_{\pi(2)} P_{\pi(1)}$. Then R_n converges weakly to P_M, that is for every $x \in H$, $\lim_{n\to\infty} \langle R_n x, y\rangle = \langle P_M x, y\rangle$ (for all $y \in H$).*

Proof. Suppose first that $\cap_{i=1}^k M_i = \{0\}$. For any $x \in H$,

$$R_{n+1} x = P_{\pi(n+1)} R_n x = \left(I - P_{M^\perp_{\pi(n+1)}}\right) R_n x = \left(I - P_{H \ominus M_{\pi(n+1)}}\right) R_n x,$$

that is

$$R_n x = R_{n+1} x + P_{H \ominus M_{\pi(n+1)}} R_n x.$$

The case when $R_n x = 0$ for some n is trivial. Assume that $R_n x \neq 0$ for all n. There can be two cases:
a)

$$\lim_{n\to\infty} \frac{\left\| P_{H\ominus M_{\pi(n+1)}} R_n x \right\|}{\|R_{n+1}x\|} = 0.$$

b) There exists $\varepsilon > 0$ such that for a subsequence $n_p \to \infty$ such that

$$\frac{\left\| P_{H\ominus M_{\pi(n_p+1)}} R_{n_p} x \right\|}{\left\|R_{n_p+1}x\right\|} > \varepsilon.$$

Consider now the last case. Since

$$\left\|R_{n_p}x\right\|^2 = \left\|R_{n_p+1}x\right\|^2 + \left\| P_{H\ominus M_{\pi(n_p+1)}} R_{n_p} x \right\|^2,$$

the last inequality implies

$$\left\|R_{n_p}x\right\|^2 \geq \left(1+\varepsilon^2\right) \left\|R_{n_p+1}x\right\|^2$$

and $\left\|R_{n_p+1}x\right\| \leq \left(1+\varepsilon^2\right)^{-1/2} \left\|R_{n_p}x\right\|$. Hence

$$\lim_{p\to\infty} \left\|R_{n_p}x\right\| = 0.$$

Since $\|R_m x\| \leq \|R_n x\|$ for $m > n$, the sequence $\|R_n x\|$ converges, that is

$$\lim_{n\infty} \|R_n x\| = 0.$$

Note that we obtained strong convergence in case b). We now turn to the case a). We show that from any subsequence of $R_n x$ it is possible to choose a subsequence which weakly converges to the zero element of H. Then we prove that $R_n x \to 0$ weakly. Pick any subsequence $\left\{R_{n_p}x\right\}$ such that $\pi(n_p) = l$ $(1 \leq l \leq k$ fixed). Without any restriction we can assume that $\pi(n_p) = 1$ for all p. Since $R_{n_p}x$ is bounded, it is possible to select a weakly convergent subsequence. For simplicity we also denote this subsequence by $\left\{R_{n_p}x\right\}$ and its weak limit by y. As $R_{n_p}x = P_1 R_{n_p}x$ we can also write that $R_{n_p}x \to P_1 y$ weakly. Hence $y = P_1 y \in M_1$. In case a) the boundedness of the sequence $\{\|R_n x\|\}$ implies that

$$\lim_{n\to\infty} \left\| P_{H\ominus M_{\pi(n+1)}} R_n x \right\| = 0.$$

As the selection π is quasi-periodic there exists an infinite sequence $\{m_p\}$ such that $m_p < K$ and $\pi(n_p + m_p) = P_2$. Then $R_{n_p+m_p}x \in M_2$ and

$$\begin{aligned} R_{n_p+m_p}x &= R_{n_p+m_p-1}x - P_{H\ominus M_{\pi(n_p+m_p)}}R_{n_p+m_p-1}x \\ &= R_{n_p+m_p-2}x - P_{H\ominus M_{\pi(n_p+m_p-1)}}R_{n_p+m_p-2}x \\ &\quad -P_{H\ominus M_{\pi(n_p+m_p)}}R_{n_p+m_p-1}x \\ &= R_{n_p}x - \sum_{j=1}^{m_p} P_{H\ominus M_{\pi(n_p+m_p-j+1)}}R_{n_p+m_p-j}x \\ &= R_{n_p}x - z_p. \end{aligned}$$

But $z_p \to 0$ for $n \to \infty$. Hence $R_{n_p+m_p}x \to y$ weakly. Also by $R_{n_p+m_p}x = P_2 R_{n_p+m_p}x \to P_2 y$ we have $y \in M_2$. Continuing this way we obtain that $y \in M_i$ for $i = 1, \ldots, k$. Thus $y = 0$ by the initial assumption. If $M = \cap_{i=1}^k M_i \neq \{0\}$, then we can reduce the problem to the case $M = \{0\}$ similarly to the proof of Theorem 7.120. ■

Finally we recall the following finite-dimensional result.

THEOREM 7.122 *(Prager, 1960). Let H be a finite-dimensional Hilbert space, $M_1, M_2, \ldots, M_k$ subspaces of H, $M = \cap_{i=1}^k M_i$, and $P_i = P_{M_i}$ $(i = 1, \ldots, k)$, $\pi : \mathbb{N} \to \{1, \ldots, k\}$ a quasi-periodic selection, and*

$$R_k = P_{\pi(k)}P_{\pi(k-1)}\cdots P_{\pi(2)}P_{\pi(1)}. \tag{7.81}$$

Then R_n converges to P_M in norm.

The best result, when π is a periodic selection, is due to Halperin [182].

THEOREM 7.123 *(Halperin, 1962). Let $M_1, M_2, \ldots, M_k$ be closed subspaces of the Hilbert space H, $M = \cap_{i=1}^k M_i$, and $P_i = P_{M_i}$ $(i = 1, \ldots, k)$. For each $x \in H$,*

$$\lim_{n\to\infty} (P_k P_{k-1} \cdots P_1)^n x = P_M x. \tag{7.82}$$

This result is often called the von Neumann-Halperin theorem. An obvious alternative form of this result, which appears in many applications, is the following

$$\lim_{n\to\infty} \left[(I - P_k)(I - P_{k-1}) \cdots (I - P_1)\right]^n x = \left(I - P_{\overline{M_1+M_2+\ldots+M_k}}\right) x \tag{7.83}$$

for $x \in H$. The von Neumann-Halperin theorem remains true, if the M_i are closed varieties (i.e., translates of subspaces) with $\cap_{i=1}^k M_i \neq \emptyset$.

Pyle [297] proved the following extension of the von Neumann theorem using the spectral decomposition [178].

THEOREM 7.124 *(Pyle, 1967). Let M_i be a linear subspace of $\mathbb{C}^n$, and let P_i be the corresponding orthogonal projection onto M_i for $i = 1, \ldots, k$. Let $M = \cap_{i=1}^k M_i$ and $P = P_M$. Furthermore let*

$$Q = P_{i_k}P_{i_{k-1}}\cdots P_{i_2}P_{i_1}P_{i_2}\cdots P_{i_{k-1}}P_{i_k},$$

where $(i_1, \ldots, i_k)$ *is any permutation of integers* $1, \ldots, k$. *Then*

$$\lim_{n\to\infty} Q^n = P_M. \tag{7.84}$$

The next result gives a closed formula for the intersection projection P_M in terms of the powers of Q and the Moore-Penrose inverse.

THEOREM 7.125 *(Pyle). Under the conditions of the previous theorem*

$$P_M = Q^\nu + \left[\left(Q^{\nu+1} - Q^\nu\right)^+ - \left(Q^\nu - Q^{\nu-1}\right)^+\right] \quad (\nu = 2, 3, \ldots). \tag{7.85}$$

Different applications such as optimization or discretization methods motivated further extensions of the von Neumann-Halperin theorem to contractions or nonexpansive operators. A linear operator $T : H \to H$ is said to be a contraction if $\|T\| \leq 1$. The orthogonal projections are clearly contractions.

Note that the relaxed Kaczmarz method contained the term $I - \mu P_i$, where P_i is an orthogonal projection. The matrix $I - \mu P_i$ is contractive in the range $0 < \mu < 2$ and projection only for $\mu = 1$.

Amemiya and Ando [13] extended the von Neumann-Halperin result to random products of contractions. Let T be a contraction in H and make use of the following criterion.

(W) $\|f_n\| \leq 1$, $\|Tf_n\| \to 1$ *imply* $(I - T) f_n \to 0$ *weakly.*

A positive semidefinite contraction, in particular, a projection, has the property (W). In fact, if T is a positive semidefinite contraction, then

$$\begin{aligned} \|(I-T) f_n\|^2 &= \left\langle (I-T)^2 f_n, f_n \right\rangle \\ &\leq \langle (I-T)(I+T) f_n, f_n \rangle = \|f_n\|^2 - \|Tf_n\|^2 \to 0 \end{aligned}$$

whenever $\|f_n\| \leq 1$ and $\|Tf_n\| \to 1$. The condition (W) has a simpler equivalent form (W'):

(W') $\|Tf\| = \|f\| \Rightarrow Tf = f$.

THEOREM 7.126 *(Amemiya, Ando, 1965). If* T_j *is a contraction with (W) or equivalently (W')* $(j = 1, \ldots, k)$*, then for any random selection* $\pi : N \to \{1, \ldots, k\}$ *the sequence*

$$S_n = T_{\pi(n)} \cdots T_{\pi(2)} T_{\pi(1)} \tag{7.86}$$

converges weakly as $n \to \infty$.

When every index j appears infinitely many times in a selection π, the limit of the sequence S_n is the projection onto the subspace of vectors invariant under all T_j $(j = 1, 2, \ldots, k)$. In fact, S_n converges weakly to $P_{\mathcal{N}(I - T_k \cdots T_2 T_1)}$.

There are several similar results (for references, see e.g., [61], [32]). Here we recall the following results of Browder [61], which show the extent of generalizations.

Let X be a real Banach space, C a closed convex subset of X. Consider the infinite sequence $\{V_j\}$ of mappings of C into C, where the V_j, in general, are nonlinear and need not be pairwise distinct. Let

$$S_n(x) = V_n V_{n-1} \dots V_2 V_1(x). \tag{7.87}$$

Let F be a nonempty subset of C which is left pointwise fixed by all the mappings V_j.

DEFINITION 7.127 *The sequence of mappings $\{V_j\}$ of C into C is an admissible sequence if the following condition holds: $0 \in C$, $V_j(0) = 0$ for all j, and there exist two continuous strictly increasing functions ϕ and ψ from $\mathbb{R}^+ = \{r \mid r \in \mathbb{R}^1, r \geq 0\}$ to $\mathbb{R}^+$ with $\phi(0) = 0$, $\psi(0) = 0$, such that for all $x \in C$ and all $j \geq 1$:*

$$\phi(\|V_j(x)\|) + \psi(\|x - V_j(x)\|) \leq \phi(\|x\|). \tag{7.88}$$

DEFINITION 7.128 *The mapping $U : C \to X$ $(C \subset X)$ is said to be a contractive mapping (in the weak sense) or a nonexpansive mapping, if for all $x, y \in C$,*

$$\|U(x) - U(y)\| \leq \|x - y\|. \tag{7.89}$$

DEFINITION 7.129 *If V is a mapping of C into X, denote by $F(V)$ the fixed point set of V in C, i.e., the set of all $x \in C$ such that $V(x) = x$.*

THEOREM 7.130 *(Browder). Let H be a Hilbert space, C a closed convex subset of H which contains 0, $\{V_j\}$ an admissible infinite sequence of nonexpansive mappings of C into C in the sense of Definition 7.127. Let $\{U_k\}$ be a finite or infinite sequence of nonexpansive mappings of C into H, and let*

$$F = \cap_k F(U_k).$$

Suppose that for all $f \in F$ and all j, $V_j(f) = f$, while $0 \in F$. Suppose also that the following recurrence condition holds:
(R) For each bounded subset C_0 of C, for each integer k, and for each $\varepsilon > 0$ there exists a positive integer $m(\varepsilon, k)$ such that from each bloc of successive integers $\{n+1, \dots, n+m(\varepsilon, k)\}$, we can choose an integer j such that for all $x \in C_0$,

$$\|V_j(x) - U_k(x)\| < \varepsilon. \tag{7.90}$$

Let $S_n(x) = V_n V_{n-1} \dots V_1(x)$ for $x \in C$. Then for each $x \in C$, the sequence $S_n(x)$ converges weakly to an element f of F.

For special cases of Theorem 7.130, see [61].

THEOREM 7.131 *(Browder). Let H be a Hilbert space, C a closed convex subset of H. Let $\{C_j\}_{j\in A}$ be a sequence (finite or infinite) of closed convex subsets of H, (where $A \subset \mathbb{N}$). Suppose that*

$$C = \cap_{j\in A} C_j.$$

Let $\pi : \mathbb{N} \to A$ be an admissible selection, and set

$$V_k = R_{C_{\pi(k)}},$$

where R_{C_j} is the nonlinear projection of H on C_j given by Definition 7.46. Then, if $S_n(x) = V_n V_{n-1} \dots V_1(x)$ for a given $x \in H$, it follows that $S_n(x)$ converges weakly to an element y_0 of C (with y_0 depending on the choice of the initial value x).

THEOREM 7.132 *(Browder). Suppose that under the hypotheses of Theorem 7.131 we have in addition that A is finite, $0 \in C$, and that for a given $\kappa > 0$, there exists a constant k_0 such that for each $x \in \overline{S(0,\kappa)} = \{x \mid \|x\| \leq \kappa\}$, we have*

$$\|x\| \leq k_0 \max_{j\in A} \left\| x - R_{C_j}(x) \right\|.$$

Then $C = \{0\}$, and for each $x \in \overline{S(0,\kappa)}$, $S_n(x)$ converges strongly to 0, uniformly over x in $\overline{S(0,\kappa)}$.

The subspaces are convex sets. Browder gives the following application of Theorem 7.132

THEOREM 7.133 *(Hildebrandt, Stummel). Let $M_1, \dots, M_k$ be closed subspaces of the Hilbert space H, $M = \cap_{i=1}^{k} M_i$, P_j the orthogonal projection of H onto M_j, P_M the orthogonal projection of H on M. Let $\pi : \mathbb{N} \to \{1, \dots, k\}$ be an admissible selection, and suppose that $M^{\perp} = \sum_{j=1}^{k} M_j^{\perp}$. If $S_n(x) = P_{\pi(n)} P_{\pi(n-1)} \cdots P_{\pi(1)} x$, then $S_n(x)$ converges in norm to $P_M(x)$.*

In fact, the generalizations of the von Neumann theorem went on different roads and rather independently (see [61], [292], [103] and [32]). The following result, whose proof is based on various matrix techniques, is an interesting late comer. Nelson and Neumann [264] developed the following concepts and result which they applied in the SOR theory of iterative methods for solving linear systems.

For $A \in \mathbb{C}^{n\times n}$ let

$$\gamma(A) = \max\{|\lambda| \mid \lambda \in \{0\} \cup \sigma(A) \setminus \{1\}\}. \tag{7.91}$$

DEFINITION 7.134 *A matrix $B \in \mathbb{C}^{n\times n}$ is called paracontracting if the spectral norm of B is bounded by unity and if*

$$0 \neq x \in \mathcal{N}^{\perp}(I - B) \Rightarrow \|Bx\|_2 < \|x\|_2.$$

If a matrix B is paracontracting, then the contraction constant

$$c(B) = \inf\left\{c \in [0,\infty) \mid \forall x \in \mathcal{N}^{\perp}(I-B),\ \|Bx\|_2 \leq c\|x\|_2\right\} \tag{7.92}$$

satisfies $0 \leq c(B) < 1$. If B is Hermitian, then $c(B) = \gamma(B)$. For any orthogonal projection P, $c(P) = 0$.

THEOREM 7.135 *(Nelson, Neumann). Let $B = B_k \cdots B_1$ be the product of k paracontracting matrices $B_i \in \mathbb{C}^{n\times n}$. Then B is paracontracting,*

$$\gamma(B) \leq c(B) < 1 \tag{7.93}$$

and

$$\left\|B^m x - P_{\mathcal{N}(I-B)}x\right\|_2 \leq c^m(B)\left\|x - P_{\mathcal{N}(I-B)}x\right\|_2 \tag{7.94}$$

hold. Furthermore

$$\mathcal{N}(I-B) = \cap_{i=1}^{k}\mathcal{N}(I-B_i) \tag{7.95}$$

and

$$c(B) \leq \left\{1 - \prod_{i=1}^{k}\left[1-c^2(B_i)\right]\prod_{i=1}^{k-1}\sin^2\theta_i\right\}^{1/2}, \tag{7.96}$$

where θ_i denotes the angle between the subspaces

$$\mathcal{N}(I-B_i) \quad \text{and} \quad \cap_{i=i+1}^{k}\mathcal{N}(I-B_i). \tag{7.97}$$

In particular, if $\gamma(B_i) = c(B_i)$ for $i = 1,\ldots,k$, then

$$\gamma(B) \leq \left\{1 - \prod_{i=1}^{k}\left[1-\gamma^2(B_i)\right]\prod_{i=1}^{k-1}\sin^2\theta_i\right\}^{1/2}. \tag{7.98}$$

The theorem extends the von Neumann-Halperin theorem in a special case and combines it with the speed estimate of Smith, Solmon and Wagner [322] to be discussed later.

3.2 Estimates for the convergence speed

We turn now to the estimates for the speed of convergence in the von Neumann-Halperin theorem.

Such estimates were given by Aronszajn [22], Smith, Solmon and Wagner [322] (see also [184]), Deutsch [102], Franchetti and Light [127], Kayalar and Weinert [221], Deutsch and Hundal [105]. The estimates make use of the concept of angle between subspaces. We follow Deutsch and Hundal [105] in the exposition of the results.

THEOREM 7.136 *(Aronszajn, 1950). For each $x \in H$, and for any integer $n \geq 1$,*

$$\|(P_2P_1)^n x - P_M x\| \leq c^{2n-1}\|x\|, \tag{7.99}$$

where $c = c(M_1, M_2)$.

Unaware of Aronszajn's result several authors devoloped similar, but weaker estimates. Youla [365] proved the bound

$$\|(P_2P_1)^n - P_M\| \leq \frac{c(M_1, M_2)^n}{1 - c(M_1, M_2)}. \tag{7.100}$$

Franchetti and Light [127] gave the upper estimate

$$\|(P_2P_1)^n - P_M\| \leq c(M_1, M_2)^n \tag{7.101}$$

while proving that the convergence speed can be arbitrarily slow if $M_1 + M_2$ is not closed. Deutsch [102] proved the following result.

THEOREM 7.137 *(Deutsch, 1984). Assume that M_1 and M_2 are closed subspaces of H. If $c(M_1, M_2) < 1$, then for each $x \in H$,*

$$\|(P_2P_1)^n x - P_M x\| \leq \frac{c(M_1, M_2)^{2n-1}}{1 - c(M_1, M_2)^2} \|x - P_2P_1x\| \quad (n = 1, 2, \ldots). \tag{7.102}$$

Finally, Kayalar and Weiner [221] proved that the Aronszajn bound is sharp.

Smith, Solmon and Wagner [322] obtained the first bounds for $k \geq 2$. They used the Friedrichs angle definition $c(M_1, M_2)$. A plane in H is defined by

$$f + M = \{f + x \mid x \in M\},$$

where M is a subspace. The angle between two closed planes is defined as the angle between the corresponding subspaces parallel to the planes. Let M be a closed subspace of H and P_M the orthogonal projection on M. The orthogonal projection R on $f + M$ is defined by

$$Rx = f + P_M(x - f) \quad (x \in H). \tag{7.103}$$

It is easy to see that for any $y \in f + M$, $x - y = (x - f) - z$ $(z \in M)$ and

$$\|x - y\| = \|(x - f) - z\| \geq \|(x - f) - P_M(x - f)\| = \|x - Rx\|.$$

Hence R is indeed the projection operator on the closed plane $f + M$.

THEOREM 7.138 *(Smith, Solmon, Wagner, 1977). For $j = 1, \ldots, k$, let R_j be the orthogonal projection on $f + M_j$, where M_j is a closed subspace of H. Define $M = \cap_{i=1}^k M_i$ and let R be the orthogonal projection on $f + M$. If $\theta_j = \alpha\left(M_j, \cap_{i=j+1}^k M_i\right)$, then for any $x \in H$, and integer $n \geq 1$,*

$$\|(R_k \cdots R_2R_1)^n x - Rx\| \leq c^n \|x - Rx\| \tag{7.104}$$

with $c = \left(1 - \prod_{j=1}^{k-1} \sin^2\theta_j\right)^{1/2}$.

Proof. For $j = 1, \ldots, k$, $R_j x = f + P_{M_j}(x - f)$. Hence for any $x \in H$, $(R_k \cdots R_1)^n x = f + (P_{M_k} \cdots P_{M_1})^n (x - f)$ and $Rx = f + P_M(x - f)$. Thus the inequality to be proved becomes

$$\|(P_{M_k} \cdots P_{M_1})^n (x - f) - P_M(x - f)\| \le c^n \|x - f - P_M(x - f)\|,$$

where $x - f \in H$ is arbitrary. Therefore, we can assume that $f = 0$ without loss of generality. Since $P_M x$ lies in M and $Q = P_{M_k} \ldots P_{M_1}$ is the identity (invariant) on M, the inequality can be written in the form

$$\|Q^n(x - P_M x)\|^2 \le \left(1 - \prod_{j=1}^{k-1} \sin^2\theta_j\right)^n \|x - P_M x\|^2. \tag{7.105}$$

For $x \in M$, both sides of the inequality are zero and hence it is true. For any $x \in H$, set $v = x - P_M x \in M^\perp$. The inequality then becomes

$$\|Q^n v\|^2 \le \left(1 - \prod_{j=1}^{k-1} \sin^2\theta_j\right)^n \|v\|^2.$$

We now show that Q is invariant on $M^\perp$. It is obvious that $P_{M_1} \cdots P_{M_k}$ is invariant (identity) on M. Let $a \in M$ and $b \in M^\perp$. Then $\langle Qb, a\rangle = \langle b, P_{M_1} \cdots P_{M_k} a\rangle = \langle b, a\rangle = 0$. Hence $Qb \in M^\perp$ and Q is invariant on $M^\perp$. Thus the inequality

$$\|Qv\|^2 \le \left(1 - \prod_{j=1}^{k-1} \sin^2\theta_j\right) \|v\|^2 \quad \left(v \in M^\perp\right) \tag{7.106}$$

implies

$$\|Q^n v\|^2 = \left\|Q\left(Q^{n-1}v\right)\right\|^2 \le c\left\|Q^{n-1}v\right\|^2 \le \cdots \le c^{n-1}\|Qv\|^2 \le c^n \|v\|^2,$$

which is the above inequality. We prove $\|Qv\|^2 \le c\|v\|^2$ by induction on k. The case $k = 1$ is evident as both sides are zero. Let $\widetilde{M} = \cap_{i=2}^k M_i$ and $\widetilde{Q} = P_{M_k} \cdots P_{M_2}$. Then $M = \widetilde{M} \cap M_1$ and $Q = \widetilde{Q} P_{M_1}$. Decompose $v \in M^\perp$ in the form $v = v_1 + v_1^\perp$, where $v_1 \in M_1$ and $v_1^\perp \in M_1^\perp$. Then $Qv = \widetilde{Q} v_1$. As $M \subset M_1$, $M^\perp \supset M_1^\perp$, $v_1 = v - v_1^\perp \in M^\perp$. Hence $v_1 \in M_1 \cap \left(M_1 \cap \widetilde{M}\right)^\perp$. Now let $v_1 = w_1 + w_1^\perp$, where $w_1 \in \widetilde{M}$ and $w_1^\perp \in \widetilde{M}^\perp$. Then $\widetilde{Q} v_1 = w_1 + \widetilde{Q} w_1^\perp$. Since $\widetilde{Q}$ is invariant on $\widetilde{M}^\perp$, $\widetilde{Q} w_1^\perp \perp w_1$. Hence

$$\left\|\widetilde{Q} v_1\right\|^2 = \|w_1\|^2 + \left\|\widetilde{Q} w_1^\perp\right\|^2.$$

As $M \subset \widetilde{M}$, $M^{\perp} \supset \widetilde{M}^{\perp}$ and $w_1 = v_1 - w_1^{\perp} \in M^{\perp}$. Hence $w_1 \in \widetilde{M} \cap \left(M_1 \cap \widetilde{M}\right)^{\perp}$. Assume that

$$\left\|\widetilde{Q}w_1^{\perp}\right\|^2 \leq \left(1 - \prod_{j=2}^{k-1} \sin^2\theta_j\right) \left\|w_1^{\perp}\right\|^2.$$

Then $\|v_1\|^2 = \|w_1\|^2 + \left\|w_1^{\perp}\right\|^2$ and the assumption imply that

$$\begin{aligned}\left\|\widetilde{Q}v_1\right\|^2 &\leq \|w_1\|^2 + \left(1 - \prod_{j=2}^{k-1} \sin^2\theta_j\right) \left\|w_1^{\perp}\right\|^2 \\ &= \left(1 - \prod_{j=2}^{k-1} \sin^2\theta_j\right) \|v_1\|^2 + \left(\prod_{j=2}^{k-1} \sin^2\theta_j\right) \|w_1\|^2.\end{aligned}$$

Now $v_1 \in M_1 \cap \left(M_1 \cap \widetilde{M}\right)^{\perp}$ and $w_1 \in \widetilde{M} \cap \left(M_1 \cap \widetilde{M}\right)^{\perp}$. Since $\alpha\left(M_1, \widetilde{M}\right) = \theta_1$ it follows that $\|w_1\|^2 = \langle v_1, w_1\rangle \leq (\cos\theta_1)\|v_1\|\|w_1\|$ and hence that $\|w_1\| \leq (\cos\theta_1)\|v_1\|$. Combining this with the last formula we obtain

$$\begin{aligned}\|Qv\|^2 = \left\|\widetilde{Q}v_1\right\|^2 &\leq \left(1 - \prod_{j=2}^{k-1} \sin^2\theta_j\right) \|v_1\|^2 + \left(\prod_{j=2}^{k-1} \sin^2\theta_j\right) \left(\cos^2\theta_1\right) \|v_1\|^2 \\ &\leq \left(1 - \prod_{j=1}^{k-1} \sin^2\theta_j\right) \|v_1\|^2.\end{aligned}$$

Using the facts that $Qv = \widetilde{Q}v_1$ and $\|v_1\| \leq \|v\|$, we get the required inequality. ∎

A related discussion of the above result can be found in Hamaker and Solmon [184]. The estimate of Smith, Solmon and Wagner was improved by Kayalar and Weinert [221] and Deutsch and Hundal [105]. Both results are extremely complicated. We quote the following somewhat simpler result [105].

THEOREM 7.139 *(Deutsch, Hundal, 1997). Let $M_1, M_2, \ldots, M_k$ be closed subspaces, $M = \cap_{i=1}^k M_i$, $\gamma_{ij} = c_0\left(M_i \cap M^{\perp}, M_j \cap M^{\perp}\right)$,*

$$\begin{aligned}\alpha &= \min\left\{\left[1 - \prod_{i=1}^{k-1}\left\{1 - c^2\left(M_i, \cap_{j=i+1}^k M_j\right)\right\}\right]^{1/2},\right. \\ &\qquad \left.\left[1 - \prod_{i=1}^{k-1}\left\{1 - c^2\left(M_{k+1-i}, \cap_{j=1}^{k-i} M_j\right)\right\}\right]^{1/2}\right\},\end{aligned}$$

and

$$\beta = \left[\alpha^2 c^2 (M_1, M_k) + \left(1 - c^2 (M_1, M_k)\right) \mu^2\right]^{1/2},$$

where

$$\begin{aligned} \mu^2 &= \min\Big\{\Big[1 - \textstyle\prod_{i=1}^{k-1}\Big\{1 - c^2\Big(M_i, \cap_{j=i+1}^{k} M_j \cap M_1\Big)\Big\}\Big], \\ &\qquad 1 - \Big[1 - c^2\Big(M_1 \cap M_k, \cap_{j=1}^{k-1} M_j\Big)\Big] \\ &\qquad\quad \times \textstyle\prod_{i=2}^{k-1}\Big[1 - c^2\Big(M_{k+1-i}, \cap_{j=1}^{k-i} M_j\Big)\Big]\Big\}. \end{aligned}$$

Then

$$\|(P_{M_k} \cdots P_{M_1})^n - P_M\| \leq \min\left\{\alpha\beta^{n-1}, \alpha^n \gamma_{1k}^{n-1}, \gamma_{12}^n \gamma_{23}^n \cdots \gamma_{k-1,k}^n, \gamma_{1k}^{n-1}\right\} \tag{7.107}$$

for $n = 1, 2, \ldots$.

Deutsch and Hundal [105] also showed that none of the previous estimates is sharp for $k \geq 3$.

References

[1] Abaffy, J., C. Broyden, and E. Spedicato: 1984, 'A class of direct methods for linear equations'. *Numerische Mathematik* **45**, 361–376.

[2] Abaffy, J. and A. Galántai: 1987, 'Conjugate Direction Methods for Linear and Nonlinear Systems of Algebraic Equations'. In: P. Rózsa (ed.): *Colloquia Mathematica Soc. János Bolyai, 50. Numerical Methods, Miskolc (Hungary) 1986.* Amsterdam, pp. 481–502.

[3] Abaffy, J., A. Galántai, and E. Spedicato: 1987, 'The local convergence of ABS methods for nonlinear algebraic equations'. *Numerische Mathematik* **51**, 429–439.

[4] Abaffy, J. and E. Spedicato: 1989, *ABS-Projection Algorithms: Mathematical Techniques for Linear and Nonlinear Algebraic Equations.* Chichester: Ellis Horwood.

[5] Abatzoglou, T.: 1978, 'The Minimum Norm Projection on C^2-Manifolds in R^n.'. *Trans. Amer. Math. Soc.* **243**, 115–122.

[6] Achieser, N. and I. Glasmann: 1954, *Theorie der Linearen Operatoren im Hilbert Raum.* Berlin: Akademie-Verlag.

[7] Adams, L. and J. Nazareth (eds.): 1996, 'Linear and Nonlinear Conjugate Gradient-Related Methods'. Philadelphia:, SIAM.

[8] Afriat, S.: 1957, 'Orthogonal and oblique projectors and the characteristics of pairs of vector spaces'. *Proc. Cambridge Phil. Soc.* **53**, 800–816.

[9] Agmon, S.: 1954, 'The Relaxation Method for Linear Inequalities'. *Canad. J. Math.* **6**, 382–392.

[10] Albert, A.: 1972, *Regression and the Moore-Penrose Pseudoinverse.* New York: Academic Press.

[11] Altman, M.: 1957a, 'An approximation process for the Gaussian least squares principle in the error theory'. *Bulletin de l'Académie Polonaise des Sciences, Cl. III* **5**(4), 371–374.

[12] Altman, M.: 1957b, 'On the approximate solution of linear algebraic equations'. *Bulletin de l'Académie Polonaise des Sciences, Cl. III* **5**(4), 365–370.

[13] Amemiya, I. and T. Ando: 1965, 'Convergence of random products of contractions in Hilbert space'. *Acta Sci. Math.* **26**, 239–244.

[14] Anderson, W.: 1971, 'Shorted operators'. *SIAM J. Appl. Math.* **20**(3), 520–525.

[15] Anderson, W. and R. Duffin: 1969, 'Series and parallel addition of matrices'. *Journal of Mathematical Analysis and Applications* **26**, 576–594.

[16] Anderson, W., E. Harner, and G. Trapp: 1985, 'Eigenvalues of the difference and product of projections'. *Linear and Multilinear Algebra* **17**, 295–299.

[17] Anderson, W. and M. Schreiber: 1972, 'The infimum of two projections'. *Acta Sci. Math.* **33**(3–4), 165–168.

[18] Anderson, W. and G. Trapp: 1975, 'Shorted operators. II'. *SIAM J. Appl. Math.* **28**(1), 60–71.

[19] Ando, T.: 1966, 'Contractive projections in L_p-spaces'. *Pacific J. Math.* **17**(3), 391–405.

[20] Andruchow, E., G. Corach, and D. Stojanoff: 1999, 'Geometry of oblique projections'. *Studia Mathematica* **137**(1), 61–79.

[21] Ansorge, R.: 1984, 'Connections Between the Cimmino-Method and the Kaczmarz-Method for the Solution of Singular and Regular Systems of Equations'. *Computing* **33**, 367–375.

[22] Aronszajn, N.: 1950, 'Theory of reproducing kernels'. *Trans. Amer. Math. Soc.* **68**, 337–404.

[23] Asplund, E.: 1973, 'Differentiability of the Metric Projection in Finite-Dimensional Euclidean Space'. *Proc. Amer. Math. Soc.* **38**(1), 218–219.

[24] Atkinson, K. and F. Potra: 1987, 'Projection and iterated projection methods for nonlinear integral equations'. *SIAM J. Numer. Anal.* **24**(6), 1352–1373.

[25] Baksalary, J. and O. Baksalary: 2000, 'Idempotency of linear combinations of two idempotent matrices'. *Linear Algebra and its Applications* **321**, 3–7.

[26] Baksalary, J. and O. Baksalary: 2002, 'Commutativity of projectors'. *Linear Algebra and its Applications* **341**, 129–142.

[27] Balakrishnan, A.: 1976, *Applied Functional Analysis*. Springer-Verlag.

[28] Ballalij, M. and H. Sadok: 1998, 'New Interpretation of Related Huang's Methods'. *Linear Algebra and its Applications* **269**, 183–195.

[29] Banerjee, K.: 1964, 'A note on idempotent matrices'. *Ann. Math. Stat.* **35**, 880–882.

[30] Bapat, R., S. K. Jain, and L. Snyder: 1997, 'Nonnegative idempotent matrices and the minus partial order'. *Linear Algebra and its Applications* **261**, 143–154.

[31] Bauer, F. and A. Householder: 1960, 'Some inequalities involving the euclidean condition of a matrix'. *Numerische Mathematik* **2**, 308–311.

[32] Bauschke, H. and J. Borwein: 1996, 'On projection algorithms for solving convex feasibility problems'. *SIAM Review* **38**(3), 367–426.

[33] Bazaraa, M., H. Sherali, and C. Shetty: 1993, *Nonlinear Programming: Theory and Algorithms.* Wiley.

[34] Bazaraa, M. and C. Shetty: 1979, *Nonlinear Programming Theory and Algorithms.* Wiley.

[35] Beiko, I., B. Bublik, and P. Zin'ko: 1983, *Metodi I Algoritmi Resheniya Zadach Optimizatsii.* Kiev: Vishcha Shkola. in Russian.

[36] Belitskii, G. and Y. Lyubich: 1988, *Matrix Norms and their Applications.* Birkhäuser Verlag.

[37] Belluce, L. and W. Kirk: 1967, 'Nonexpansive Mappings and Fixed-Points in Banach Spaces'. *Illinois J. Math.* **11**, 474–479.

[38] Belousov, E.: 1977, *Vvedenie V Vipuklii Analiz I Tselochislennoe Programmirovanie.* Moscow University. in Russian.

[39] Ben-Israel, A.: 1967, 'On the geometry of subspaces in Euclidean n-spaces'. *SIAM J. Appl. Math.* **15**(5), 1184–1198.

[40] Ben-Israel, A. and T. Greville: 1974, *Generalized Inverses: Theory and Applications.* John Wiley and Sons.

[41] Benzi, M. and C. Meyer: 1995, 'A direct projection method for sparse linear systems'. *SIAM J. Sci. Comput.* **16**(5), 1159–1176.

[42] Berkson, E.: 1963, 'Some metrics on the subspaces of a Banach space'. *Pacific. J. Math.* **13**, 7–22.

[43] Berman, A. and R. Plemmons: 1994, *Nonnegative Matrices in the Matematical Sciences.* SIAM.

[44] Bertsekas, D.: 1976, 'On the Goldstein-Levitin-Polyak Gradient Projection Method'. *IEEE Transactions on Automatic Control* **AC-21**(2), 174–184.

[45] Bertsekas, D.: 1982, 'Projected Newton Methods for Optimization Problems with Simple Constraints'. *SIAM J. Control and Optimization* **20**(2), 221–246.

[46] Bhatia, R.: 1997, *Matrix Analysis.* New York: Springer-Verlag.

[47] Björck, A. and T. Elfving: 1979, 'Accelerated projection methods for computing pseudoinverse solutions of systems of linear equations'. *BIT* **19**, 145–163.

[48] Björck, A. and G. Golub: 1973, 'Numerical methods for computing angles between linear subspaces'. *Mathematics of Computation* **27**(123), 579–594.

[49] Bodewig, E.: 1959, *Matrix Calculus.* Amsterdam: North-Holland, second edition.

[50] Bosznay, A.: 1986, 'A Remark on the Alternating Algorithm'. *Periodica Mathematica Hungarica* **17**(3), 241–244.

[51] Bosznay, A. and B. Garay: 1986, 'On norms of projections'. *Acta Sci. Math.* **50**(1–2), 87–92.

[52] Boullion, T. and P. Odell: 1971, *Generalized Inverse Matrices.* John Wiley and Sons.

[53] Bramley, R. and A. Sameh: 1992, 'Row projection methods for large nonsymmetric linear systems'. *SIAM J. Sci. Stat. Comput.* **13**(1), 168–193.

[54] Bregman, L.: 1965, 'Nakhozhdeniye Obshchei Tochki Vipuklikh Mnozhestv Metodom Posledovatel'nogo Proektirovaniya'. *Dokl. Akad. Nauk. SSSR* **162**(3), 487–490. in Russian.

[55] Bregman, L.: 1967, 'Relaksatsionnii Metod Nakhozhdeniya Obshei Tochki Vipuklikh Mnozhestv I Ego Primeneniye Dlya Resheniya Zadach Vipuklogo Programmirovaniya'. *Zhurnal Vichislitel'noi Matematiki i Matematicheskoi Fiziki* **7**(3), 620–631. in Russian.

[56] Brezinski, C.: 1997a, 'Projection methods for linear systems'. *Journal of Computational and Applied Mathematics* **77**, 35–51.

[57] Brezinski, C.: 1997b, *Projection Methods for Systems of Equations*. Elsevier.

[58] Brezinski, C.: 1999, 'Multiparameter descent methods'. *Linear Algebra and its Applications* **296**, 113–141.

[59] Brezinski, C.: 2000, 'Acceleration procedures for matrix iterative methods'. *Numerical Algorithms* **25**, 63–73.

[60] Browder, F.: 1958, 'On some Approximation Methods for Solutions of the Dirichlet Problem for Linear Elliptic Equations of Arbitrary Order'. *Journal of Mathematics and Mechanics* **7**(1), 69–80.

[61] Browder, F.: 1967a, 'Convergence theorems for sequences of nonlinear operators in Banach spaces'. *Math. Zeitschr.* **100**, 201–225.

[62] Browder, F.: 1967b, 'Nonlinear Mapping of Nonxpansive and Accretive Type in Banach Spaces'. *Bull. Amer. Math. Soc.* **73**, 875–882.

[63] Browder, F. and W. Petryshyn: 1966, 'The Solution by Iteration of Nonlinear Functional Equations in Banach Spaces'. *Bull. Amer. Math. Soc.* **72**, 571–575.

[64] Browder, F. and W. Petryshyn: 1967, 'Construction of Fixed Points of Nonlinear Mappings in Hilbert Space'. *Journal of Mathematical Analysis and Applications* **20**, 197–228.

[65] Brown, P.: 1987, 'A local convergence theory for combined inexact-Newton/finite-difference projection methods'. *SIAM J. Numer. Anal.* **24**(2), 407–434.

[66] Broyden, C.: 1985, 'On the numerical stability of Huang's and related methods'. *JOTA* **47**, 401–412.

[67] Broyden, C.: 1993, 'Block Conjugate Gradient Methods'. *Optimization Methods and Software* **2**, 1–17.

[68] Broyden, C.: 1996, 'A New Taxonomy of Conjugate Gradient Methods'. *Computers and Mathematics with Applications* **31**(4/5), 7–17.

[69] Broyden, C.: 1997, 'The Gram-Schmidt Method - a Hierarchy of Algorithms'. In: A. Sydow (ed.): *15th IMACS World Congress on Scientific Computation, Modelling and Applied Mathematics, Vol. 2, Numerical Mathematics*. Berlin, pp. 545–550.

[70] Bruck, R.: 1973, 'Properties of Fixed-Point Sets of Nonexpansive Mappings in Banach Spaces'. *Trans. Amer. Math. Soc.* **179**, 251–262.

[71] Buckholtz, D.: 1997, 'Inverting the difference of Hilbert space projections'. *Am. Math. Monthly* **104**, 60–61.

[72] Buckholtz, D.: 1999, 'Hilbert space idempotents and involutions'. *Proc. Am. Math. Soc.* **128**(5), 1415–1418.

[73] Calamai, P. and J. Moré: 1987, 'Projected Gradient Methods for Linearly Constrained Problems'. *Mathematical Programming* **39**, 93–116.

[74] Campbell, S.: 1976, 'Differentiation of the Drazin Inverse'. *SIAM J. Appl. Math.* **30**(4), 703–707.

[75] Campbell, S. and C. Meyer: 1991, *Generalized Inverses of Linear Transformations.* New York: Dover.

[76] Carlson, D.: 1975, 'Matrix Decompositions Involving the Schur Complement'. *SIAM J. Appl. Math.* **28**(3), 577–587.

[77] Censor, Y.: 1981, 'Row-action methods for huge and sparse systems and their applications'. *SIAM Review* **23**(4), 444–466.

[78] Censor, Y.: 1988, 'Parallel application of block-iterative methods in medical imaging and radiation therapy'. *Mathematical Programming* **42**(307–325).

[79] Censor, Y., N. Cohen, T. Kutscher (Kotzer), and J. Shamir: 2002, 'Summed square distance error reduction by simultaneous multiprojections and applications'. *Applied Mathematics and Computation* **126**, 157–179.

[80] Censor, Y., P. Eggermont, and D. Gordon: 1983, 'Strong underrelaxation in Kaczmarz's method for inconsistent system'. *Numerische Mathematik* **41**, 83–92.

[81] Censor, Y. and T. Elfving: 1982, 'New Methods for Linear Inequalities'. *Linear Algebra and its Applications* **42**, 199–211.

[82] Censor, Y. and A. Lent: 1981, 'An Iterative Row-Action Method for Interval Convex Programming'. *JOTA* **34**(3), 321–353.

[83] Censor, Y. and S. Zenios: 1997, *Parallel Optimization: Theory, Algorithms and Applications.* Oxford University Press.

[84] Chatelin, F.: 1993, *Eigenvalues of Matrices.* John Wiley and Sons.

[85] Chipman, J. and M. Rao: 1964, 'Projections, generalized inverses, and quadratic forms'. *Journal of Mathematical Analysis and Applications* **9**(1), 1–11.

[86] Chu, K.: 1986, 'Generalization of the Bauer-Fike Theorem'. *Numerische Mathematik* **49**, 685–691.

[87] Chu, M., R. Funderlic, and G. Golub: 1995, 'A Rank-One Reduction Formula and its Applications to Matrix Factorizations'. *SIAM Review* **37**, 512–530.

[88] Cimmino, G.: 1938, 'Calcolo Approssimato Per Le Soluzioni Dei Sistemi Di Equazioni Lineari'. *La Ricerca Scientifica* **II**(1), 326–333.

[89] Corach, G., A. Maestripieri, and D. Stojanoff: 2001, 'Oblique projections and Schur complements'. *Acta. Sci. Math. (Szeged)* **67**, 337–356.

[90] Corach, G., A. Maestripieri, and D. Stojanoff: 2002, 'Generalized Schur complements and oblique projections'. *Linear Algebra and its Applications* **341**(259–272).

[91] Crombez, G.: 1991, 'Image recovery by convex combinations of projections'. *Journal of Mathematical Analysis and Applications* **155**, 413–419.

[92] Csikós, B.: 2000, 'On the Principal Angles Between the Osculating K-Planes of a Curve'. *Acta Math. Hungar.* **86**(3), 261–273.

[93] Davis, C. and W. Kahan: 1969, 'Some new bounds on perturbation of subspaces'. *Bull. Am. Math. Soc.* **75**(4), 863–868.

[94] Davis, C. and W. Kahan: 1970, 'The rotation of eigenvectors by a perturbation. III'. *SIAM J. Numer. Anal.* **7**(1), 1–46.

[95] Davis, P.: 1975, *Interpolation and Approximation.* New York: Dover.

[96] Dax, A.: 1990, 'The convergence of linear stationary iterative processes for solving singular unstructured systems of linear equations'. *SIAM Review* **32**(4), 611–635.

[97] de Oliveira, G.: 1967, 'Matrices whose sum is the identity matrix'. *Gazeta de Matematica* (105-108), 24–25.

[98] De Pierro, A. and A. Iusem: 1985, 'A Simultaneous Projections Method for Linear Inequalities'. *Linear Algebra and its Applications* **64**, 243–253.

[99] De Pierro, A. and A. Iusem: 1986, 'A Relaxed Version of Bregman's Method for Convex Programming'. *JOTA* **5**(3), 421–440.

[100] DeMarr, R.: 1974, 'Nonnegative idempotent matrices'. *Proceedings of the American Mathematical Society* **45**(2), 185–188.

[101] Dennis, J. and R. Schnabel: 1983, *Numerical Methods for Unconstrained Optimization and Nonlinear Equations.* Englewood Cliffs, New Jersey: Prentice-Hall, Inc.

[102] Deutsch, F.: 1984, 'Rate of convergence of the method of alternating projections'. In: *International Series of Numerical Mathematics*, Vol. 72. Basel, pp. 96–107.

[103] Deutsch, F.: 1992, 'The Method of Alternating Orthogonal Projections'. In: S. P. Singh (ed.): *Approximation Theory, Spline Functions and Applications.* pp. 105–121.

[104] Deutsch, F.: 1995, 'The angle between subspaces of a Hilbert space'. In: S. Singh (ed.): *Approximation Theory, Wavelets and Applications.* pp. 107–130.

[105] Deutsch, F. and H. Hundal: 1997, 'The rate of convergence for the method of alternating projections. II'. *Journal of Mathematical Analysis and Applications* **205**, 381–405.

[106] Diniz-Ehrhardt, M. and J. Martínez: 1996, 'Successive projection methods for the solution of overdetermined nonlinear systems'. In: G. Di Pillo and F. Gianessi (eds.): *Nonlinear Optimization and Applications.* New York, pp. 75–84.

[107] Djokovic, D.: 1969, 'Note on two problems of matrices'. *Am. Math. Monthly* **76**, 652–654.

[108] Dokovic, D.: 1991, 'Unitary Similarity of Projectors'. *Aequationes Mathematicae* **42**, 220–224.

[109] Du, D.-Z. and X.-S. Zhang: 1986, 'A Convergence Theorem of Rosen's Gradient Projection Method'. *Mathematical Programming* **36**, 135–144.

[110] Du, D.-Z. and X.-S. Zhang: 1989, 'Global Convergence of Rosen's Gradient Projection Method'. *Mathematical Programming* **44**, 357–366.

[111] Duff, I.: 1991, 'Parallel algorithms for general sparse systems'. In: E. Spedicato (ed.): *Computer Algorithms for Solving Linear Algebraic Equations*. pp. 277–297.

[112] Dunn, J.: 1981, 'Global and Asymptotic Convergence Rate Estimates for a Class of Projected Gradient Processes'. *SIAM J. Control and Optimization* **19**(3), 368–400.

[113] Edwards, R.: 1965, *Functional Analysis, Theory and Applications*. New York: Holt, Rinehart and Winston.

[114] Egerváry, E.: 1960, 'On Rank-Diminishing Operations and their Applications to the Solution of Linear Equations'. *ZAMP* **11**, 376–386.

[115] Egerváry, J.: 1956, 'Az inverz mátrix általánosítása'. *A Magyar Tudományos Akadémia Matematikai Kutató Intézetének Közleményei* **1**, 315–324. in Hungarian.

[116] Elfving, T.: 1980, 'Block-iterative methods for consistent and inconsistent linear equations'. *Numerische Mathematik* **35**, 1–12.

[117] Elfving, T.: 1998, 'A stationary iterative pseudoinverse algorithm'. *BIT* **38**(2), 275–282.

[118] Faddejew, D. and W. Faddejewa: 1973, *Numerischen Methoden der Linearen Algebra*. Berlin: VEB Deutscher Verlag der Wissenschaften.

[119] Farahat, H.: 1966, 'The Semigroup of Doubly-Stochastic Matrices'. *Proc. Glasgow Math. Ass.* **7**, 178–183.

[120] Federer, H.: 1969, *Geometric Measure Theory*. Springer-Verlag.

[121] Fiedler, M. and V. Pták: 1962, 'On Matrices with Nonpositive Off-Diagonal Elements and Positive Principal Minors'. *Czech. Math. J.* **12**, 619–633.

[122] Fitzpatrick, S. and R. Phelps: 1982, 'Differentiability of the Metric Projection in Hilbert Space'. *Trans. Amer. Math. Soc.* **270**(2), 483–501.

[123] Flam, S. and J. Zowe: 1990, 'Relaxed Outer Projections, Weighted Averages and Convex Feasibility'. *BIT* **30**, 289–300.

[124] Fletcher, R.: 1997, 'Dense Factors of Sparse Matrices'. In: M. Buhman and A. Iserles (eds.): *Approximation Theory and Optimization*. pp. 145–166.

[125] Flor, P.: 1969, 'On Groups of Non-Negative Matrices'. *Compositio Mathematica* **21**(4), 376–382.

[126] Forsythe, G.: 1953, 'Solving linear algebraic equations can be interesting'. *Bull. Amer. Math. Soc.* **59**, 299–329.

[127] Franchetti, C. and W. Light: 1986, 'On the von Neumann alternating algorithm in Hilbert space'. *Journal of Mathematical Analysis and Applications* **114**, 305–314.

[128] Friedman, A.: 1970, *Foundations of Modern Analysis.* New York: Dover.

[129] Frommer, A.: 1986a, 'Comparison of Brown's and Newton's Method in the Monotone Case'. *Numerische Mathematik* **52**, 511–521.

[130] Frommer, A.: 1986b, 'Monotonie und Einschließung Beim Brown-Verfahren'. Ph.D. thesis, Inst. für Angewandte Mathematik, University of Karlsruhe.

[131] Frommer, A.: 1992a, 'Verallgemeinerte Diagonaldominanz Bei Nichtlinearen Funktionen.'. *ZAMM* **72**(9), 431–444.

[132] Frommer, A.: 1992b, 'Verallgemeinerte Diagonaldominanz Bei Nichtlinearen Funktionen II: Anwendung Auf Asynchrone Iterationsverahren und Beispiele'. *ZAMM* **72**(11), 591–605.

[133] Galántai, A., 'Perturbation Bounds for Triangular and Full Rank Factorizations'. *Computers and Mathematics with Applications.* submitted.

[134] Galántai, A.: 1990, 'Block ABS Methods for Nonlinear Systems of Algebraic Equations'. In: E. Allgower and K. Georg (eds.): *Computational Solution of Nonlinear Systems of Equations (AMS-SIAM Summer Seminar on the Computational Solution of Nonlinear Systems of Equations, Fort Collins, Colorado, 1988), Lectures in Applied Mathematics, 26.* Providence, RI, pp. 181–187.

[135] Galántai, A.: 1991a, 'Analysis of error propagation in the ABS class for linear systems'. *Ann. Inst. Statist. Math.* **43**, 597–603.

[136] Galántai, A.: 1991b, 'Convergence Theorems for the Nonlinear ABS-Methods'. In: D. Greenspan and P. Rózsa (eds.): *Colloquia Mathematica Soc. János Bolyai, 59. Numerical Methods, Miskolc (Hungary) 1990.* Amsterdam, pp. 65–79.

[137] Galántai, A.: 1994, 'ABS Methods for the Solution of Linear and Nonlinear Systems of Algebraic Equations'. Habilitation dissertation, Technical University of Budapest.

[138] Galántai, A.: 1995, 'The Global Convergence of the ABS Methods for a Class of Nonlinear Problems'. *Optimization Methods and Software* **4**, 283–295.

[139] Galántai, A.: 1996, 'A Monotonicity Property of M-Matrices'. *Publ. Math. Debrecen* **44**(3-4), 265–268.

[140] Galántai, A.: 1997a, 'Hybrid Quasi-Newton ABS Methods for Nonlinear Equations'. In: F. Cobos, J. Gómez, and F. Mateos (eds.): *EAMA97 Meeting on Matrix Analysis and Applications, Sevilla, September 10-12, 1997.* pp. 221–228.

[141] Galántai, A.: 1997b, 'The Local Convergence of Quasi-Newton ABS Methods'. *Publ. Univ. of Miskolc, Series D. Natural Sciences* **37., Mathematics**, 31–38.

[142] Galántai, A.: 1997-99, 'Rank Reduction and Conjugation'. *Acta Technica Acad. Sci. Hung.* **108**(1–2), 107–130.

[143] Galántai, A.: 1998, 'A Note on Projector Norms'. *Publ. Univ. of Miskolc, Series D. Natural Sciences* **38., Mathematics**, 41–49.

[144] Galántai, A.: 1999a, 'The Geometry of LU Decomposability'. *Publ. Univ. of Miskolc, Series D. Natural Sciences* **40, Mathematics**, 21–24.

[145] Galántai, A.: 1999b, 'Parallel ABS Projection Methods for Linear and Nonlinear Systems with Block Arrowhead Structure'. *Computers and Mathematics with Applications* **38**, 11–17.

[146] Galántai, A.: 2000, 'Componentwise Perturbation Bounds for the LU, LDU and LDL^T Decompositions'. *Mathematical Notes, Miskolc* **1**, 109–118.

[147] Galántai, A.: 2001, 'Rank Reduction, Factorization and Conjugation'. *Linear and Multilinear Algebra* **49**, 195–207.

[148] Galántai, A.: 2003, 'Perturbations of Triangular Matrix Factorizations'. *Linear and Multilinear Algebra* **51**, 175–198.

[149] Galántai, A. and A. Jeney: 1996, 'Quasi-Newton ABS Methods for Solving Nonlinear Algebraic Systems of Equations'. *JOTA* **89**(3), 561–573.

[150] Gantmacher, F.: 1959, *The Theory of Matrices*, Vol. I-II. New York: Chelsea.

[151] García-Palomares, U.: 1999, 'Preconditioning projection methods solving algebraic linear systems'. *Numerical Algorithms* **21**, 157–164.

[152] Gastinel, N.: 1958, 'Procédé itératif pour la résolution numérique d'un systéme d'équations linéaires'. *C.R. Acad. Sci. Paris* **246**, 2571–2574.

[153] Gastinel, N.: 1960, 'Matrices du second degré et normes générales en analyse numérique linéaire'. Ph.D. thesis, Université de Grenoble.

[154] Gastinel, N.: 1963, 'Sur-décomposition de normes générales et procédés itératifs'. *Numerische Mathematik* **5**, 142–151.

[155] Gastinel, N.: 1972, *Lineare numerische Analysis.* Berlin: VEB Deutscher Verlag der Wissenschaften.

[156] Gastinel, N. and J. Joly: 1970, 'Condition numbers and general projection method'. *Linear Algebra and its Applications* **3**, 185–224.

[157] Gentle, J.: 1998, *Numerical Linear Algebra for Applications in Statistics.* Springer-Verlag.

[158] Gluck, H.: 1967, 'Higher Curvatures of Curves in Euclidean Space, II'. *Amer. Math. Monthly* **74**, 1049–1056.

[159] Gochberg, I. and I. Feldman: 1974, *Faltungsgleichungen und Projektionverfahren Zu Ihrer Lösung.* Berlin: Akademie-Verlag.

[160] Gohberg, I. and M. Krein: 1969, *Introduction to the Theory of Nonselfadjoint Operators.* Providence, RI: AMS.

[161] Gohberg, I., P. Lancaster, and L. Rodman: 1986, *Invariant Subspaces of Matrices with Applications.* Wiley.

[162] Goldstein, A.: 1964, 'Convex Programming in Hilbert Space'. *Bull. Amer. Math. Soc.* **70**, 709–710.

[163] Golub, G. and D. O'Leary: 1989, 'Some History of the Conjugate Gradient and Lanczos Algorithms: 1948-1976'. *SIAM Review* **31**(1), 50–102.

[164] Golub, G. and V. Pereyra: 1973, 'The Differentiation of Pseudo-Inverses and Nonlinear Least Squares Problems Whose Variables Separate'. *SIAM J. Numer. Anal.* **10**, 413–432.

[165] Golub, G. and C. Van Loan: 1983, *Matrix Computations.* Baltimore: The John Hopkins University Press.

[166] Gonzaga, C. and H. Lara: 1997, 'A note on properties of condition numbers'. *Linear Algebra and its Applications* **261**, 269–273.

[167] Graybill, F. and G. Marsaglia: 1957, 'Idempotent matrices and quadratic forms in the general linear hypothesis'. *Ann. Math. Stat.* **28**, 678–686.

[168] Groß, J.: 1999, 'On the product of orthogonal projectors'. *Linear Algebra and its Applications* **289**, 141–150.

[169] Groß, J. and G. Trenkler: 1997, 'Generalized and hypergeneralized projectors'. *Linear Algebra and its Applications* **264**, 463–474.

[170] Groß, J. and G. Trenkler: 1998, 'On the product of oblique projectors'. *Linear and Multilinear Algebra* **44**, 247–259.

[171] Groß, J. and G. Trenkler: 1999, 'Nonsingularity of the difference of two oblique projectors'. *SIAM J. Matrix Anal. Appl.* **21**(2), 390–395.

[172] Groß, J., G. Trenkler, and S. Troschke: 1999, 'On semi-orthogonality and a special class of matrices'. *Linear Algebra and its Applications* **289**, 169–182.

[173] Gurin, L., B. Polyak, and E. Raik: 1967, 'Metodi Proektsii Dlya Otiskaniya Obshei Tochki Vipuklikh Mnozhestv'. *Zhurnal Vichislitel'noi Matematiki i Matematicheskoi Fiziki* **7**(6), 1211–1228. in Russian.

[174] Gustafson, K.: 1999, 'The geometrical meaning of the Kantorovich-Wielandt inequalities'. *Linear Algebra and its Applications* **296**, 143–151.

[175] Guttman, L.: 1946, 'Enlargement methods for computing the inverse matrix'. *Ann. Math. Statist.* **17**, 336–343.

[176] Hageman, L. and D. Young: 1981, *Applied Iterative Methods.* New York: Academic Press.

[177] Halmos, P.: 1951, *Introduction to Hilbert Space.* New York: Chelsea.

[178] Halmos, P.: 1958, *Finite Dimensional Vector Spaces.* Van Nostrand, second edition.

[179] Halmos, P.: 1969, 'Two subspaces'. *Trans. Am. Math. Soc.* **144**, 381–389.

[180] Halmos, P.: 1970, 'Finite-dimensional Hilbert spaces'. *Am. Math. Monthly* **77**(5), 457–464.

[181] Halmos, P.: 1982, *A Hilbert Space Problem Book.* Springer-Verlag, second edition.

[182] Halperin, I.: 1962, 'The product of projection operators'. *Acta Sci. Math.* **23**, 96–99.

[183] Halpern, B.: 1967, 'Fixed points of nonexpanding maps'. *Bull. Amer. Math. Soc.* **73**, 957–961.

[184] Hamaker, C. and D. Solmon: 1978, 'The angles between the null spaces of X rays'. *Journal of Mathematical Analysis and Applications* **62**, 1–23.

[185] Han, S.: 1988, 'A successive projection method'. *Mathematical Programming* **40**, 1–14.

[186] Hanke, M.: 1994, 'Iterative Consistency: A Concept for the Solution of Singular Systems of Linear Equations'. *SIAM J. Matrix. Anal. Appl.* **15**(2), 569–577.

[187] Hanson, R. and C. Lawson: 1969, 'Extensions and applications of the Householder algorithm for solving linear least squares problems'. *Mathematics of Computation* **23**, 787–812.

[188] Hart, W. and T. Motzkin: 1956, 'A composite Newton-Raphson gradient method for the solution of systems of equations'. *Pacific J.Math.* **6**, 691–707.

[189] Hartwig, R. and M. Putcha: 1990, 'When is a matrix a sum of idempotents?'. *Linear and Multilinear Algebra* **26**, 279–286.

[190] Hartwig, R. and G. Styan: 1987, 'Partially ordered idempotent matrices'. In: T. Pukkila and S. Puntanen (eds.): *Proc. Second International Tampere Conference in Statistics.* pp. 361–383.

[191] Harville, D.: 1997, *Matrix Algebra From a Statistician's Perspective.* Springer-Verlag.

[192] He, B. and J. Stoer: 1992, 'Solution of Projection Problems over Polytopes'. *Numerische Mathematik* **61**, 73–90.

[193] Hegedûs, C.: 1991, 'Generating Conjugate Directions for Arbitrary Matrices by Matrix Equations, I-II.'. *Comp. Math. Applic.* **21**, 71–94.

[194] Hegedûs, C.: 2001, 'Short Proofs for the Pseudoinverse in Linear Algebra'. *Annales Univ. Sci. Budapest* **44**, 115–121.

[195] Herman, G., A. Lent, and P. Lutz: 1978, 'Relaxation Methods for Image Reconstruction'. *Comm. ACM.* **21**, 152–158.

[196] Hestenes, M.: 1973, 'Iterative Methods for Solving Linear Equations'. *JOTA* **11**(4), 323–359. reprint of the original NAML Report No. 52-9, 1951.

[197] Hestenes, M. and E. Stiefel: 1952, 'Methods of Conjugate Gradients for Solving Linear Systems'. *Journal of Research of the National Bureau of Standards* **49**(6), 409–436.

[198] Higham, N.: 1996, *Accuracy and Stability of Numerical Algorithms.* Philadelphia: SIAM.

[199] Hiriart-Urruty, J.-B.: 1982, 'At What Points is the Projection Mapping Differentiable?'. *Amer. Math. Monthly* **89**, 456–458.

[200] Hiriart-Urruty, J.-B. and C. Lemaréchal: 1993a, *Convex Analysis and Minimization Algorithms I. Fundamentals*. Springer-Verlag.

[201] Hiriart-Urruty, J.-B. and C. Lemaréchal: 1993b, *Convex Analysis and Minimization Algorithms II. Advanced Theory and Bundle Methods*. Springer-Verlag.

[202] Holmes, R.: 1973, 'Smoothness of Certain Metric Projections on Hilbert Space'. *Trans. Amer. Math. Soc.* **183**, 87–100.

[203] Horn, R. and C. Johnson: 1985, *Matrix Analysis*. Cambridge University Press.

[204] Horn, R. and C. Johnson: 1991, *Topics in Matrix Analysis*. Cambridge University Press.

[205] Householder, A.: 1953, *Principles of Numerical Analysis*. New York: McGraw-Hill.

[206] Householder, A.: 1964, *The Theory of Matrices in Numerical Analysis*. Blaisdell Publishing Company.

[207] Householder, A. and F. Bauer: 1960, 'On certain iterative methods for solving linear systems'. *Numerische Mathematik* **2**, 55–59.

[208] Huard, P.: 1975, 'Optimization Algorithms and Point-to-Set-Maps'. *Mathematical Programming* **8**, 308–331.

[209] Ikebe, Y., T. Inagaki, and S. Miyamoto: 1987, 'The monotonicity theorem, Cauchy's interlace theorem, and the Courant-Fischer theorem'. *Am. Math. Monthly* **94**, 352–354.

[210] Ikramov, K.: 1996, 'A Canonical Form for Projectors under Unitary Similarity'. *Comp. Maths. Math. Phys.* **36**, 279–281.

[211] Ikramov, K.: 1999, 'An algebraic proof of a result by Gonzaga and Lara'. *Linear Algebra and its Applications* **299**, 191–194.

[212] Ipsen, I. and C. Meyer: 1995, 'The angle between complementary subspaces'. *Am. Math. Monthly* **102**, 904–911.

[213] Iusem, A. and A. De Pierro: 1986, 'Convergence Results for an Accelerated Nonlinear Cimmino Algorithm'. *Numerische Mathematik* **49**, 367–378.

[214] Ivanov, V.: 1986, *Metodi Vichislenii Na EVM*. Kiev: Naukova Dumka. in Russian.

[215] Kaczmarz, S.: 1937, 'Angenäherte Auflösung von Systemen linearer Gleichungen'. *Bulletin International de l'Académie Polonaise des Sciences et des Lettres. Classe des Sciences Mathématiques et Naturelles. Série A. Sciences Mathématiques* pp. 355–357.

[216] Kamath, C. and A. Sameh: 1988/89, 'A projection method for solving nonsymmetric linear systems on multiprocessors'. *Parallel Computing* **9**, 291–312.

[217] Kantorovich, L. and G. Akilov: 1964, *Functional Analysis in Normed Spaces*. Pergamon Press.

[218] Karmanov, V.: 1974, 'Otsenki Skhodimosti Iteratsionnikh Metodov Minimizatsii'. *Zh. Vychisl. Mat. Mat. Fiz.* **14**(1), 3–14. in Russian.

[219] Karmanov, V.: 1989, *Mathematical Programming.* Moscow: Mir.

[220] Kato, T.: 1966, *Perturbation Theory for Linear Operators.* Springer-Verlag.

[221] Kayalar, S. and H. Weinert: 1988, 'Error bounds for the method of alternating projections'. *Math. Control Signals Systems* **1**, 43–59.

[222] Kelley, C.: 1999, *Iterative Methods for Optimization.* Philadelphia: SIAM.

[223] Kestelman, H.: 1966, 'A generalization of Cochran's theorem'. *Am. Math. Monthly* **75**, 301–303.

[224] Kirk, W.: 1965, 'A Fixed Point Theorem for Mappings Which Do Not Increase Distances'. *Amer. Math. Monthly* **72**, 1004–1006.

[225] Kiwiel, K. and B. Lopuch: 1997, 'Surrogate projection methods for finding fixed points of firmly nonexpansive mappings'. *SIAM J. Optimization* **7**(4), 1084–1102.

[226] Kotzer, T., N. Cohen, and J. Shamir: 1997, 'A Projection-Based Algorithm for Consistent and Inconsistent Constraints'. *SIAM J. Optimization* **7**(2), 527–546.

[227] Kovarik, Z.: 1977, 'Similarity and interpolation between projectors'. *Acta Sci. Math.* **39**, 341–351.

[228] Kovarik, Z.: 1980, 'Manifolds of frames of projectors'. *Linear Algebra and its Applications* **31**, 151–158.

[229] Krasnosel'skii, M.: 1955, 'Dva Zamechaniya O Metode Posledovatel'nikh Priblizhenii'. *Uspekhi Matematicheskikh Nauk* **10**(1-63), 123–127. In Russian.

[230] Krasnosel'skii, M., G. Vainikko, P. Zabreiko, Y. Rutitskii, and V. Stetsenko: 1972, *Approximate Solution of Operator Equations.* Groningen: Wolters-Nordhoff.

[231] Laurie, C., B. Mathes, and H. Radjavi: 1994, 'Sums of idempotents'. *Linear Algebra and its Applications* **208/209**, 175–197.

[232] Levitin, E. and B. Polyak: 1965, 'Metodi Minimizatsii Pri Nalichii Ogranichenii'. *Zh. Vychisl. Mat. Mat. Fiz.* **6**(5), 787–823. in Russian.

[233] Lewin, M.: 1991, 'Nonnegative idempotent matrices and their graphs'. *SIAM J. Matrix Anal. Appl.* **12**(1), 166–171.

[234] Li, X. and Y. Wei: 2001, 'An improvement on the perturbation of the group inverse and oblique projection'. *Linear Algebra and its Applications* **338**, 53–66.

[235] Liu, C.: 1995, 'An acceleration scheme for row projection methods'. *Journal of Computational and Applied Mathematics* **57**, 363–391.

[236] Liusternik, L. and V. Sobolev: 1961, *Elements of Functional Analysis.* New York: Frederic Ungar.

[237] Ljashko, I., V. Makarov, and A. Skorobogat'ko: 1977, *Metodi Vichislenii.* Kiev: Vishcha Skola. in Russian.

[238] Lou, G. and S. Han: 1988, 'A Parallel Projection Method for Solving Generalized Linear Least-Squares Problems'. *Numerische Mathematik* **53**, 255–264.

[239] Loynes, R.: 1966, 'On idempotent matrices'. *Ann. Math. Stat.* **37**, 295–296.

[240] Luchka, A. Y.: 1993, *Proektsionno-Iterativnye Metody.* Kiev: Naukova Dumka. in Russian.

[241] Maess, G.: 1979, 'Iterative Lösung Linearer Gleichungssysteme'. *Nova Acta Leopoldina* **52**(238), 1–79.

[242] Mäkeläinen, T. and G. Styan: 1976, 'A decomposition of an idempotent matrix where nonnegativity implies idempotence and none of the matrices need be symmetric'. *Shankhyā: The Indian Journal of Statistics* **38**, 400–403.

[243] Marek, I. and K. Zitny: 1983, 1986, *Matrix Analysis for Applied Sciences Vol. 1-2*, Vol. 60, 84 of *Teubner–Texte Zur Mathematik.* Leipzig: Teubner.

[244] Marsaglia, G. and G. Styan: 1974, 'Equalities and Inequalities for Ranks of Matrices'. *Linear and Multilinear Algebra* **2**, 269–292.

[245] Marshall, A. and I. Olkin: 1973, 'Norms and inequalities for condition numbers, III'. *Linear Algebra and its Applications* **7**, 291–300.

[246] Marshall, A. and I. Olkin: 1979, *Inequalities: Theory of Majorization and Its Applications.* New York: Academic Press.

[247] Martínez, J.: 1985, 'The projection method for solving nonlinear systems of equations under the "most violated constraint" control'. *Comp. and Maths. with Appls.* **11**(10), 987–993.

[248] Martínez, J.: 1986a, 'The method of successive orthogonal projections for solving nonlinear simultaneous equations'. *Calcolo* **23**, 93–104.

[249] Martínez, J.: 1986b, 'Solution of nonlinear systems of equations by an optimal projection method'. *Computing* **37**, 59–70.

[250] McCormick, G. and R. Tapia: 1972, 'The Gradient Projection Method under Mild Differentiability Conditions'. *SIAM J. Control* **10**(1), 93–98.

[251] McCormick, S.: 1975, 'An iterative procedure for the solution of constrained nonlinear equations with application to optimization problems'. *Numerische Mathematik* **23**, 371–385.

[252] McCormick, S.: 1977, 'The methods of Kaczmarz and row orthogonalization for solving linear equations and least squares problems in Hilbert space'. *Indiana University Mathematics Journal* **26**(6), 1137–1150.

[253] McCormick, S.: 1992, *Multilevel Projection Methods for Partial Differential Equations.* Philadelphia: SIAM.

[254] Meany, R.: 1969, 'A matrix inequality'. *SIAM J. Numer. Anal.* **6**(1), 104–107.

[255] Meyer, C. and R. Plemmons: 1977, 'Convergent Powers of a Matrix with Applications to Iterative Methods for Singular Linear Systems'. *SIAM J. Numer. Anal.* **14**(2), 699–705.

[256] Meyn, K.: 1983, 'Solution of underdetermined nonlinear equations by stationary iteration methods'. *Numerische Mathematik* **42**, 161–172.

[257] Mitrinovic, D.: 1970, *Analytic Inequalities.* New York: Springer-Verlag.

[258] Mizel, V. and M. Rao: 1962, 'Nonsymmetric projections in Hilbert space'. *Pacific. J. Math.* **12**, 343–357.

[259] Morgenstern, D.: 1956, 'Begründung Des Alternierenden Verfahrens Durch Orthogonalprojektion'. *Z. angew. Math. Mech.* **36**(7/8), 255–256.

[260] Motte, D.: 1988, 'The numerical computation of the minimal projection'. *IMA J. Numer. Anal.* **8**, 219–230.

[261] Motzkin, T. and I. Schoenberg: 1954, 'The Relaxation Method for Linear Inequalities'. *Canad. J. Math.* **6**, 393–404.

[262] Nakano, H.: 1953, *Spectral Theory in the Hilbert Space.* Tokyo: Japan Society for the Promotion of Science.

[263] Nashed, M. (ed.): 1976, 'Generalized Inverses and Applications'. New York:, Academic Press.

[264] Nelson, S. and M. Neumann: 1987, 'Generalizations of the Projection Method with Application to SOR Theory for Hermitian Positive Semidefinite Linear Systems'. *Numer. Math.* **51**, 123–141.

[265] Neumann, J. v.: 1937, 'Some matrix-inequalities and metrization of matric-space'. *Tomsk Univ. Rev.* **1**, 286–300. in Collected Works, Vol. IV., Pergamon Press, 1962, pp. 205-219.

[266] Neumann, J. v.: 1950, 'On Rings of Operators, Reduction Theory'. *Annals of Math.* **50**, 401–485.

[267] Nicolaides, R.: 1974, 'On a Geometrical Aspect of SOR and the Theory of Consistent Ordering for Positive Definite Matrices'. *Numerische Mathematik* **23**, 99–104.

[268] O'Leary, D.: 1995, 'Why Broyden's nonsymmetric method terminates on linear equations'. *SIAM J. Optimization* **5**(2), 231–235.

[269] Opial, Z.: 1967, 'Weak Convergence of the Sequence of Successive Approximations for Nonexpansive Mappings'. *Bull. Amer. Math. Soc.* **73**, 591–597.

[270] Ortega, J.: 1972, *Numerical Analysis: A Second Course.* New York: Academic Press.

[271] Ortega, J. and W. Rheinboldt: 1970, *Iterative Solution of Nonlinear Equations in Several Variables.* Academic Press.

[272] Ouellette, D.: 1981, 'Schur complements and statistics'. *Linear Algebra and its Applications* **36**, 187–295.

[273] Paige, C. and M. Saunders: 1981, 'Towards a Generalized Singular Value Decomposition'. *SIAM J. Numer. Anal.* **18**(3), 398–405.

[274] Paige, C. and M. Wei: 1994, 'History and generality of the CS decomposition'. *Linear Algebra and its Applications* **208/209**, 303–326.

[275] Pasternak-Winiarski, Z.: 1998, 'On the dependence of the orthogonal projector on deformations of the scalar product'. *Studia Mathematica* **128**(1), 1–17.

[276] Peters, W.: 1976, 'Lösung Linearer Gleichungssysteme Durch Projektion Auf Schnittraume Von Hyperebenen und Berechnung Einer Verallgemeinerten Inversen'. *Beitrage zur Numerischen Mathematik* **5**, 129–146.

[277] Petryshyn, W.: 1967, 'Projection methods in nonlinear numerical functional analysis'. *Journal of Mathematics and Mechanics* **17**(4), 353–372.

[278] Petryshyn, W.: 1968, 'On the iteration, projection and projection-iteration methods in the solution of nonlinear functional equations'. *Journal of Mathematical Analysis and Applications* **21**, 575–607.

[279] Phelps, R.: 1985, 'Metric Projections and the Gradient Projection Method in Banach Spaces'. *SIAM J. Control and Optimization* **23**(6), 973–977.

[280] Phelps, R.: 1986, 'The Gradient Projection Method Using Curry's Steplength'. *SIAM J. Control. Optim.* **24**(4), 692–699.

[281] Pierra, G.: 1984, 'Decomposition Through Formalization in a Product Space'. *Mathematical Programming* **28**, 96–115.

[282] Pietrzykowski, T.: 1960, 'Projection method'. Technical Report A 8, Polskiej Akademii Nauk, Warsaw. Zaklad Aparatow Matematycznych PAN.

[283] Polak, E.: 1971, *Computational Methods in Optimization.* Academic Press.

[284] Pólya, G. and G. Szegõ: 1972, *Problems and Theorems in Analysis I.* Springer-Verlag.

[285] Polyak, B.: 1967, 'A General Method for Solving Extremum Problems'. *Soviet Math. Dokl.* **8**(3), 593–597.

[286] Polyak, B.: 1969, 'Minimizacija Negladkikh Funkcionalov'. *Zh. Vychisl. Mat. Mat. Fiz.* **9**(3), 509–521. in Russian.

[287] Polyak, B.: 1971, 'Skhodimost' Metodov Vozmozhnikh Napravlenii V Ekstremal'nikh Zadachakh'. *Zh. Vychisl. Mat. Mat. Fiz.* **11**(4), 855–869. in Russian.

[288] Ponstein, J.: 1967, 'Seven Kinds of Convexity'. *SIAM Review* **9**(1), 115–119.

[289] Popa, C.: 1995, 'Least-Squares Solution of Overdetermined Inconsistent Linear Systems Using Kaczmar's Relaxation'. *Intern. J. Computer Math.* **55**, 79–89.

[290] Popa, C.: 1999, 'Block-Projections Algorithms with Blocks Containing Mutually Orthogonal Rows and Columns'. *BIT* **39**(2), 323–338.

[291] Popa, C.: 2000, 'A Fast Kaczmarz-Like Solver for Linear Least Squares Problems'. *ZAMM* **80**(S3), S811–S812.

[292] Prager, M.: 1960, 'On a Principle of Convergence in a Hilbert Space'. *Czech. Math. J.* **10**, 271–282. in Russian.

[293] Prößdorf, S. and B. Silberman: 1977, *Projektionsverfahren und Die Naherungsweise Lösung Singularer Gleichungen.* Leipzig: Teubner.

[294] Pshenichny, B. and Y. Danilin: 1982, *Numerical Methods in Extremal Problems.* Moscow: Mir.

[295] Pták, V.: 1985, 'Extremal operators and oblique projections'. *Casopis Pest. Mat.* **110**, 343–350.

[296] Pyle, L.: 1964, 'Generalized inverse computations using the gradient projection method'. *JACM* **11**(4), 422–428.

[297] Pyle, L.: 1967, 'A generalized inverse ϵ-algorithm for constructing intersection projection matrices, with applications'. *Numerische Mathematik* **10**, 86–102.

[298] Pyle, L.: 1972, 'The Generalized Inverse in Linear Programming. Basic Structure'. *SIAM J. Appl. Math.* **22**(3), 335–355.

[299] Pyle, L. and R. Cline: 1973, 'The Generalized Inverse in Linear Programming–Interior Gradient Projection Methods'. *SIAM J. Appl. Math.* **24**(4), 511–534.

[300] Rao, C. and S. Mitra: 1971, *Generalized Inverse of Matrices and its Applications.* John Wiley and Sons.

[301] Riesz, F. and B. Sz.-Nagy: 1955, *Functional Analysis.* New York: Frederick Unger.

[302] Rockafellar, R.: 1970, *Convex Analysis.* Princeton, New Jersey: Princeton University Press.

[303] Rockafellar, R. and R. J.-B. Wets: 1998, *Variational Analysis.* Springer-Verlag.

[304] Rosen, J.: 1960, 'The gradient projection method for nonlinear programming. Part I. Linear constraints'. *J. Soc. Indust. Appl. Math.* **8**(1), 181–217.

[305] Rosen, J.: 1961, 'The gradient projection method for nonlinear programming. Part II. Nonlinear constraints'. *J. Soc. Indust. Appl. Math.* **9**(4), 514–532.

[306] Saad, Y.: 1996, *Iterative Methods for Sparse Linear Systems.* PWS Publishing Company.

[307] Saad, Y. and H. van der Vorst: 2000, 'Iterative solution of linear systems in the 20th century'. *Journal of Computational and Applied Mathematics* **123**, 1–33.

[308] Sargent, R. and B. Murtagh: 1973, 'Projection methods for non-linear programming'. *Mathematical Programming* **4**, 245–268.

[309] Scales, L.: 1985, *Introduction to Non-Linear Optimization.* New York: Springer-Verlag.

[310] Schatten, R.: 1947, 'On projections with bound 1'. *Annals of Mathematics* **48**(2), 321–325.

[311] Schatten, R.: 1960, *Norm Ideals of Completely Continuous Operators.* Springer-Verlag.

[312] Schmidt, J. and U. Patzke: 1981, 'Iterative Nachverbesserung mit Fehlereingrenzung der Cholesky-Faktoren Von Stieltjes-Matrizen'. *Journ. für die Reine und Angewandte Mathematik* **327**, 81–92.

[313] Schott, D.: 1991, 'A general iterative scheme with applications to convex optimization and related fields'. *Optimization* **22**(6), 885–902.

[314] Schröder, J.: 1980, *Operator Inequalities.* New York: Academic Press.

[315] Schwarz, S.: 1964, 'On the Structure of the Semigroup of Stochastic Matrices'. *Magyar. Tud. Akad. Mat. Kut. Int. Közl.* **9**, 297–311.

[316] Schwarz, S.: 1967, 'A Note on the Structure of the Semigroup of Doubly-Stochastic Matrices'. *Mat. Casopis* **17**(4), 308–316.

[317] Scolnik, H., N. Echebest, M. Guardarucci, and M. Vacchino: 2002, 'A class of optimized row projection methods for solving large nonsymmetric linear systems'. *Applied Numerical Mathematics* **41**, 499–513.

[318] Sebestyén, Z.: 1990, 'Characterization of subprojection suboperators'. *Acta Math. Hun.* **56**(1–2), 115–119.

[319] Sloboda, F.: 1978, 'A parallel projection method for linear algebraic systems'. *Aplikace Matematiky* **23**(3), 185–198.

[320] Sloboda, F.: 1991, 'A projection method of the Cimmino type for linear algebraic systems'. *Parallel Computing* **17**, 435–442.

[321] Sloboda, F. and F. Sgallari: 1995, 'On iterative solution of linear equations arising in BVPs of ODEs'. *Linear Algebra and its Applications* **225**, 195–205.

[322] Smith, K., D. Solmon, and S. Wagner: 1977, 'Practical and mathematical aspects of the problem of reconstructing objects from radiographs'. *Bull. Am. Math. Soc.* **83**(6), 1227–1270.

[323] Sommerville, D.: 1958, *An Introduction to the Geometry of N Dimensions.* New York: Dover.

[324] Stampfli, J.: 1964, 'Sums of projections'. *Duke Mathematical Journal* **31**(3), 455–461.

[325] Stewart, G.: 1973, 'Conjugate direction methods for solving systems of linear equations'. *Numerische Mathematik* **21**, 285–297.

[326] Stewart, G.: 1977, 'On the perturbation of pseudo-inverses, projections and linear least squares problems'. *SIAM Review* **19**(4), 634–662.

[327] Stewart, G.: 1982, 'Computing the CS decomposition of a partioned orthonormal matrix'. *Numerische Mathematik* **40**(2), 297–306.

[328] Stewart, G.: 1989, 'On scaled projections and pseudoinverses'. *Linear Algebra and its Applications* **112**, 189–193.

[329] Stewart, G.: 1993, 'On the Early History of the Singular Value Decomposition'. *SIAM Review* **35**(4), 551–566.

[330] Stewart, G.: 1998, *Matrix Algorithms*, Vol. I: Basic Decompositions. Philadelphia: SIAM.

[331] Stewart, G. and J. Sun: 1990, *Matrix Perturbation Theory.* Academic Press.

[332] Stoer, J. and C. Witzgall: 1970, *Convexity and Optimization in Finite Dimensions.* Springer-Verlag.

[333] Sun, J.-G.: 1992, 'Componentwise Perturbation Bounds for some Matrix Decompositions'. *BIT* **32**, 702–714.

[334] Takane, Y. and H. Yanai: 1999, 'On oblique projectors'. *Linear Algebra and its Applications* **289**, 297–310.

[335] Taldikin, A.: 1982, *Elementi Prikladnogo Funktsional'nogo Analiza.* Moscow: Visshaya Shkola. in Russian.

[336] Tan, W.: 1975, 'Some matrix results and extensions of Cochran's theorem'. *SIAM J. Appl. Math.* **28**(3), 547–554.

[337] Tanabe, K.: 1971, 'Projection method for solving a singular system of linear equations and its applications'. *Numerische Mathematik* **17**, 203–214.

[338] Tanabe, K.: 1974, 'Characterization of linear stationary iterative processes for solving a singular system of linear equations'. *Numerische Mathematik* **22**, 349–359.

[339] Tewarson, R.: 1969, 'Projection Methods for Solving Sparse Linear Systems'. *The Computer Journal* **12**, 77–80.

[340] Tian, Y. and G. Styan: 2001, 'Rank equalities for idempotent and involutory matrices'. *Linear Algebra and its Applications* **335**, 101–117.

[341] Tiel, J. V.: 1984, *Convex Analysis.* Wiley.

[342] Tompkins, C.: 1955, 'Projection methods in calculation'. In: H. Antosiewicz (ed.): *Proc. Second Symposium on Linear Programming.* Washington, D.C., pp. 425–448.

[343] Topkis, D. and A. Veinott: 1967, 'On the Convergence of some Feasible Direction Algorithms for Nonlinear Programming'. *J. SIAM Control* **5**(2), 268–279.

[344] Trenkler, G.: 1994, 'Characterizations of oblique and orthogonal projectors'. In: *Proceedings of the International Conference on Linear Statistical Inference LINSTAT'93.* pp. 255–270.

[345] Trenogin, V., B. Pisarevskii, and T. Soboleva: 1984, *Zadachi I Uprazhnenija Po Funkcional'nomu Analizu.* Moscow: Nauka.

[346] Trummer, M.: 1981, 'Reconstructing Pictures from Projections: On the Convergence of the ART Algorithm with Relaxation'. *Computing* **26**, 189–195.

[347] Trummer, M.: 1983, 'SMART-an Algorithm for Reconstructing Pictures from Projections'. *Z. Angew. Math. Phys.* **34**, 746–753.

[348] Trummer, M.: 1984, 'A Note on the ART Relaxation'. *Computing* **33**, 349–352.

[349] Turner, K.: 1991, 'Computing projections for the Karmarkar algorithm'. *Linear Algebra and its Applications* **152**, 141–154.

[350] Valeev, K.: 1986, *Rashchepleniye Spektra Matric.* Kiev: Vishscha Shkola. in Russian.

[351] Vasil'chenko, G. and A. Svetlanov: 1980, 'Proektsionnyj algoritm resheniya sistem linejnykh algebraicheskikh uravnenij bol'shoj razmernosti'. *Zh. Vychisl. Mat. Mat. Fiz.* **20**(1), 3–10. in Russian.

[352] Voyevodin, V.: 1983, *Linear Algebra.* Moscow: Mir.

[353] Wainwright, R. and R. Keller: 1977, 'Algorithms for Projection Methods for Solving Linear Systems of Equations'. *Comp. Maths. with Appls.* **3**, 235–245.

[354] Watkins, D.: 1991, *Fundamentals of Matrix Computations.* Wiley and Sons.

[355] Wedin, P.: 1983, 'On angles between subspaces of a finite dimensional inner product space'. In: B. Kagstrom and A. Ruhe (eds.): *Matrix Pencils.* pp. 263–285.

[356] Wei, Y.: 1999, 'On the perturbation of the group inverse and oblique projection'. *Appl. Math. Comp.* **98**, 29–42.

[357] Whitney, T. and R. Meany: 1967, 'Two algorithms related to the method of steepest descent'. *SIAM J. Numer. Anal.* **4**(1), 109–118.

[358] Wiener, N.: 1955, 'On the Factorization of Matrices'. *Commentarii Mathematici Helvetici* **29**, 97–111.

[359] Wimmer, H.: 1999a, 'Canonical angles of unitary spaces and perturbations of direct complements'. *Linear Algebra and its Applications* **287**, 373–379.

[360] Wimmer, H.: 1999b, 'Lipschitz continuity of oblique projections'. *Proc. Am. Math. Soc.* **128**(3), 873–876.

[361] Windisch, G.: 1989, *M-Matrices in Numerical Analysis.* Leipzig: Teubner.

[362] Winter, B.: 1972, 'Problem 5842'. *Am. Math. Monthly* **79**, 307.

[363] Witsch, K.: 1977, 'Konvergenzaussagen Für Projektionsverfahren Bei Linearen Operatoren'. *Numerische Mathematik* **27**, 339–354.

[364] Yeh, L.: 1995, 'A Note on Wielandt's Inequality'. *Appl. Math. Lett.* **8**, 29–31.

[365] Youla, D.: 1978, 'Generalized image restoration by the method of alternating orthogonal projections'. *IEEE Transactions on Circuits and Systems* **25**(9), 694–702.

[366] Young, D.: 1971, *Iterative Solution of Large Linear Systems.* New York: Academic Press.

[367] Zarontonello, E.: 1971, 'Projections on Convex Sets in Hilbert Space and Spectral Theory'. In: E. Zarontonello (ed.): *Contributions to Nonlinear Functional Analysis.* New York-London, pp. 237–424.

[368] Zassenhaus, H.: 1964, 'Angles of inclination in correlation theory'. *Am. Math. Monthly* **71**, 218–219.

[369] Zoutendijk, G.: 1960, *Methods of Feasible Directions.* Amsterdam: Elsevier.

Index